AF179627

# Medical Geology: Effects of Geological Environments on Human Health

# Medical Geology: Effects of Geological Environments on Human Health

Editor: Justin Richmond

MURPHY & MOORE
www.murphy-moorepublishing.com

www.murphy-moorepublishing.com

MURPHY & MOORE

**Cataloging-in-Publication Data**

Medical geology : effects of geological environments on human health / edited by Justin Richmond.
p. cm.
Includes bibliographical references and index.
ISBN 978-1-63987-779-9
1. Medical geography. 2. Geology--Health aspects. 3. Environmental health. 4. Public health.
5. Environmental geology. I. Richmond, Justin.
RA792 .E87 2023
614.42--dc23

© Murphy & Moore Publishing, 2023

Murphy & Moore Publishing
1 Rockefeller Plaza,
New York City,
NY 10020, USA

ISBN 978-1-63987-779-9

This book contains information obtained from authentic and highly regarded sources. Copyright for all individual chapters remain with the respective authors as indicated. All chapters are published with permission under the Creative Commons Attribution License or equivalent. A wide variety of references are listed. Permission and sources are indicated; for detailed attributions, please refer to the permissions page and list of contributors. Reasonable efforts have been made to publish reliable data and information, but the authors, editors and publisher cannot assume any responsibility for the validity of all materials or the consequences of their use.

**Trademark Notice:** Registered trademark of products or corporate names are used only for explanation and identification without intent to infringe.

# Contents

# Preface

I am honored to present to you this unique book which encompasses the most up-to-date data in the field. I was extremely pleased to get this opportunity of editing the work of experts from across the globe. I have also written papers in this field and researched the various aspects revolving around the progress of the discipline. I have tried to unify my knowledge along with that of stalwarts from every corner of the world, to produce a text which not only benefits the readers but also facilitates the growth of the field.

Geology is a field of natural science that involves the study of the Earth and other astronomical objects, characteristics and composition of these objects and how they change over time. Medical geology is a new interdisciplinary field that investigates the relationship between natural geological factors and their impact on animal and human health. It also attempts to develop an understanding of how the environment influences the geographical distribution of health issues in both animals and humans. Naturally high quantities of minerals, elements, microbes, or other compounds may also be found in geologic materials including soils, volcanic emissions, rocks and dusts, which can improve or harm human health. They may also consist of impurities from human sources such as chemicals, pathogen contaminants or minerals. There are several diseases which can be caused due to geologic materials such as cardiovascular disease and iodine deficiency. This book elucidates the concepts and innovative models around prospective developments in the field of medical geology. Those in search of information to further their knowledge on the effects of geological environments on human health will be greatly assisted by it.

Finally, I would like to thank all the contributing authors for their valuable time and contributions. This book would not have been possible without their efforts. I would also like to thank my friends and family for their constant support.

**Editor**

# 1

# A Longitudinal Study of Association between Heavy Metals and Itchy Eyes, Coughing in Chronic Cough Patients: Related with Non-Immunoglobulin E Mediated Mechanism

**Thao Thi Thu Nguyen [1], Tomomi Higashi [2], Yasuhiro Kambayashi [1], Enoch Olando Anyenda [1], Yoshimasa Michigami [3], Johsuke Hara [4], Masaki Fujimura [4,5], Hiromasa Tsujiguchi [1], Masami Kitaoka [1], Hiroki Asakura [1], Daisuke Hori [1], Yuri Hibino [1], Tadashi Konoshita [6] and Hiroyuki Nakamura [1,*]**

Academic Editor: Paul B. Tchounwou

1 Department of Environmental Health and Preventive Medicine, Graduate School of Medical Sciences, Kanazawa University, 13-1 Takara-machi, Kanazawa 920-8640, Japan; toi_fs@yahoo.com (T.T.T.N.); ykamba@med.kanazawa-u.ac.jp (Y.K.); enolany@yahoo.com (E.O.A.); t-hiromasa@med.kanazawa-u.ac.jp (H.T.); m-ymzk16@stu.kanazawa-u.ac.jp (M.K.); heros@stu.kanazawa-u.ac.jp (H.A.); hori_d@mbr.nifty.com (D.H.); hibino@staff.kanazawa-u.ac.jp (Y.H.)

2 Department of Hygiene, Graduate School of Medical Sciences, Kanazawa University, 13-1 Takara-machi, Kanazawa 920-8640, Japan; tomotomo@med.kanazawa-u.ac.jp

3 Environment Preservation Center, Kanazawa University, Kakuma-machi, Kanazawa 920-1192, Japan; mitigami@t.kanazawa-u.ac.jp

4 Respiratory Medicine, Cellular Transplantation Biology, Graduate School of Medical Sciences, Kanazawa University, 13-1 Takara-machi, Kanazawa 920-8640, Japan; hara0728@gmail.com (J.H.); fujimura@nanao.hosp.go.jp (M.F.)

5 Respiratory Medicine, National Hospital Organization Nanao Hospital, 3-1 Mattou-machi Yabe, Nanao, Ishikawa 926-8531, Japan

6 Third Department of Internal Medicine, Fukui University School of Medicine, Eiheiji, Fukui 910-1193, Japan; konosita@u-fukui.ac.jp

* Correspondence: hiro-n@po.incl.ne.jp.

**Abstract:** The association between heavy metals exposure and respiratory diseases or allergic sensitization showing high serum immunoglobulin E (IgE) has been suggested. However, previous findings have been inconsistent and the mechanisms responsible remain unclear. We evaluated heavy metal exposure and its association with coughing, itchy eyes in chronic cough patients with different IgE levels. Ninety outpatients in Kanazawa University Hospital were recruited between January–June 2011. Subjects whose total IgE measured by radioimmunosorbent test were asked to record their daily symptoms. We collected daily total suspended particles (TSP) from which concentrations of calcium (Ca), cadmium (Cd), chromium (Cr), iron (Fe), manganese (Mn), nickel (Ni), and lead (Pb) were determined then divided into high and low level groups. Generalized estimating equations were applied to compute the relationship between concentrations of these metals and symptoms. All metals at high levels were significantly associated with itchy eyes compared with low levels, with exception of Ca, the six others were significant in patients with IgE $< 250$ IU/mL. Cd, Fe, Mn had association with coughing (odds ratio-OR (95% confidence interval-CI): 1.13 (1.03, 1.24), 1.22 (1.05, 1.42), and 1.13 (1.01, 1.27), respectively), this relationship remained significant for Cd (OR (95% CI): 1.14 (1.03, 1.27)) and Mn (OR (95% CI): 1.15 (1.00, 1.31)) in patients with lower IgE. Our findings demonstrate the relationship between aerial heavy metals and itchy eyes, coughing in chronic cough patients, suggesting these symptoms may be due to a non-IgE mediated mechanism.

**Keywords:** heavy metals; coughing; itchy eyes; chronic cough; total serum IgE level

## 1. Introduction

Due to economic and industrial growth, air pollution such as particulate matter (PM) and gaseous emissions have become an important environmental issue throughout the world. PM is a heterogeneous mixture of biologically active components in the air, and heavy metals and polycyclic aromatic hydrocarbons have been identified as the predominate toxic constituents [1–3]. Airborne heavy metals originate from various sources involving diesel exhaust emissions (iron (Fe), manganese (Mn), lead (Pb), cadmium (Cd), calcium (Ca), and nickel (Ni)) and non-exhaust emissions such as the abrasion of brakes and tires (Fe and copper (Cu)), and are regarded as urban air pollutants that have negative effects on health [4,5].

Epidemiological and experimental studies have provided evidence for the adverse effects of PM exposure, particularly heavy metals, on respiratory diseases and allergic sensitization. For example, exposure to ambient PM has been identified as a high risk factor for increasing the incidence of atopic diseases and allergic sensitization in children [6]. Heavy traffic exposure, one source of heavy metals, has associated with increases in the incidence and prevalence of childhood asthma and wheeze as well as allergic sensitization, bronchial hyperresponsiveness and respiratory symptoms in children [7,8]. A previous study reported that the concentration of Fe in $PM_{10}$ was associated with allergic sensitization at the birth address as well as the current address and may also increase the risk of asthma and allergy in schoolchildren [9]. Increases in ambient Ni concentrations significantly increased wheeze symptoms in young children in New York City [10]. Higher levels of Fe and Cd have been detected in the exhaled breath condensates of chronic obstructive pulmonary disease patients than in those of non-smoking subjects [11]. However, this relationship is weak or inconsistent and the underlying mechanisms for the adverse effects of pollutants on human health currently remain unclear.

Chronic cough is the fifth most common condition encountered by primary care physicians [12]. In addition to eosinophilic bronchitis, gastro-esophageal reflux disease, postnasal drip syndrome, and rhinosinusitis, chronic obstructive pulmonary disease, pulmonary fibrosis, bronchiectasis, and bronchial asthma have been identified as some of the most common causes of chronic cough worldwide. Cough-variant asthma and atopic cough, two asthma-related conditions, also accompany chronic cough [13]. In Japan, coughing is the most common reason for patients (11.7% of all) to visit clinics [14], and cough-variant asthma as well as atopic cough have been identified as the main causes of isolated chronic non-productive cough [15]. Previous studies in the United Kingdom and Australia identified the most common cause of coughing as rhinitis, followed by asthma [16,17].

Immunoglobulin E (IgE) plays an important role in inflammation. It has interdependent and independent roles in the complex immune responses that manifest clinically as allergic disorders [18]. Inhaled environmental exposure and its influences on T-helper (Th) cell and serum IgE have been demonstrated [19,20]. Possible mechanism of the effects of heavy metals on allergic inflammation has been suggested in which heavy metals suppress Th1 development by inhibiting interferon-$\gamma$ (IFN-$\gamma$) expression and promote Th2 development by enhancing interleukin 4 expression, increase IgE as well as IgE-dependent basophil-mediated inflammation [21]. Several experimental studies on heavy metals have implicated Fe exposure in alterations in humoral and cell-mediated immunity and the development of allergic conditions, including the production of serum IgE [22,23], and Cd at low concentrations has been suggested to inhibit the synthesis of IgE by human B lymphocytes [24]. In contrast, an epidemiological study on an urban Canadian population of pregnant woman and their children found no relationship between metals and elevated serum IgE [25]. Furthermore, there is currently no evidence for average total serum IgE levels in patients with chronic respiratory diseases and allergic sensitization whose total serum IgE levels may be normal or high. Hence the effects of heavy metals on IgE-mediated allergic responses as well as the roles of IgE in airway inflammation have not yet been elucidated in detail.

Information provided by longitudinal studies on the relationship between heavy metal exposure and respiratory diseases is still limited. The involvement of IgE mechanisms in the symptoms of patients with chronic airway inflammation remains unclear. Therefore, we conducted this study to investigate the relationship between heavy metal concentrations and the appearance of symptoms including itchy eyes and coughing in chronic cough patients with different total serum IgE levels.

## 2. Experimental Section

### 2.1. Participant Recruitment

Ninety-nine patients diagnosed with at least one of the following: Bronchial asthma, atopic cough, and cough-variant asthma, were recruited at Kanazawa University Hospital between 4 January 2011 and 30 June 2011. All subjects gave their informed consent for inclusion before they participated in the study. The study was conducted in accordance with the principles of the Declaration of Helsinki of the World Medical Association, and the protocol was approved by the Medical Ethics Committee of Kanazawa University (Project Identification Code: 981). We used Japan Asthma Prevention and Management Guidelines 2011 to diagnose asthma [26]. The criteria of the Japanese Cough Research Society was used to diagnose cough-variant asthma [27]. Atopic cough was diagnosed using reported criteria, which are based on bronchodilator-resistant cough and the resolution of coughing with the use of histamine H1 antagonists and/or inhaled corticosteroids (ICS) [28]. Detailed recruiting methods have been published elsewhere [29]. We excluded five current smokers, three patients with an unknown smoking status, and one patient whose total serum IgE level was not obtained. Among the remaining 90 patients, 87 had a recorded frequency of coughing, while 76 had a recorded level of itchy eyes (Figure 1).

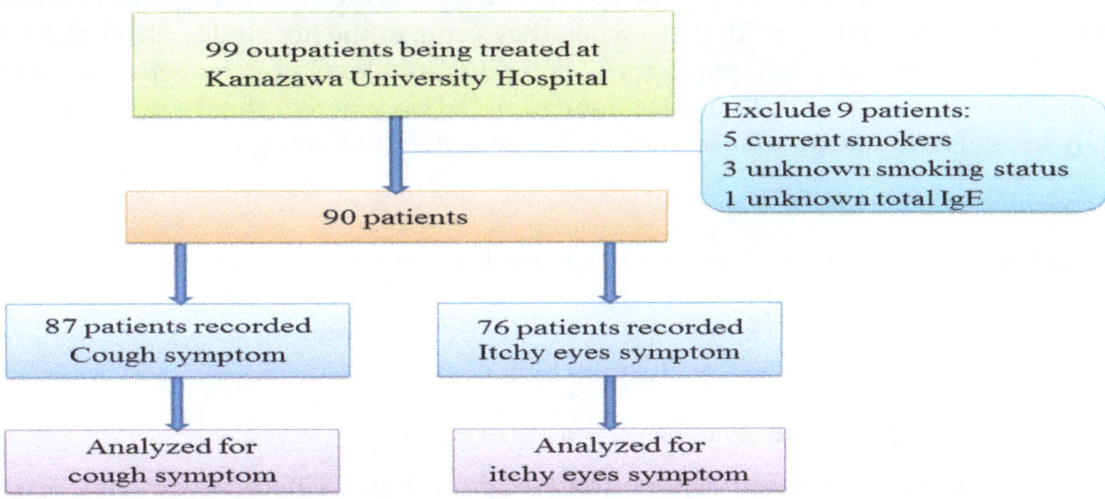


**Figure 1.** Flow chart showing description of patients in the study.

Among the 76 patients with recorded itchy eye symptoms, 59 had a total serum IgE level less than 250 IU/mL, while that of 17 patients was more than or equal to 250 IU/mL (Table 1). Forty-three, 18, and 29 patients were diagnosed with bronchial asthma, cough-variant asthma, and atopic cough, respectively. Among the 87 patients with recorded cough symptoms, 65 had a total IgE level less than 250 IU/mL, while that of 22 patients was more than or equal to 250 IU/mL. Fifty-four, 18, and 30 patients were diagnosed with bronchial asthma, cough-variant asthma, and atopic cough, respectively. Total serum IgE levels were significantly lower in atopic cough patients than in non-atopic cough patients. No significant differences were observed in gender, age, BMI, smoking status, or the onset or duration of disease between the groups of patients with total IgE ⩾ 250 IU/mL and < 250 IU/mL.

**Table 1.** Characteristics of the patients.

| Characteristics | Patient-Recorded Itchy Eyes Symptoms ($n$ = 76) | | Patient-Recorded Cough Symptoms ($n$ = 87) | |
|---|---|---|---|---|
| | **IgE < 250 IU/m ($n$ = 59)** | **IgE ⩾ 250 IU/mL ($n$ = 17)** | **IgE < 250 IU/mL ($n$ = 65)** | **IgE ⩾ 250 IU/Ml ($n$ = 22)** |
| Gender (Male/Female) | 18/41 | 13/4 | 20/45 | 8/14 |
| Smoking status | | | | |
| Non-Smoker | 44 | 13 | 47 | 14 |
| Ex-Smoker | 15 | 4 | 18 | 8 |
| Age (years) (Mean ± SD) | 63.8 ± 11.9 | 56.8 ± 15.4 | 64.8 ± 11.2 | 61.6 ± 16.3 |
| BMI (kg/m$^2$) (Mean ± SD) | 22.9 ± 3.8 | 23.2 ± 2.6 | 22.6 ± 3.6 | 22.8 ± 2.5 |
| Duration of disease (years) | 15.0 ± 8.8 | 18.1 ± 8.1 | 15.7 ± 10.8 | 18.8 ± 10.0 |
| Onset (age) | 49.0 ± 13.3 | 38.9 ± 18.8 | 64.3 ± 11.2 | 61.1 ± 16.2 |
| BA patients | 31 | 12 | 37 | 17 |
| CVA patients | 13 | 5 | 13 | 5 |
| AC patients [a,b] | 26 | 3 | 27 | 3 |

SD: standard deviation, BA: Bronchial asthma, CVA: Cough variant asthma, AC: atopic cough. [a] significantly different at 0.05 level between two groups of patients with recorded itchy eyes. [b] significantly different at 0.05 level between two groups of patients with recorded cough.

### 2.2. Health Outcomes

Patients recorded their symptoms in an allergy diary every morning, afternoon, evening, and night from the beginning to the last day of the study. Patients started recording the frequency of coughing from 4 January 2011, while the level of itchy eyes was started on 1 February 2011. We made every effort to avoid missing data due to patients forgetting to record their symptoms in their diaries by asking them to bring their diaries with them when they came to the hospital to check their symptoms or have consultations. The average frequency of coughing and level of itchy eyes was calculated for each patient. The presence of symptoms was defined based on whether the frequency of coughing or level of itchy eyes of each patient on each day was higher than the average.

### 2.3. Measurement of Total Serum IgE Levels

At the initiation of the study, total serum IgE levels were measured from a blood sample taken from the peripheral blood vessels of all patients using a radioimmunosorbent test. Patients were divided into two groups with total serum IgE < 250 IU/mL and total serum IgE ⩾ 250 IU/mL based on criteria used at Kanazawa University Hospital [30].

### 2.4. Measurement of Heavy Metals in Total Suspended Particulates

Hydrochloric acid and nitric acid were purchased from Kanto Chemical Co., Inc. (Tokyo, Japan) for atomic absorption spectrometry. A borosilicate glass fiber filter coated with fluorocarbon (T60A20, 8 × 10 in, Pallflex (Putnam, CT, USA)) was used. Total suspended particulates (TSP) were collected on the borosilicate glass fiber filter coated with fluorocarbon using a high-volume air sampler (120SL, Kimoto Electric Co., Ltd., Osaka, Japan) at a flow rate of 1000 L, located at Kanazawa University (136.7 degrees east longitude, 36.6 degrees north latitude). After 24 h of collecting ambient dust (start at 13:30 to 13:30 on the next day), the collected filter was removed and replaced with a blank filter. The collected filter was kept for 16 h–48 h in a desiccator in order to remove water and then stored at −30 °C before the extraction and analysis of heavy metals. The method of measurement used has been described previously [31]. A quarter of the filter sample was cut into small pieces and placed in a beaker. Fifteen milliliters of hydrochloric acid and 5 mL of nitric acid were added to the filter and heated on a hot plate for 2 h. The supernatant was filtered through filter paper. Five milliliters of hydrochloric acid and 1.5 mL of nitric acid were added to the remaining filter and heated on an electric hot plate for 15 min. The supernatant was filtered through filter paper, added to the filtrated solution, and then evaporated to dryness on the electric hot plate. After cooling, 9 mL of Milli-Q water and 1 mL of nitric acid were added and heated on the electric hot plate in order to dissolve metals for 15 min. After the dilution of metals, Cd, Co, Cr, Cu, Mn, Ni, and Pb were measured using a graphite atomic absorption spectrometer (model: AA-6800, Shimadzu Corporation, Kyoto, Japan), while Ca and Fe were measured using a flame atomic absorption spectrometer (model: AA-7000, Shimadzu Corporation, Kyoto, Japan). Metal data were not obtained between March 2nd and 16th due to the failure of the high-volume air sampler. Cu and Co were excluded from data for statistical analyses due to missing data for more than 20 days.

### *2.5. Statistical Analysis*

In order to assess the effects of heavy metals on the appearance of symptoms over the 6-month follow-up, we applied the generalized estimating equation (GEE) model to adjust the relationship between changes in the concentrations of heavy metals and symptoms in the same patient as well as to determine the effects of these metals on the symptoms of patients with different total serum IgE levels. We categorized the concentration of each metal into two levels (high and low levels) using the median of each one. Since patients recorded the frequency of coughing and level of itchy eyes at different periods, we used the median concentration of heavy metals in these periods. The odds ratios of high levels of each metal with 95% confidence intervals were calculated relative to the low levels. The characteristics of patients (BMI, age, and gender), meteorological factors (temperature, humidity, and precipitation), air pollutants (sulfur dioxide ($SO_2$)), and pollen count (cedar and cypress pollen) were used as potential confounders. Data were statistically analyzed using the SPSS software program for MS Windows, version 19.0 (SPSS, Inc., New York, NY, USA). The significance of differences was set at $p < 0.05$ for all analyses.

## 3. Results

### *3.1. Heavy Metal Concentrations*

During the study period, information on the concentrations of metals was not obtained for 15 days. The range of metal concentrations was very large. The metal with the highest concentration was Fe (mean 503.23 ng/m$^3$). The mean Fe concentration was more than 2-fold higher than that of Ca (mean 238.53 ng/m$^3$), and at least 13-fold higher than that of Mn (mean 37.54 ng/m$^3$) and the others. The average concentrations of metals from high to low were Fe > Ca > Mn > Pb > Ni > Cr > Cd (Table 2).

**Table 2.** Description of heavy metals from 4 January 2011 to 30 June 2011.

| Heavy Metals | Days of Observation | Mean | IQR | Minimum | Maximum |
|---|---|---|---|---|---|
| Ca (ng/m$^3$) | 163 | 238.53 ± 232.06 | 184.817 | 41.29 | 1777.49 |
| Cd (ng/m$^3$) | 161 | 0.26 ± 0.42 | 0.122 | 0.00 | 3.44 |
| Cr (ng/m$^3$) | 163 | 2.03 ± 2.30 | 1.279 | 0.00 | 17.54 |
| Fe (ng/m$^3$) | 163 | 503.23 ± 920.15 | 247.813 | 15.50 | 6926.79 |
| Mn (ng/m$^3$) | 163 | 37.54 ± 254.06 | 8.326 | 0.00 | 3211.78 |
| Ni (ng/m$^3$) | 162 | 3.31 ± 4.33 | 1.866 | 0.00 | 38.33 |
| Pb (ng/m$^3$) | 163 | 5.98 ± 6.25 | 3.956 | 0.00 | 37.33 |

IQR: interquartile range.

Correlations were observed between most of the metals examined, except for between Mn, Ca, and Cd (Supplementary Table S1). The strongest correlation was observed between Mn and Fe (0.628), whereas the weakest was between Ca and Cd (0.172).

Although the concentrations of heavy metals changed every day in a large range (Figure 2), except for Fe, the concentration of which was significantly high in May and Spring, no significant differences were observed between the concentrations of each metal due to the month or season.

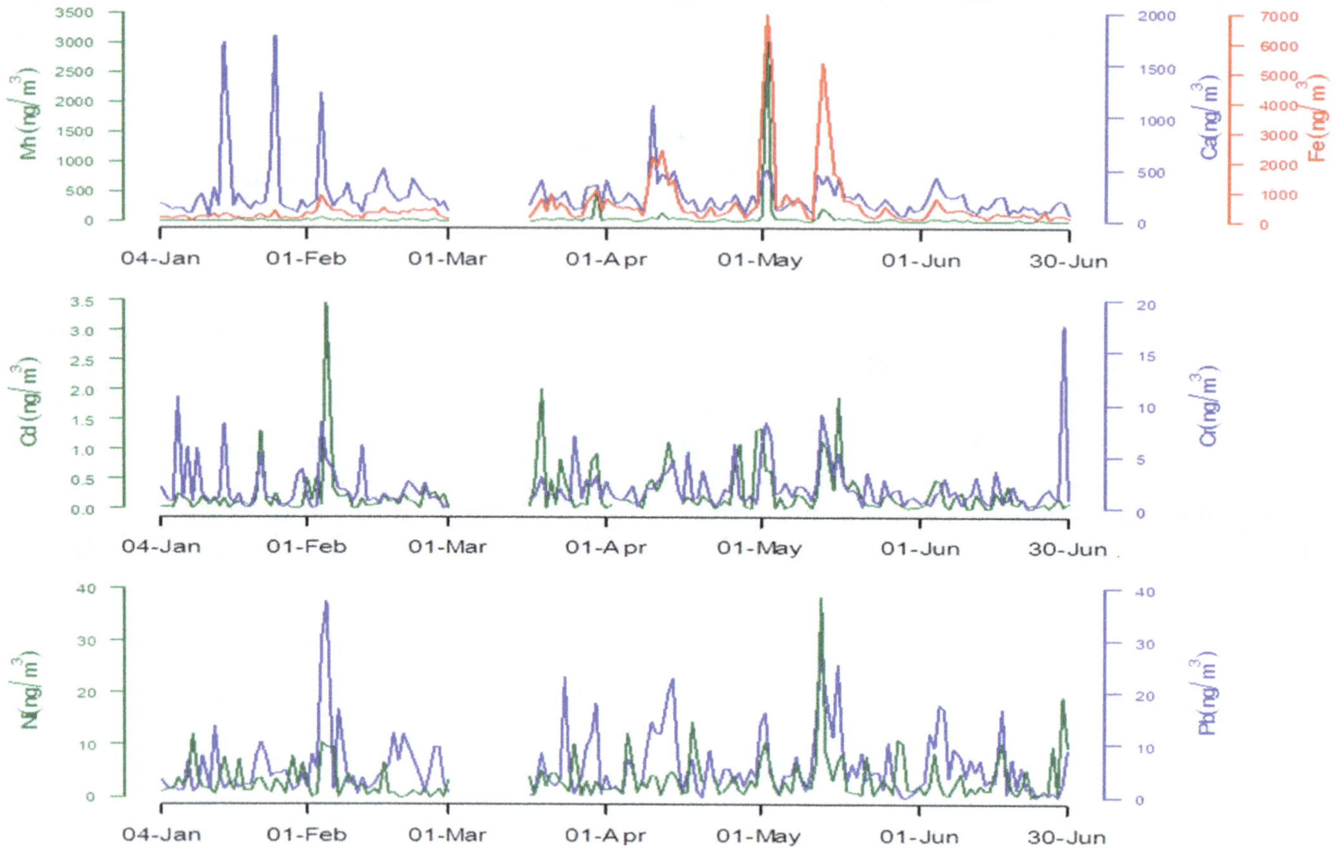


**Figure 2.** Daily concentrations of heavy metals between 4 January 2011 and 30 June 2011.

### *3.2. Relationship between Heavy Metals and Itchy Eyes and Coughing*

Prior to adjusting with potential confounders, no relationship was observed between Ni and itchy eyes. Six other metals at high levels correlated with the appearance of itchy eyes, with ORs for Fe, Mn, Mn, Ca, Cd, Cr, and Pb of 1.83 (95% CI: 1.40, 2.40), 1.77 (95% CI: 1.43, 2.18), 1.49 (95% CI: 1.20, 1.84), 1.34 (95% CI: 1.15, 1.57), 1.29 (95% CI: 1.11, 1.49), and 1.25 (95% CI: 1.05, 1.50), respectively. However, after using covariates (age, BMI, gender, temperature, humidity, precipitation, $SO_2$, cedar pollen, and cypress pollen) for adjustments, we found that all metals at high levels correlated with itchy eyes with ORs for Mn, Fe, Cd, Ni, Ca, Pb, and Cr of 1.60 (95% CI: 1.31, 1.95), 1.56 (95% CI: 1.20, 2.02), 1.27 (95% CI: 1.09, 1.47), 1.23 (95% CI: 1.02, 1.49), 1.22 (95% CI: 1.02, 1.45), 1.20 (95% CI: 1.01, 1.41), and 1.17 (95% CI: 1.03, 1.33), respectively (Table 3).

We also evaluated the relationship between heavy metal levels and the occurrence of coughing. Although all metals were positively related to coughing, four out of seven metals correlated with coughing: Fe, Mn, Cd, and Cr with ORs of 1.26 (95% CI: 1.07, 1.48), 1.17 (95% CI: 1.03, 1.33), 1.15 (95% CI: 1.04, 1.26), and 1.12 (95% CI: 1.01, 1.25), respectively. After adjustments for age, BMI, gender, temperature, humidity, precipitation, $SO_2$, cedar pollen, and cypress pollen, the relationship between Cr levels and coughing was not significant. However, correlations remained with Fe (OR (95% CI): 1.22 (1.05, 1.42)), Mn (OR (95% CI): 1.13, (1.01, 1.27)), and Cd (OR (95% CI): 1.13 (1.03, 1.24)) (Table 3).

**Table 3.** Adjusted odds ratio and 95% confidence intervals for itchy eyes and cough symptom associated with level of heavy metals' concentrations.

| Metals | Itchy Eyes Symtom | | | | Cough Symptom | | | |
|---|---|---|---|---|---|---|---|---|
| | OR (95% CI) [a] | p-Value [a] | OR (95% CI) [b] | p-Value [b] | OR (95% CI) [a] | p-Value [a] | OR (95% CI) [b] | p-Value [b] |
| Ca | 1.49 (1.20–1.84) | <0.001 | 1.22 (1.02–1.45) | 0.026 | 1.11 (0.99–1.25) | 0.084 | 1.03 (0.93–1.13) | 0.663 |
| Cd | 1.34 (1.15–1.57) | <0.001 | 1.27 (1.09–1.47) | 0.002 | 1.15 (1.04–1.26) | 0.005 | 1.13 (1.03–1.24) | 0.012 |
| Cr | 1.29 (1.11–1.49) | 0.001 | 1.17 (1.03–1.33) | 0.019 | 1.12 (1.01–1.25) | 0.027 | 1.10 (0.99–1.02) | 0.053 |
| Fe | 1.83 (1.40–2.40) | <0.001 | 1.56 (1.20–2.02) | 0.001 | 1.26 (1.07–1.48) | 0.007 | 1.22 (1.05–1.42) | 0.010 |
| Mn | 1.77 (1.43–2.18) | <0.001 | 1.60 (1.31–1.95) | <0.001 | 1.17 (1.03–1.33) | 0.015 | 1.13 (1.01–1.27) | 0.030 |
| Ni | 1.18 (0.99–1.41) | 0.068 | 1.23 (1.02–1.49) | 0.031 | 1.01 (0.92–1.11) | 0.886 | 1.02 (0.93–1.13) | 0.663 |
| Pb | 1.25 (1.05–1.50) | 0.015 | 1.20 (1.01–1.41) | 0.038 | 1.05 (0.97–1.14) | 0.227 | 1.04 (0.96v1.13) | 0.376 |

[a] Only heavy metal without adjustment. [b] With adjustment for age, BMI, gender, temperature, humidity, precipitation, $SO_2$, cedar pollen, cypress pollen.

### *3.3. Relationship between Heavy Metals and Itchy Eyes, Coughing, and Serum IgE Levels*

We determined whether symptoms similarly occurred in patients with different total serum IgE levels. Ca was the only metal that did not correlate with the appearance of itchy eyes in patients with total serum IgE levels less than 250 IU/mL. The relationship between heavy metal concentrations and itchy eyes were significant with ORs for Mn, Fe, Cd, Ni, Pb, Cr of 1.65 (95% CI: 1.32, 2.05), 1.61 (95% CI: 1.22, 2.12), 1.30 (95% CI: 1.12, 1.51), 1.23 (95% CI: 1.00, 1.52), 1.21 (95% CI: 1.03, 1.42), and 1.18 (95% CI: 1.03, 1.35), respectively (Table 4). In another analysis, two out of seven metals correlated with the appearance of coughing in patients with total serum IgE levels less than 250 IU/mL, including Mn and Cd with ORs of 1.15 (95% CI: 1.00, 1.31) and 1.14 (95% CI: 1.03, 1.27), respectively. Although Fe also positively correlated with coughing in patients with lower total serum IgE levels (an OR of 1.22 with a *p*-value of 0.027), this relationship did not account for the difference observed in the effects of Fe on coughing in patients with high total IgE levels (an OR of 1.27 and $p$-value > 0.05). No correlations were observed between heavy metal concentrations and symptoms in patients with total serum IgE $\geqslant$ 250 IU/mL (Table 4).

**Table 4.** Adjusted odds ratio and 95% confidence intervals for itchy eyes and cough symptom associated with level of heavy metals' concentration at groups of patients with different total serum IgE levels.

| Metals | IgE < 250 IU/mL | | | | IgE ⩾ 250 IU/mL | | | |
|---|---|---|---|---|---|---|---|---|
| | Itchy Eyes Symtom | | Cough Symptom | | Itchy Eyes Symtom | | Cough Symptom | |
| | OR (95% CI) | p-Value | OR (95% CI) | p-Value | OR (95% CI) | p-Value | OR (95% CI) | p-Value |
| Ca | 1.16 (0.96–1.40) | 0.118 | 1.06 (0.95–1.19) | 0.315 | 1.56 (0.99–2.47) | 0.056 | 0.91 (0.75–1.12) | 0.388 |
| Cd | 1.30 (1.12–1.51) | 0.001 | 1.14 (1.03–1.27) | 0.014 | 1.13 (0.72–1.79) | 0.599 | 1.08 (0.88–1.32) | 0.458 |
| Cr | 1.18 (1.03–1.35) | 0.015 | 1.08 (0.97–1.21) | 0.178 | 1.11 (0.73–1.69) | 0.620 | 1.16 (0.97–1.39) | 0.100 |
| Fe | 1.61 (1.22–2.12) | 0.001 | 1.22 (1.02–1.45) | 0.027 | 1.34 (0.73–2.48) | 0.347 | 1.27 (0.89–1.81) | 0.195 |
| Mn | 1.65 (1.32–2.05) | <0.001 | 1.15 (1.00–1.31) | 0.046 | 1.39 (0.92–2.10) | 0.122 | 1.11 (0.87–1.41) | 0.414 |
| Ni | 1.23 (1.00–1.52) | 0.048 | 1.03 (0.92–1.17) | 0.599 | 1.21 (0.70–2.07) | 0.494 | 1.00 (0.84–1.20) | 0.992 |
| Pb | 1.21 (1.03–1.42) | 0.019 | 1.07 (0.97–1.17) | 0.194 | 1.14 (0.64–2.05) | 0.657 | 0.97 (0.81–1.16) | 0.703 |

Adjusted odd ratio and 95% confidence intervals for age, BMI, gender, temperature, humidity, precipitation, $SO_2$, cedar pollen, cypress pollen.

## 4. Discussion

To the best of our knowledge, this is the first epidemiological study to provide evidence for a relationship between airborne heavy metals and coughing or itchy eyes in chronic cough patients with different total serum IgE levels. The results of the present study indicate that patients with lower IgE levels are more likely to be affected by heavy metals and also that the relationship between heavy metal exposure and airway inflammation may involve a non-IgE mediated mechanism.

Some environmental studies in large industrial cities in Asia such as Beijing, Guangzhou, Chengdu (China), and Seoul (Korea) reported markedly higher concentrations of PM and heavy metals than those of the present study [32–34]. Differences have also been noted in the relationships between metals as well as the order of concentrations reported. These studies showed that the metal with the highest concentration was Pb, an element evoking the greatest concern in terms of environmental heavy metal pollution [35,36], with at least a 2-fold higher concentration than that of Mn, which had the second highest concentration. A recent study showed the metal with the highest concentration was Fe, which has been associated with brake wear in the proximity of busy roads [37], and also that the concentration of Mn was more than 2-fold higher than that of Pb. In addition, a comparison with other studies from similar prefectures located along the coastline facing the Sea of Japan (Tottori) [38] and Pacific Ocean (Nagoya) [39] revealed that the concentrations of heavy metals in Kanazawa in this study followed the same order. These results suggest that the dominant source of metals in Kanazawa, an urban city without large industrial zones is traffic emission from direct tailpipe emission, brake and tire abrasion, and the resuspension of road way dust [40,41].

We clearly showed that all of the metals examined correlated with the presence of itchy eyes after adjusting with covariates. Furthermore, Cd, Cr, Fe, and Mn also correlated with coughing, and this relationship remained significant for Cd, Fe, and Mn after adjusting with possible confounders. A previous study on New York children for 24 months did not find a relationship between coughing and Fe in single models [10]. Other indirect evidence also demonstrated that a higher concentration of transition metals in $PM_{2.5}$ may be responsible for more severe inflammation [42], and short-term reductions in air pollution may improve lung function in children [43]. Children and adolescents with residential proximity to a major roadway more frequently had asthma-related symptoms accompanied by itchy-watery eyes than responders who resided 200–500 meters from a major road [44]. Although Fe was also the main coarse fraction of $PM_{10}$ in the Netherlands with the highest average concentration, a correlation was not observed after adjusting for asthma symptoms and rhinitis but only for allergic sensitization at birth [9]. Exposure to Mn in mine workers has been associated with the appearance of more respiratory symptoms as well as a higher prevalence of impaired lung function [45]. Mn has been suggested as the cause of the high frequency of work-related symptoms such as ocular symptoms and dry cough in welder workers [46]. Ocular surface symptoms have been reported in taxi drivers and traffic controllers due to exposure to air pollution [47]. Our results clearly showed correlations between all heavy metals at high concentrations and itchy eyes, one of the common symptoms of a clinical diagnosis of allergic rhinitis [48], and also between Cd, Fe, and Mn and coughing in multiple-pollutant models. Recent studies have indicated that a relationship exists between asthma as well as chronic respiratory diseases with allergic rhinitis and the recommended reasons of prevalence, links with asthma along with the impact of quality of life and work [49]. The deterioration of upper or lower respiratory, ocular, or skin symptoms has been reported in 33% of asthma patients on Asian Dust day [50]. These findings suggest the presence of a strong relationship between asthma and allergic rhinitis, as supported by our results on metal exposure and the appearance of itchy eyes in patients with respiratory diseases. In the present study, we adjusted for coexisting factors related to allergies such as pollen including Japanese cedar pollen and Japanese cypress pollen. Japanese cedar pollinosis is the most common pollinosis in Japan [51,52], and the prevalence of allergic rhinitis, particularly cedar pollinosis, is increasing [53]. Furthermore, a significantly higher number of patients exhibited itchy eyes and coughing during Asian dust days with pollen [29]. The findings of another study suggest that high levels of anthropogenic air pollution including Fe and Pb are transported with Asian dust [54]. However, it currently remains unclear whether the aggravation of symptoms depends on pollen, metals, or Asian dust. Therefore, we used pollen as a potential confounder in order to better adjust the relationship between heavy metals and symptoms in chronic cough patients.

We clearly demonstrated that correlations existed between six out of the seven heavy metals (except for Ca) at high concentrations and the presence of itchy eyes in patients with total serum IgE levels less than 250 IU/mL. Furthermore, Cd and Mn positively correlated with coughing in patients with total serum IgE levels less than 250 IU/mL. The findings of a current study also showed that none of the metals tested correlated with the appearance of itchy eyes and coughing in patients with total serum IgE levels more than or equal to 250 IU/mL. Several possible mechanisms have been proposed for IgE-dependent allergic inflammation pathology [55,56]. A strong correlation was previously reported between serum IgE levels and asthma and its phenotype [57], suggesting the development of airway hyperresponsiveness and inflammation through the induction of Th2-type cytokine production by IgE in mice [58]. Previous studies have suggested that heavy metals exposure such as lead or cadmium, alters immune system components and is associated with increased production of IgE [59–61] However, other studies have suggested effector mechanisms that are independent of IgE may also contribute to the pathology of allergic inflammation. In chronic asthma mouse model, mast cells can substantially influence features of chronic allergic inflammation and tissue remodeling, independently of mast cell signaling through either IgE-Fc$\varepsilon$RI (high-affinity Fc receptor for IgE) or antigen-IgG-Fc$\gamma$RIII [62]. Different mast cell population show different patterns of expression of receptors for pathogen-associated molecular patterns and activation of mast cells can induce these cells to secrete distinct patterns of cytokines or chemokines [63–66]. Thus mast cells have potential to drive important features of allergic inflammation independently of IgE. In human, evidence for non-IgE-mediated and irritant-induced work-related rhinitis has recently been increasing [67–69]; however, the mechanisms responsible remain largely unclear [69]. Furthermore, Ni has been suggested to act as an allergen that induces an adaptive immune response related to inflammation independently of IgE [55]. Moreover, a previous study on occupational asthma and occupational rhinitis patients caused by new low-molecular-weight molecules including chemicals and metals demonstrated that the prevalence of patients with IgE-mediated mechanisms was markedly lower than the total of patients [70]. These findings suggest the role of non-IgE mediated or other mechanisms in the influence of these factors on asthma and rhinitis patients. In addition, non-IgE-mediated related rhinitis and asthma have recently been described in several case reports, indirect epidemiological studies, and surveillance and experimental research [69]. No epidemiological study has directly shown the effects of total serum IgE levels on the relationship between coughing or itchy eyes and heavy metal concentrations in the air, and the mechanisms underlying IgE-induced airway inflammation still need to be elucidated.

The present longitudinal cohort study with a detailed heavy metal concentration exposure assessment for 6 continuous months with patients recording their symptoms in a diary during the study period in addition to the adjustment of the effects of metals on each patient with possible confounders including patient characteristics (age, gender, and BMI), weather (temperature, humidity, and precipitation), air pollution ($SO_2$), and pollen (cedar pollen and cypress pollen) are the important strengths of our study. However, there were some limitations. First, we only obtained data on metal concentrations in TSP at Kanazawa University-Takaramachi Campus; therefore it is possible that data at the observation did not always reflective the concentration in the study area and individual exposures. Second, the findings for metals may be applicable to traffic-related constituents especially largely relate to non-exhaust emission of traffic [9]; however, it is not clear to what extent other sources could be in case metals in the air in some Japan coastal prefectures increase by others events (Asian Dust days) [38,39,54,71]. We only measured the total serum IgE levels of 90 patients at the initiation of the study without re-checking after six months to determine whether changes occurred in order to confirm the relationship between heavy metal concentrations and symptoms related to total IgE levels. Further studies with a larger number of patients whose total serum IgE levels are known are needed in order to more clearly determine the relationship between itchy eyes and coughing in patients with chronic cough and heavy metal concentrations.

## 5. Conclusions

The results of the present study provide evidence for the effects of aerial heavy metals on itchy eyes and coughing in patients with chronic cough, particularly due to different total serum IgE levels. The present results suggest correlations between not only the non-exhaust emissions of traffic and general health, but also asthma and rhinitis. The relationship observed between heavy metal concentrations and the symptoms of our patients with total serum IgE levels less than 250 IU/mL is an important result and indicates the relevance of the impact of heavy metals on airway inflammation through a non-IgE-mediated mechanism.

**Acknowledgments:** We thank all the patients for their cooperation as well as all the field workers and laboratory personnel involved for their efforts. This study was supported by JSPS KAKENHI Grant Numbers 25460166 and 26340048.

**Author Contributions:** Thao Thi Thu Nguyen performed epidemiologic study and statistical analysis and drafted the manuscript. Hiroyuki Nakamura designed the study. Enoch Olando Anyenda, Johsuke Hara, Masaki Fujimura, Hiromasa Tsujiguchi, Masami Kitaoka, Hiroki Asakura, Daisuke Hori, Yuri Hibino and Tadashi Konoshita helped to contributed to the data interpretation and manuscript writing. Yasuhiro Kambayashi was responsibility for measurement of heavy metals concentration. Tomomi Higashi obtained the patients data and helped drafting the manuscript. Thao Thi Thu Nguyen, Yasuhiro Kambayashi and Hiroyuki Nakamura have primary responsibility for final content. All authors read and approved the final manuscript.

## Abbreviations

| | |
|---|---|
| IgE | Immunoglobulin E |
| TSP | total suspended particles |
| PM | particulate matter |
| Ca | calcium |
| Cd | cadmium |
| Co | cobalt |
| Cu | copper |
| Cr | chromium |
| Fe | iron |
| Mn | manganese |
| Ni | nickel |
| Pb | lead |
| BMI | body mass index |
| $SO_2$ | sulfur dioxide |
| OR | Odds ratio |
| CI | Confidence interval |
| SD | Standard deviation |
| SPSS | Statistical Package for Social Sciences |
| GEE | generalized estimating equation |

## References

1. Bae, S.; Pan, X.-C.; Kim, S.-Y.; Park, K.; Kim, Y.-H.; Kim, H.; Hong, Y.-C. Exposures to particulate matter and polycyclic aromatic hydrocarbons and oxidative stress in schoolchildren. *Environ. Health Perspect.* **2010**, *118*, 579–583. [CrossRef] [PubMed]
2. Wei, Y.; Han, I.-K.; Shao, M.; Hu, M.; Zhang, J.; Tang, X. $PM_{2.5}$ constituents and oxidative DNA damage in humans. *Environ. Sci. Technol.* **2009**, *43*, 4757–4762. [CrossRef] [PubMed]
3. Wang, T.; Feng, W.; Kuang, D.; Deng, Q.; Zhang, W.; Wang, S.; He, M.; Zhang, X.; Wu, T.; Guo, H. The effects of heavy metals and their interactions with polycyclic aromatic hydrocarbons on the oxidative stress among coke-oven workers. *Environ. Res.* **2015**, *140*, 405–413. [CrossRef] [PubMed]
4. Eeftens, M.; Beelen, R.; De Hoogh, K.; Bellander, T.; Cesaroni, G.; Cirach, M.; Declercq, C.; Dėdelė, A.; Dons, E.; de-Nazelle, A.; *et al.* Development of land use regression models for $PM_{2.5}$, $PM_{2.5}$ absorbance, $PM_{10}$ and PM coarse in 20 European study areas; Results of the ESCAPE project. *Environ. Sci. Technol.* **2012**, *46*, 11195–11205. [CrossRef] [PubMed]

5. De Hoogh, K.; Wang, M.; Adam, M.; Badaloni, C.; Beelen, R.; Birk, M.; Cesaroni, G.; Cirach, M.; Declercq, C.; Dėdelė, A.; *et al.* Development of land use regression models for particle composition in twenty study areas in Europe. *Environ. Sci. Technol.* **2013**, *47*, 5778–5786. [CrossRef] [PubMed]
6. Morgenstern, V.; Zutavern, A.; Cyrys, J.; Brockow, I.; Koletzko, S.; Krämer, U.; Behrendt, H.; Herbarth, O.; von Berg, A.; Bauer, C.P.; *et al.* Atopic diseases, allergic sensitization, and exposure to traffic-related air pollution in children. *Am. J. Respir. Crit. Care. Med.* **2008**, *177*, 1331–1337. [CrossRef] [PubMed]
7. Gasana, J.; Dillikar, D.; Mendy, A.; Forno, E.; Ramos Vieira, E. Motor vehicle air pollution and asthma in children: A meta-analysis. *Environ. Res.* **2012**, *117*, 36–45. [CrossRef] [PubMed]
8. Janssen, N.A.H.; Brunekreef, B.; Van Vliet, P.; Aarts, F.; Meliefste, K.; Harssema, H.; Fischer, P. The relationship between air pollution from heavy traffic and allergic sensitization, bronchial hyperresponsiveness, and respiratory symptoms in Dutch schoolchildren. *Environ. Health Perspect.* **2003**, *111*, 1512–1518. [CrossRef] [PubMed]
9. Gehring, U.; Beelen, R.; Eeftens, M.; Hoek, G.; De Hoogh, K.; De Jongste, J.C.; Keuken, M.; Koppelman, G.H.; Meliefste, K.; Oldenwening, M.; *et al.* Particulate matter composition and respiratory health: The PIAMA Birth Cohort study. *Epidemiology* **2015**, *26*, 300–309. [CrossRef] [PubMed]
10. Patel, M.M.; Hoepner, L.; Garfinkel, R.; Chillrud, S.; Reyes, A.; Quinn, J.W.; Perera, F.; Miller, R.L. Ambient metals, elemental carbon, and wheeze and cough in New York city children through 24 months of age. *Am. J. Respir. Crit. Care. Med.* **2009**, *180*, 1107–1113. [CrossRef] [PubMed]
11. Mutti, A.; Corradi, M.; Goldoni, M.; Vettori, M.V.; Bernard, A.; Apostoli, P. Exhaled metallic elements and serum pneumoproteins in asymptomatic smokers and patients with COPD or asthma. *Chest* **2006**, *129*, 1288–1297. [CrossRef] [PubMed]
12. Braman, S.S.; Corrao, W.M. Chronic cough: Diagnosis and treatment. *Prim. Care* **1985**, *12*, 217–225. [PubMed]
13. Chung, K.F.; Pavord, I.D. Prevalence, pathogenesis, and causes of chronic cough. *Lancet* **2008**, *371*, 1364–1374. [CrossRef]
14. Yamada, T.; Yoshimura, M.; Nago, N.; Asai, Y.; Koga, Y.; Inoue, Y.; Hamasaki, K.; Mise, J.; Lamberts, H.; Okkes, I.M. What is the common diseases and common health problems?—The use of ICPC in the community-based project. *JpN J. Prim. Care* **2000**, *23*, 80–89.
15. Fujimura, M. Cough variant asthma and related disorders: Atopic cough. *Arerugi no Rinsho* **1996**, *16*, 22–27.
16. Fujimura, M. Eosinophilic bronchitis is an important cause of chronic cough. *Am. J. Respir Crit. Care Med.* **2000**, *161*, 1764–1765. [PubMed]
17. Carney, I.K.; Gibson, P.G.; Murree-Allen, K.; Saltos, N.; Olson, L.G.; Hensley, M.J. A systematic evaluation of mechanisms in chronic cough. *Am. J. Respir Crit. Care Med.* **1997**, *156*, 211–216. [CrossRef] [PubMed]
18. Galli, S.J.; Tsai, M. IgE and mast cells in allergic disease. *Nat. Med.* **2012**, *18*, 693–704. [CrossRef] [PubMed]
19. Liu, J.; Ballaney, M.; Al-Alem, U.; Quan, C.; Jin, X.; Perera, F.; Chen, L.C.; Miller, R.I. Combined inhaled diesel exhaust particles and allergen exposure alter methylation of T helper genes and IgE production *in vivo*. *Toxicol. Sci.* **2008**, *102*, 76–81. [CrossRef] [PubMed]
20. Herr, C.E.W.; Ghosh, R.; Dostal, M.; Skokanova, V.; Ashwood, P.; Lipsett, M.; Joad, J.P.; Pinkerton, K.E.; Yap, P.-S.; Frost, J.D.; *et al.* Exposure to air pollution in critical prenatal time windows and IgE levels in newborns. *Pediatr. Allergy. Immunol.* **2011**, *22*, 75–84. [CrossRef] [PubMed]
21. Yang, S.-N.; Hsieh, C.-C.; Kuo, H.-F.; Lee, M.-S.; Huang, M.-Y.; Kuo, C.-H.; Hung, C.-H. The effects of environmental toxins on allergic inflammation. *Allergy Asthma Immunol. Res.* **2014**, *6*, 478–484. [CrossRef] [PubMed]
22. Tong, S.; von Schirnding, Y.E.; Prapamontol, T. Environmental lead exposure: A public health problem of global dimensions. *Bull. World Health Organ.* **2000**, *78*, 1068–1077. [PubMed]
23. Mishra, K.P. Lead exposure and its impact on immune system: A review. *Toxicol. Vitr.* **2009**, *23*, 969–972. [CrossRef] [PubMed]
24. Marth, E.; Jelovcan, S.; Kleinhappl, B.; Gutschi, A.; Barth, S. The effect of heavy metals on the immune system at low concentrations. *Int. J. Occup. Med. Environ. Health* **2001**, *14*, 375–386. [PubMed]
25. Ashley-martin, J.; Levy, A.R.; Arbuckle, T.E.; Platt, R.W.; Marshall, J.S.; Dodds, L. Maternal exposure to metals and persistent pollutants and cord blood immune system biomarkers. *Environ. Health* **2015**. [CrossRef] [PubMed]
26. Ohta, K.; Yamaguchi, M.; Akiyama, K.; Adachi, M.; Ichinose, M.; Takahashi, K.; Nishimuta, T.; Morikawa, A.; Nishima, S. Japanese guideline for adult asthma. *Allergol. Int.* **2011**, *60*, 115–145. [CrossRef] [PubMed]

27. Kohno, S.; Ishida, T.; Uchida, Y.; Kishimoto, H.; Sasaki, H.; Shioya, T.; Tokuyama, K.; Niimi, A.; Nishi, K.; Fujimura, M.; *et al.* The Japanese Respiratory Society guidelines for management of cough. *Respirology* **2006**, *11*, S135–S186. [PubMed]
28. Fujimura, M.; Sakamoto, S.; Kamio, Y.; Matsuda, T. Effects of methacholine induced bronchoconstriction and procaterol induced bronchodilation on cough receptor sensitivity to inhaled capsaicin and tartaric acid. *Thorax* **1992**, *47*, 441–445. [CrossRef] [PubMed]
29. Higashi, T.; Kambayashi, Y.; Ohkura, N.; Fujimura, M.; Nakanishi, S.; Yoshizaki, T.; Saijoh, K.; Hayakawa, K.; Kobayashi, F.; Michigami, Y.; *et al.* Exacerbation of daily cough and allergic symptoms in adult patients with chronic cough by Asian dust: A hospital-based study in Kanazawa. *Atmos. Environ.* **2014**, *97*, 537–543. [CrossRef]
30. Ikeda, Y.; Makino, S. Measurement of total IgE in sera from normal subjects and allergic patients by chemiluminescent immunoassay. *Arerugi* **1994**, *43*, 134–141. [PubMed]
31. Higashi, T.; Kambayashi, Y.; Ohkura, N.; Fujimura, M.; Nakai, S.; Honda, Y.; Saijoh, K.; Hayakawa, K.; Kobayashi, F.; Michigami, Y.; *et al.* Effects of Asian dust on daily cough occurrence in patients with chronic cough: A panel study. *Atmos. Environ.* **2014**, *92*, 506–513. [CrossRef]
32. Yang, F.; Tan, J.; Zhao, Q.; Du, Z.; He, K.; Ma, Y.; Duan, F.; Chen, G. Characteristics of $PM_{2.5}$ speciation in representative megacities and across China. *Atmos. Chem. Phys.* **2011**, *11*, 5207–5219. [CrossRef]
33. Li, Y.; Zhang, Z.; Liu, H.; Zhou, H.; Fan, Z.; Lin, M.; Wu, D.; Xia, B. Characteristics, sources and health risk assessment of toxic heavy metals in $PM_{2.5}$ at a megacity of southwest China. *Environ. Geochem. Health* **2015**. [CrossRef] [PubMed]
34. Park, E.; Kim, D.; Park, K. Monitoring of ambient particles and heavy metals in a residential area of Seoul, Korea. *Environ. Monit. Assess.* **2008**, *137*, 441–449. [CrossRef] [PubMed]
35. Laraqui, C.H.; Caubet, A.; Laraqui, O.; Rahhali, A.E.; Curtes, J.P.; Verger, C. Health risks study in a pottery environment in Morocco. *Sante* **2000**, *10*, 249–254. [PubMed]
36. Wieloch, M.; Kamiński, P.; Ossowska, A.; Koim-Puchowska, B.; Stuczyński, T.; Kuligowska-Prusińska, M.; Dymek, G.; Mańkowska, A.; Odrowąż-Sypniewska, G. Do toxic heavy metals affect antioxidant defense mechanisms in humans? *Ecotoxicol. Environ. Safety* **2012**, *78*, 195–205. [CrossRef] [PubMed]
37. Majestic, B.J.; Anbar, A.D.; Herckes, P. Elemental and iron isotopic composition of aerosols collected in a parking structure. *Sci. Total Environ.* **2009**, *407*, 5104–5109. [CrossRef] [PubMed]
38. Watanabe, M.; Yamasaki, A.; Burioka, N.; Kurai, J.; Yoneda, K.; Yoshida, A.; Igishi, T.; Fukuoka, Y.; Nakamoto, M.; Takeuchi, H.; *et al.* Correlation between Asian dust storms and worsening asthma in western Japan. *Allergol. Int.* **2011**, *60*, 267–275. [CrossRef] [PubMed]
39. Okuda, T.; Fujimori, E.; Hatoya, K.; Takada, H.; Kumata, H.; Nakajima, F.; Hatakeyama, S.; Uchida, M.; Tanaka, S.; He, K.; *et al.* Rapid and simple determination of multi-elements in aerosol samples collected on quartz fiber filters by using EDXRF coupled with fundamental parameter quantification technique. *Aerosol. Air Qual. Res.* **2013**, *13*, 1864–1876. [CrossRef]
40. Lough, G.C.; Schauer, J.J.; Park, J.-S.; Shafer, M.M.; DeMinter, J.T.; Weinstein, J.P. Emissions of metals associated with motor vehicle roadways. *Environ. Sci. Technol* **2005**, *39*, 826–836. [CrossRef] [PubMed]
41. Salma, I.; Maenhaut, W. Changes in elemental composition and mass of atmospheric aerosol pollution between 1996 and 2002 in a Central European city. *Environ. Pollut.* **2006**, *143*, 479–488. [CrossRef] [PubMed]
42. Schaumann, F.; Borm, P.J.A.; Herbrich, A.; Knoch, J.; Pitz, M.; Schins, R.P.F.; Luettig, B.; Hohlfeld, J.M.; Heinrich, J.; Krug, N. Metal-rich ambient particles (particulate matter 2.5) cause airway inflammation in healthy subjects. *Am. J. Respir. Crit. Care Med.* **2004**, *170*, 898–903. [CrossRef] [PubMed]
43. Frye, C.; Hoelscher, B.; Cyrys, J.; Wjst, M.; Wichmann, H.-E.; Heinrich, J. Association of lung function with declining ambient air pollution. *Environ. Health Perspect.* **2003**, *111*, 383–387. [CrossRef] [PubMed]
44. Porebski, G.; Woźniak, M.; Czarnobilska, E. Residential proximity to major roadways is associated with increased prevalence of allergic respiratory symptoms in children. *Ann. Agric. Environ. Med.* **2014**, *21*, 760–766. [CrossRef] [PubMed]
45. Boojar, M.M.A.; Goodarzi, F. A longitudinal follow-up of pulmonary function and respiratory symptoms in workers exposed to manganese. *J. Occup. Environ. Med.* **2002**, *44*, 282–290. [CrossRef] [PubMed]
46. Hedmer, M.; Karlsson, J.-E.; Andersson, U.; Jacobsson, H.; Nielsen, J.; Tinnerberg, H. Exposure to respirable dust and manganese and prevalence of airways symptoms, among Swedish mild steel welders in the manufacturing industry. *Int. Arch. Occup. Environ. Health* **2014**, *87*, 623–634. [CrossRef] [PubMed]

47. Torricelli, A.A.M.; Novaes, P.; Matsuda, M.; Braga, A.; Saldiva, P.H.N.; Alves, M.R.; Monteiro, M.L.R. Correlation between signs and symptoms of ocular surface dysfunction and tear osmolarity with ambient levels of air pollution in a large metropolitan area. *Cornea* **2013**, *32*, 11–15. [CrossRef] [PubMed]
48. Seidman, M.D.; Gurgel, R.K.; Lin, S.Y.; Schwartz, S.R.; Baroody, F.M.; Bonner, J.R.S.; Dawson, D.E.; Dykewicz, M.S.; Hackell, J.M.; Han, J.K.; *et al.* Clinical practice guideline: Allergic rhinitis. *Otolaryngol. Head Neck Surg.* **2015**, *152*, 1–43. [CrossRef] [PubMed]
49. Bousquet, J.; Khaltaev, N.; Cruz, A.A.; Denburg, J.; Fokkens, W.J.; Togias, A.; Zuberbier, T.; Canonica, G.W.; Van Weel, C.; Agache, I.; *et al.* Allergic rhinitis and its impact on asthma (ARIA) 2008. *Allergy* **2008**, *63*, 8–160. [CrossRef] [PubMed]
50. Watanabe, M.; Igishi, T.; Burioka, N.; Yamasaki, A.; Kurai, J.; Takeuchi, H.; Sako, T.; Yoshida, A.; Yoneda, K.; Fukuoka, Y.; *et al.* Pollen augments the influence of desert dust on symptoms of adult asthma patients. *Allergol. Int.* **2011**, *60*, 517–524. [CrossRef] [PubMed]
51. Nakamura, H.; Miyagawa, K.; Ogino, K.; Endo, T.; Imai, T.; Ozasa, K.; Motohashi, Y.; Matsuzaki, I.; Sasahara, S.; Hatta, K.; *et al.* High contribution contrast between the genes of eosinophil peroxidase and IL-4 receptor alpha-chain in Japanese cedar pollinosis. *J. Allergy Clin. Immunol.* **2003**, *112*, 1127–1131. [CrossRef] [PubMed]
52. Saito, Y. Japanese cedar pollinosis: Discovery, nomenclature, and epidemiological trends. *Proc. Jpn. Acad. Ser. B Phys. Biol. Sci.* **2014**, *90*, 203–210. [CrossRef] [PubMed]
53. Okamoto, Y.; Horiguchi, S.; Yamamoto, H.; Yonekura, S.; Hanazawa, T. Present situation of cedar pollinosis in Japan and its immune responses. *Allergol. Int.* **2009**, *58*, 155–162. [CrossRef] [PubMed]
54. Coulibaly, S.; Minami, H.; Abe, M.; Hasei, T.; Sera, N.; Yamamoto, S.; Funasaka, K.; Asakawa, D.; Watanabe, M.; Honda, N.; *et al.* Seasonal fluctuations in air pollution in Dazaifu, Japan, and effect of long-range transport from mainland East Asia. *Biol. Pharm. Bull.* **2015**, *38*, 1395–1403. [CrossRef] [PubMed]
55. Galli, S.J.; Tsai, M.; Piliponsky, A.M. The development of allergic inflammation. *Nature* **2008**, *454*, 445–454. [CrossRef] [PubMed]
56. Gould, H.J.; Sutton, B.J. IgE in allergy and asthma today. *Nat. Rev. Immunol.* **2008**, *8*, 205–217. [CrossRef] [PubMed]
57. Platts-Mills, T.A. The role of immunoglobulin E in allergy and asthma. *Am. J. Respir. Crit. Care Med.* **2001**, *164*, 1–5. [CrossRef] [PubMed]
58. Mizutani, N.; Goshima, H.; Nabe, T.; Yoshino, S. Complement C3a-induced IL-17 plays a critical role in an IgE-mediated late-phase asthmatic response and airway hyperresponsiveness via neutrophilic inflammation in mice. *J. Immunol.* **2012**, *188*, 5694–5705. [CrossRef] [PubMed]
59. Heo, Y.; Parsons, P.J.; Lawrence, D.A. Lead differentially modifies cytokine production *in vitro* and *in vivo*. *Toxicol. Appl. Pharmacol.* **1996**, *138*, 149–157. [CrossRef] [PubMed]
60. Lutz, P.M.; Wilson, T.J.; Ireland, J.; Jones, A.L.; Gorman, J.S.; Gale, N.L.; Johnson, J.C.; Hewett, J.E. Elevated immunoglobulin E (IgE) levels in children with exposure to environmental lead. *Toxicology* **1999**, *134*, 63–78. [CrossRef]
61. Sarasua, S.M.; Vogt, R.F.; Henderson, L.O.; Jones, P.A.; Lybarger, J.A. Serum immunoglobulins and lymphocyte subset distributions in children and adults living in communities assessed for lead and cadmium exposure. *J. Toxicol. Environ. Health* **2000**, *60*, 1–15.
62. Yu, M.; Tsai, M.; Tam, S.-Y.; Jones, C.; Zehnder, J.; Galli, S.J. Mast cells can promote the development of multiple features of chronic asthma in mice. *J. Clin. Investig.* **2006**, *116*, 1633–1641. [CrossRef] [PubMed]
63. Galli, S.J.; Tsai, M. Mast cells in allergy and infection: Versatile effector and regulatory cells in innate and adaptive immunity. *Eur. J. Immunol.* **2010**, *40*, 1843–1851. [CrossRef] [PubMed]
64. Gilfillan, A.M.; Austin, S.J.; Metcalfe, D.D. Mast cell biology: Introduction and overview. *Adv. Exp. Med. Biol.* **2011**, *716*, 2–12. [PubMed]
65. Hofmann, A.M.; Abraham, S.N. New roles for mast cells in pathogen defense and allergic disease. *Discov. Med.* **2010**, *9*, 79–83. [PubMed]
66. Moon, T.C.; St Laurent, C.D.; Morris, K.E.; Marcet, C.; Yoshimura, T.; Sekar, Y.; Befus, A.D. Advances in mast cell biology: New understanding of heterogeneity and function. *Mucosal. Immunol.* **2010**, *3*, 111–128. [CrossRef] [PubMed]
67. Pala, G.; Pignatti, P.; Moscato, G. The use of fractional exhaled nitric oxide in investigation of work-related cough in a hairdresser. *Am. J. Ind. Med.* **2011**, *54*, 565–568. [CrossRef] [PubMed]

68. Díaz Angulo, S.; Szram, J.; Welch, J.; Cannon, J.; Cullinan, P. Occupational asthma in antibiotic manufacturing workers: Case reports and systematic review. *J. Allergy* **2011**. [CrossRef] [PubMed]
69. Siracusa, A.; Folletti, I.; Moscato, G. Non-IgE-mediated and irritant-induced work-related rhinitis. *Curr. Opin. Allergy Clin. Immunol.* **2013**, *13*, 159–166. [CrossRef] [PubMed]
70. Pralong, J.A.; Cartier, A.; Vandenplas, O.; Labrecque, M. Occupational asthma: New low-molecular-weight causal agents, 2000–2010. *J. Allergy* **2012**. [CrossRef] [PubMed]
71. Watanabe, M.; Kurai, J.; Igishi, T.; Yamasaki, A.; Burioka, N.; Takeuchi, H.; Sako, T.; Touge, H.; Nakamoto, M.; Hasegawa, Y.; *et al.* Influence of Asian desert dust on lower respiratory tract symptoms in patients with asthma over 4 years. *Yonago Acta Med.* **2012**, *55*, 41–48. [PubMed]

# Cadmium Chloride Induces DNA Damage and Apoptosis of Human Liver Carcinoma Cells via Oxidative Stress

**Anthony Skipper [1], Jennifer N. Sims [2], Clement G. Yedjou [1] and Paul B. Tchounwou [1,*]**

Academic Editor: William Toscano

[1] Molecular Toxicology Research Laboratory, NIH-Center for Environmental Health, College of Science, Engineering and Technology, Jackson State University, 1400 Lynch Street, Box 18540, Jackson, MS 39217, USA; anthony.c.skipper@gmail.com (A.S.); clement.yedjou@jsums.edu (C.G.Y.)

[2] Department of Pathology, Beth Israel Deaconess Medical Center, Harvard Medical School, 330 Bookline Avenue, Boston, MA 02215, USA; jnsims@bidmc.harvard.edu

* Correspondence: paul.b.tchounwou@jsums.edu.

**Abstract:** Cadmium is a heavy metal that has been shown to cause its toxicity in humans and animals. Many documented studies have shown that cadmium produces various genotoxic effects such as DNA damage and chromosomal aberrations. Ailments such as bone disease, renal damage, and several forms of cancer are attributed to overexposure to cadmium. Although there have been numerous studies examining the effects of cadmium in animal models and a few case studies involving communities where cadmium contamination has occurred, its molecular mechanisms of action are not fully elucidated. In this research, we hypothesized that oxidative stress plays a key role in cadmium chloride-induced toxicity, DNA damage, and apoptosis of human liver carcinoma ($HepG_2$) cells. To test our hypothesis, cell viability was determined by MTT assay. Lipid hydroperoxide content stress was estimated by lipid peroxidation assay. Genotoxic damage was tested by the means of alkaline single cell gel electrophoresis (Comet) assay. Cell apoptosis was measured by flow cytometry assessment (Annexin-V/PI assay). The result of MTT assay indicated that cadmium chloride induces toxicity to $HepG_2$ cells in a concentration-dependent manner, showing a 48 hr-$LD_{50}$ of 3.6 μg/mL. Data generated from lipid peroxidation assay resulted in a significant ($p < 0.05$) increase of hydroperoxide production, specifically at the highest concentration tested. Data obtained from the Comet assay indicated that cadmium chloride causes DNA damage in $HepG_2$ cells in a concentration-dependent manner. A strong concentration-response relationship ($p < 0.05$) was recorded between annexin V positive cells and cadmium chloride exposure. In summary, these *in vitro* studies provide clear evidence that cadmium chloride induces oxidative stress, DNA damage, and programmed cell death in human liver carcinoma ($HepG_2$) cells.

**Keywords:** cadmium chloride; $HepG_2$ cells; cytotoxicity; oxidative stress; DNA damage; apoptosis

## 1. Introduction

Cadmium is one of the naturally occurring heavy metals. However, it is often used in industry, and exerts toxic human health effects. It is classified as a human carcinogen by the International Agency for Research on Cancer and belongs to the group I carcinogens [1]. Cadmium intoxication in humans usually occurs through inhalation (cigarette smoke) and ingestion (consumption of contaminated water and food). Acute intoxication of cadmium may lead to liver, lung, and testis damages [2] while chronic intoxication may result in obstruction of pulmonary disease, disturbance of metabolism, disregulation of blood pressure, obstruction of kidney function, structure of bones and immune system [1,3,4].

Although the mechanism of cadmium induced toxicity is poorly understood, it has been reported that cadmium causes damage to cells through the generation of reactive oxygen species [5]. Studies using two-dimensional gel electrophoresis have shown that several stress response systems are expressed in response to cadmium exposure, including those for heat shock, oxidative stress, stringent response, cold shock, and Son of Sevenless (SOS) [6–8]. *In vivo* studies have shown that cadmium modulates male reproduction in a mice model at a concentration of 1 mg/kg body weight [9]. However, cadmium is a weak mutagen when compared with other carcinogenic metals [10]. Previous reports revealed that cadmium affects signal transduction pathways; inducing inositol polyphosphate formation, increasing cytosolic free calcium levels in various cell types [11], and blocking calcium channels [12,13]. A line of evidence shows that cadmium alters antioxidant defense mechanisms and increases generation of reactive oxygen species (ROS) including superoxide anion and hydrogen peroxide [14–16]. Hence, the present investigation was designed to prove that oxidative stress plays a key role in cadmium chloride-induced DNA damage and apoptosis of human liver carcinoma ($HepG_2$) cells.

## 2. Materials and Methods

### 2.1. Chemicals and Test Media

DMEM-F12 containing 2.5 mM L-glutamine, 15 mM HEPES, 0.5 mM sodium pyruvate, and 1200 mg/L sodium bicarbonate, was supplied by American Type Culture Collection-ATCC (Manassas, VA, USA), and was used as the growth medium. Costar Company (Cambridge, MA, USA) was the source for obtaining the ninety six-well plates, while Sigma Chemical Company (St. Louis, MO, USA) provided reagents such as fetal bovine serum (FBS), penicillin G and streptomycin, phosphate buffered saline (PBS), G418 and MTT assay kit.

### 2.2. Cell/Tissue Culture

Human liver carcinoma (HepG2) cells obtained from ATCC were conserved in liquid nitrogen. During experimentation their containers/vials were gently shaken for 2 min in a water bath at 37 °C, and the content of each vial was transferred to a 25 $cm^2$ tissue culture flask in which DMEM-F12 medium containing 10% (v/v) fetal bovine serum (FBS), 0.4 mg/mL G418, and 1% (w/v) penicillin/streptomycin, was added up to a total volume of 10 mL. The cells were examined using an inverted tissue culture microscope, and incubated for 24 h in a humidified 5% $CO_2$ incubator at 37 °C. The Trypan blue exclusion test (Life Technologies, Carlsbad, CA, USA) was performed to determine the cell viability based on the number of live cells counted, using a hemocytometer.

### 2.3. Assessment of Cell Viability by MTT Assay

$HepG_2$ cells were cultured in enriched DMEM-F12 medium as described above, and 180 μL aliquots cell suspension ($5 \times 10^5$/mL) were pipetted and placed 96-well polystyrene tissue culture plates, followed by the addition of 20 μL aliquots of stock solutions to make-up six replicates of final cadmium chloride concentrations of 1, 2, 3, 4, and 5 μg/mL. Control cells received 20 μL of distilled water. After chemical treatment, HepG2 cells were incubated for 48 h in a humidified 5% $CO_2$ incubator at 37 °C. After incubation, the MTT assay for cell viability was performed as previously described [17,18].

### 2.4. Assessment of Oxidative Stress by Lipid Hydroperoxide Assay

To test the hypothesis that oxidative stress plays a key role in cadmium chloride-induced toxicity to $HepG_2$ cells, lipid hydroperoxide assay (Calbiochem-Novabiochem, San Diego, CA, USA) was performed and the production level of hydroperoxide content was estimated in untreated and treated cells. This experiment was conducted according to the manufacturer's instructions (Calbiochem-Novabiochem) [19,20], with few modifications as previously described in our laboratory [21–23].

### 2.5. *Assessment of DNA Damage by Comet Assay*

The Comet assay was carried out by the method previously described by Collins and his collaborators [24,25] with some modifications [26]. Briefly, $1 \times 10^6$ cells/mL were treated with either media or cadmium chloride (0, 1, 2, 3, 4, and 5 μg/mL) respectively and incubated in a 5% $CO_2$ at 37 °C for 48 h. After incubation, the cells were centrifuged, washed with cold PBS, and $1 \times 10^5$ cells/mL counted from the pool of untreated and treated cells were used for performing the comet assay as previously described [26,27].

### 2.6. *Assessment of Apoptosis by Annexin V/PI Assay*

Annexin V FITC/PI assay was performed as described previously [28] to evaluate the apoptotic effect of cadmium chloride to human liver carcinoma ($HepG_2$) cells. Briefly, 2 mL of cells ($1 \times 10^6$ cells/mL) were added to each well of 6 plates and treated with 1, 2, 3, 4, and 5 μg/mL of cadmium chloride for 48 h. Control well plates were also made without cadmium chloride. After 48 h of incubation, $1 \times 10^6$ cells/mL were counted and washed in PBS, re-suspended in binding buffer (10 mM Hepes/NaOH pH 7.4, 140 mM NaCl, 25mM $CaCl_2$), and stained with FITC-conjugated annexin V (Pharmingen, Becton Dickinson Co., San Diego, CA, USA). After staining, the cells were incubated for 15 min in the dark at room temperature. Cells were re-washed with binding buffer and analyzed by flow cytometry (FACS Calibur; Becton-Dickinson) using CellQuest software.

## 3. Results and Discussion

### 3.1. *Inhibition of Cell Viability by Cadmium Chloride*

The results of the cytotoxicity of cadmium chloride to human liver carcinoma ($HepG_2$) cells are presented in Figure 1. Data obtained from this assay demonstrated a strong concentration-response relationship with regard to the cytotoxic of cadmium chloride in $HepG_2$ cells. As indicated in this Figure, there was a gradual decrease in the viability of $HepG_2$ cells, with increasing concentrations of cadmium chloride that resulted in a 48 hr-$LD_{50}$ of 3.6 μg/mL. $HepG_2$ cells exposed to cadmium chloride concentrations of 1, 2, 3, 4, and 5 μg/mL showed significant mortalities ($p < 0.05$) compared to control cells, according to ANOVA Dunnett's test (Figure 1). Cadmium, a potent toxic metal, has a high potential to accumulate in the environment. It is very harmful to the environment and to human beings. The toxicity of cadmium as an industrial pollutant and a food contaminant, and as one of the major components in cigarette smoke is well established [29]. Cadmium can cause a number of lesions in many organs, such as the kidney, the testis, the lung, the liver, the brain, the bone, the blood system [30]. However, the mechanism of toxicity of cadmium is not yet clear. In the present study, we showed that cadmium chloride at the concentrations of 1, 2, 3, 4, and 5 μg/mL significant caused cell mortalities ($p < 0.05$) compared to the control. Consistent with our report, recent study indicated that cadmium chloride is highly cytotoxic to human lens epithelial cells [31]. *In vitro* studies also indicated that cadmium induces cytotoxic effects and free radical-dependent DNA damage in bacteria [32,33], which causes single-strand DNA damage and disrupts the synthesis of nucleic acids and proteins [34]. Published studies reported that acute ingestion of cadmium causes health adverse effects such as abdominal pain, burning sensation, nausea, vomiting, salivation, muscle cramps, vertigo, shock, loss of consciousness and convulsions usually appear within 15 to 30 min [35]. Acute cadmium ingestion can also cause gastrointestinal tract erosion, pulmonary, hepatic or renal injury and coma, depending on the route of poisoning [35,36]. Cadmium can cause a number of lesions in many organs, such as the kidney, the testis, the lung, the liver, the brain, the bone, the blood system [30]. Human exposure to cadmium is possible through a number of several sources including employment in primary metal industries, eating contaminated food, smoking cigarettes, and working in cadmium-contaminated work places, with smoking being a major contributor [1,37]. Other sources of cadmium include emissions from industrial activities, including mining, smelting, and manufacturing of batteries, pigments, stabilizers, and alloys [38]. Cadmium is also present in trace amounts in certain foods such as leafy vegetables, potatoes, grains and seeds, liver and kidney, and crustaceans and mollusks [39].

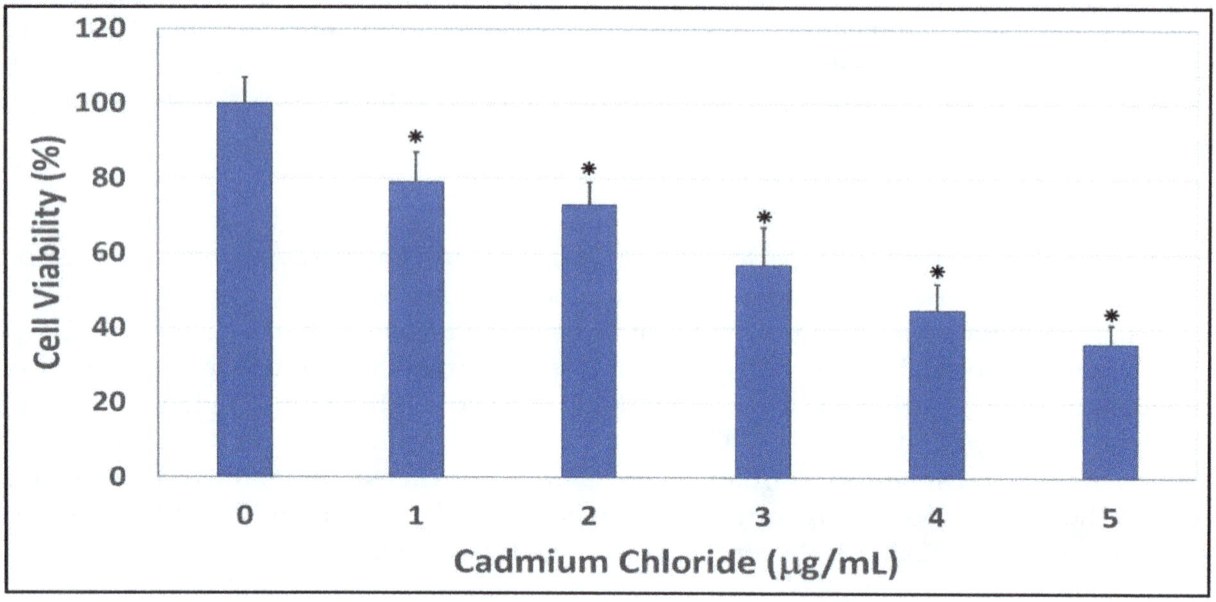


**Figure 1.** Cytotoxic effect of cadmium chloride on human liver carcinoma ($HepG_2$) cells. Cells were cultured with increasing concentrations (1, 2, 3, 4, and 5 μg/mL) of cadmium chloride for 48 h as indicated in the Materials and Methods. Cell viability was determined based on the MTT assay. Each point represents a mean ± SD of three experiments with six replicates per concentration. * Significantly different ($p < 0.05$) from the control, according to the Dunnett's test.

### *3.2. Induction of Lipid Hydroperoxide by Cadmium Chloride*

To test whether oxidative stress plays a key role in cadmium chloride-induced toxicity to $HepG_2$ cells, we performed a lipid hydroperoxide assay. As shown in Figure 2, our results indicated that the treatment of $HepG_2$ cells with cadmium chloride resulted in a significant increase of lipid hydroperoxide levels, a major degradation product of unsaturated phospholipids and glycolipids. Similar results have been recorded by Bashandy and collaborators [40]. Our result is also in agreement with a previous report indicating that cadmium induces formation of superoxide ion and hydrogen peroxide in HeLa human tumor cells and bovine aorta endothelial cells [41]. Consistent with our result, previous reports also indicated that lead and cadmium-induced tissue damages through oxidative stress [42]. Cadmium stimulates the formation of metallothioneins and reactive oxygen species, thus causing oxidative damage to erythrocytes and various tissues resulting in loss of the membrane functions [43]. Another study indicated that chronic exposure to cadmium increased lipid peroxidation and caused inhibition of superoxide dismutase (SOD) activity showing oxidative damage in liver, kidney, and testes [44]. A previous scientific report showed that treatment with cadmium causes a more pronounced reduction in intracellular glutathione levels and a significantly higher free radical accumulation in progenitors [45]. Other reports indicated that cadmium decreases intracellular glutathione content and activities of cellular antioxidant enzymes, superoxide dismutase, peroxidase and catalase, leading to the accumulation of ROS and an increase in intracellular oxidative stress in cadmium exposed CRL-1439 normal rat liver kidney cells [46,47].

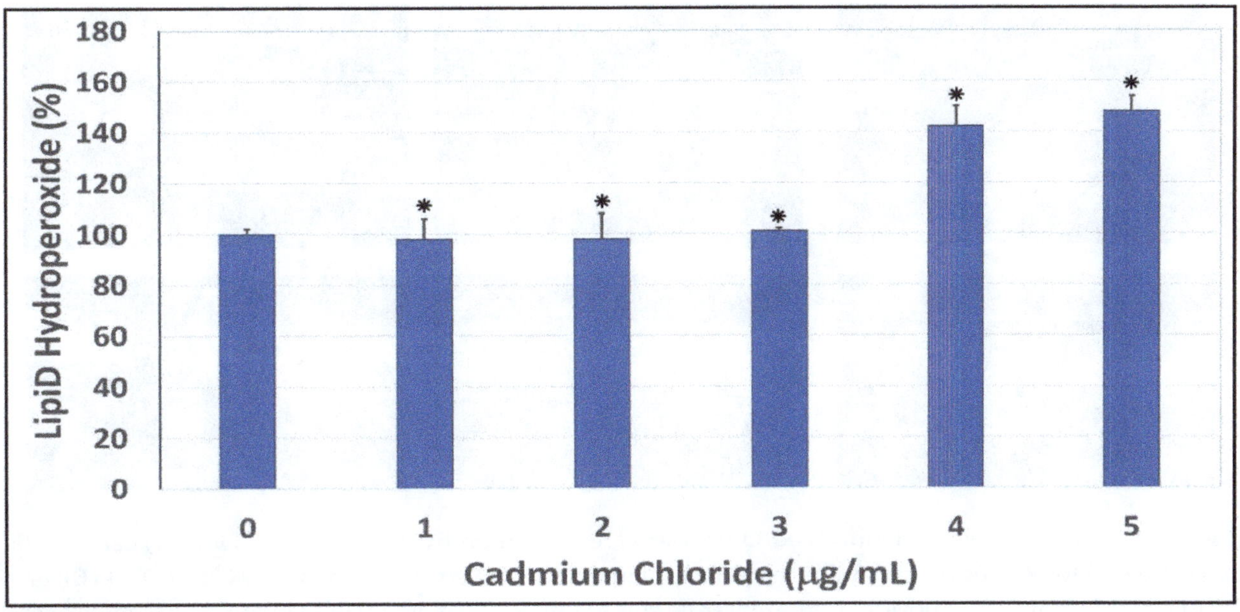


**Figure 2.** Cadmium chloride-induced lipid peroxidation in human liver carcinoma ($HepG_2$) cells. Cells were incubated for 48 h with increasing concentrations of cadmium chloride (1, 2, 3, 4, and 5 μg/mL). Lipid hydroperoxide levels were determined as described in Materials and Methods. * Significantly different ($p < 0.05$) from the control, according to the Dunnett's test. Data are representative of three independent experiments.

### *3.3. Induction of DNA Damage by Cadmium Chloride*

To evaluate the ability of cadmium to trigger genotoxic damage in hepatocytes, $HepG_2$ cells were treated with different concentrations of cadmium chloride, in the range 0–5 μg/mL for 48 h, and the degree of DNA damage was quantified by the means of LAI's Comet Assay Analysis System software (Loates Associates, Inc. Westminster, MD, USA) after staining with SYBR Green. Our results showed that cadmium chloride at the concentrations of 1, 2, 3, 4 and 5 μg/mL causes DNA single strand breaks in $HepG_2$ cells and there is a gradual concentration-response relationship. The representative comet assay images of control and cadmium chloride-treated $HepG_2$ cells are presented in Figure 3. This Figure showed a significant increase in the percentage of DNA damage and length of comet tail in a concentration-dependent manner in human liver carcinoma cells exposed to cadmium chloride. The percentages of DNA cleavage were as follows: (1.7 ± 1.2)%, (2.9 ± 1.6)%, (21.9 ± 5.5)%, (37.5 ± 4.7)%, (52.4 ± 11.4)%, and (66.5 ± 18.3)% for 0, 1, 2, 3, 4 and 5 μg/mL of cadmium chloride, respectively (Figure 4). Similar trend was observed in the mean length of comet tail. The tail lengths of DNA comets were all longer in cadmium chloride-treated cells compared to the control ($p < 0.05$) (Figure 4).

Several possible mechanisms may be involved in the induction of DNA damages. With the currently available data, cadmium chloride seems to have direct genotoxic activity to $HepG_2$ cells at concentrations relevant tested. Studies have provided evidence that reactive oxygen species (ROS) are involved in DNA damage induced by carcinogenic metal ion [48]. It has been shown that cadmium enhances lipid peroxidation in cultured cells and animals [49,50]. Low concentrations of cadmium binds to proteins, decreases DNA repair [51], activates protein degradation, up-regulates cytokines and proto-oncogenes such as c-*fos*, c-*jun*, and c-*myc* [52], and induces expression of several genes including metallothioneins [53], heme oxygenases, glutathione transferases, heat-shock proteins, acute-phase reactants, and DNA polymerase β [54].

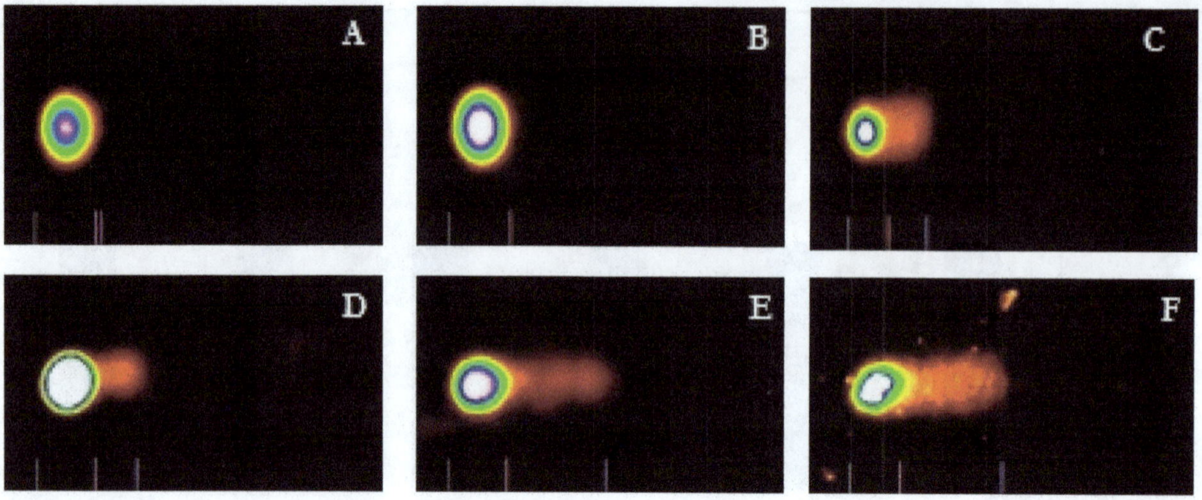


**Figure 3.** Cadmium chloride induced DNA damage in human liver carcinoma ($HepG_2$) cells. Cells were treated for 48 hours with medium (**A**) supplemented with solvent or 1 (**B**); 2 (**C**); 3 (**D**); 4 (**E**); and 5 (**F**) µg/mL cadmium chloride. Representative comet images were analyzed using LAI's Comet Assay Analysis System software (Loates Associates, Inc. Westminster, MD, USA).

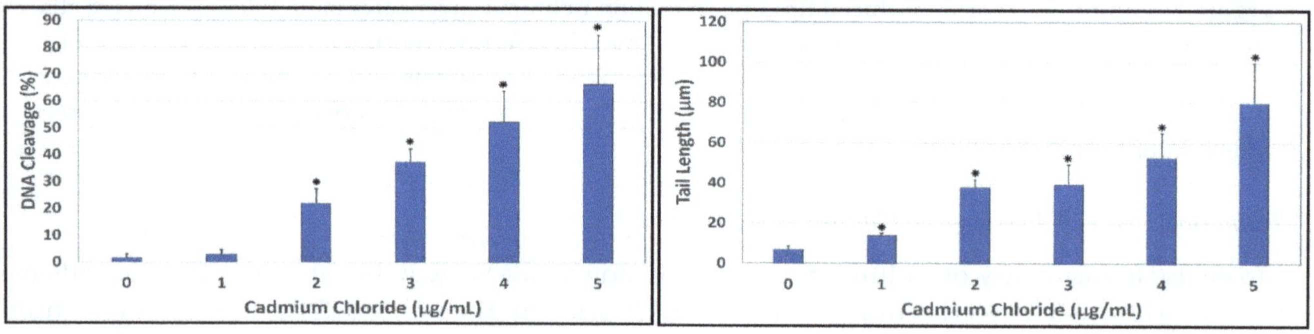


**Figure 4.** Comet assay of $HepG_2$ cells showing the percentages of DNA cleavage (**Left**) and comet tail lengths (**Right**) as a function of cadmium chloride concentrations. Each point represents mean ± SD of three independent experiments. * Significantly different ($p < 0.05$) from the control, according to the Dunnett's test.

### *3.4. Induction of Apoptosis by Cadmium Chloride*

To gain insight into the mechanism of cadmium chloride-induced apoptosis, we examined the modulation of phosphatidylserine externalization in $HepG_2$ cells. We observed that cadmium chloride induces cellular apoptosis of $HepG_2$ cells in a concentration-dependent manner, showing a gradual increase of annexin positive cells in cadmium chloride-treated cells compared to the control (Figure 5). Figure 6 shows the percentages of both annexin V and PI positive cells were (10.3 ± 3.2)%, (14.4 ± 5.6)%, (21.4 ± 4.6)%, (30.5 ± 2.8)%, (43.2 ± 7.5)%, and (52.5 ± 9.4)% in 0, 1, 2, 3, 4, and 5 µg/mL cadmium chloride, respectively.

The effect of cadmium chloride was more pronounced at 5 µg/mL ($p < 0.05$) compared to the control cells. We observed that the percentage of annexin positive cells increased gradually ($p < 0.05$) with increasing cadmium chloride concentrations and reached a maximum of (52.5 ± 9.4)% cell death upon 48 h of exposure. Cadmium can affect cell proliferation and differentiation, cell cycle progression, DNA synthesis and repair, apoptosis and other cellular activities [55,56]. Recently, several reports have shown that cadmium can induce apoptosis of many tissues and cells both *in vivo* and *in vitro* including cells of the respiratory system, the testis, the kidney, the liver, and the immune system [57–61].

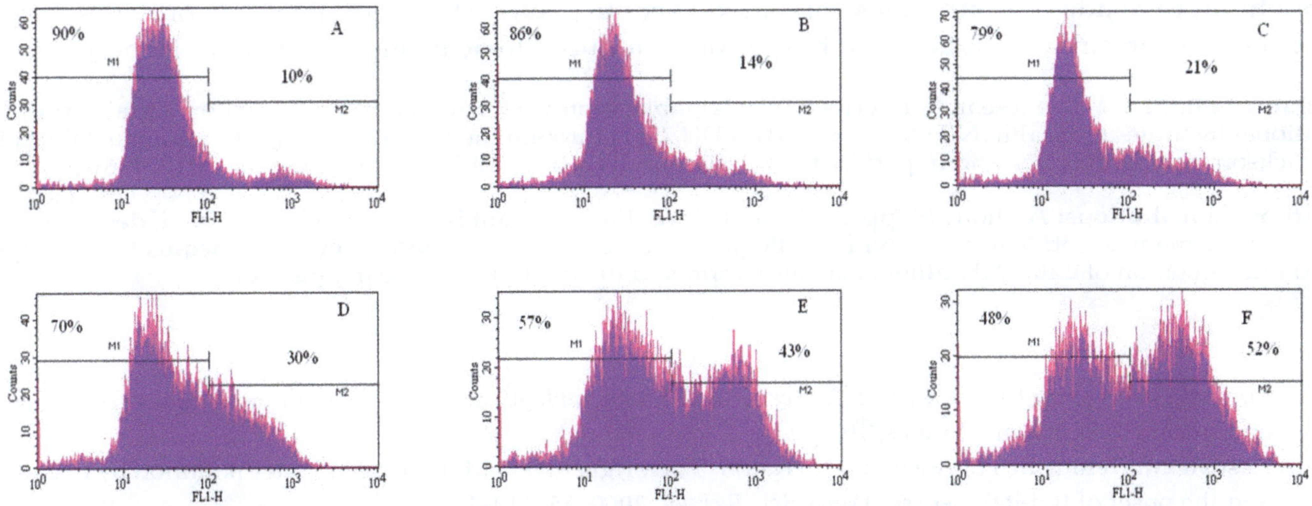


**Figure 5.** Representative flow cytometry analysis data from annexin V/PI assay. The histograms show a comparison of the distribution of annexin V/PI negative cells (M1) and annexin V/PI positive cells (M2) after 48 h exposure to cadmium chloride. **A** = control; **B** = 1 µg/mL; **C** = 2 µg/mL; **D** = 3 µg/mL; **E** = 4 µg/mL; and **F** = 5 µg/mL.

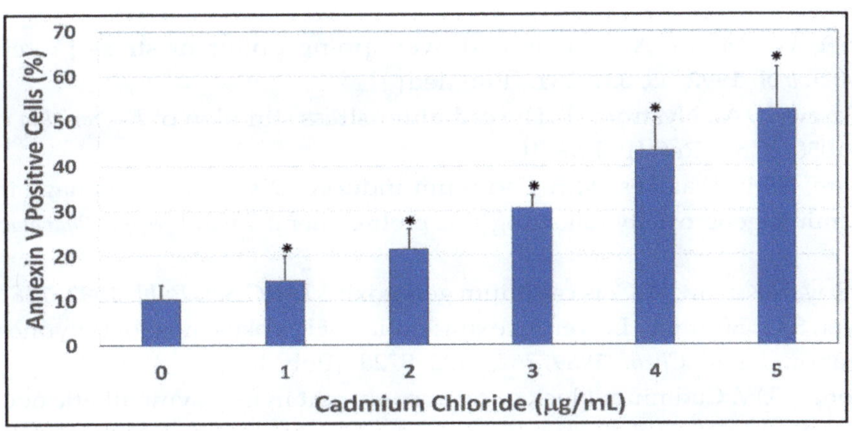


**Figure 6.** Annexin V and PI positive cells. Cells were exposed to different concentrations of cadmium chloride as described in the Materials and Methods. * Significantly different ($p < 0.05$) from the control, according to the Dunnett's test.

In addition to the apoptotic effects of cadmium, other reports have indicated that cadmium chloride decreases the viability of HepG2 cells and increases lactate dehydrogenase leakage, DNA damage, malondialdehyde, and antioxidant enzymes activities [62,63]. Additionally, a significant decrease in ATP production and increase in ROS levels in cadmium chloride-treated HepG2 cells was observed at all concentrations tested. A significant decrease in GSH/GSSG ratio was also found. These effects were reported to be attenuated when cells were co-exposed to *N*-acetylcysteine (NAC) and cadmium chloride [62,63].

## 4. Conclusions

The present *in vitro* study demonstrated that cadmium chloride exposure gradually decreases the viability of $HepG_2$ cells; increases lipid hydroperoxide levels resulting from reactive oxygen species formation; induces DNA damage and triggers apoptosis of $HepG_2$ cells through phosphatidylserine externalization. In summary, these findings suggest that oxidative stress plays a role in cadmium chloride-induced cyto/genotoxicity and apoptosis of $HepG_2$ cells, especially at higher level of exposure (4 and 5 µg/mL) where $CdCl_2$-treated cells show a statistically significant difference in lipid hydroperoxide concentrations compared to control cells. This study therefore provides insight into the mechanism underlying cadmium chloride-induced toxicity and apoptosis of $HepG_2$ cells.

**Acknowledgments:** The research described in this publication was made possible in part by a grant from the National Institutes of Health (NIH/NIMHD-G12MD007581) through the RCMI-Center for Environmental Health at Jackson State University; and in part by the Mississippi INBRE grant (NIH/NIGMS-P20GM103476).

**Author Contributions:** Anthony Skipper, Clement G. Yedjou, and Paul B. Tchounwou conceived, designed, and drafted the manuscript. Jennifer N. Sims participated in the implementation of the study, acquisition, analysis and interpretation of data. All authors read and approved the final draft of the manuscript.

## References

1. *International Agency for Research on Cancer (IARC) Monographs-Cadmium*; International Agency for Research on Cancer (IARC): Lyon, France, 1993.
2. Kasuya, M.; Teranishi, H.; Aoshima, K.; Katoh, T.; Horiguchi, H.; Morikawa, Y. Water pollution by cadmium and the onset of Itai-itai disease. *Water Sci. Technol.* **2000**, *25*, 149–156.
3. Jarup, L.; Berglund, M.; Elinder, C.; Nordberg, G.; Vahteram, M. Health effects of cadmium exposure-a review of the literature and a risk estimate. *Scand. J. Work Environ. Health* **1998**, *1*, 1–52.
4. Jin, T.; Lu, J.; Nordberg, M. Toxicokinetics and biochemistry of cadmium with special emphasis on the role of metallothioneins. *Neurotoxicology* **1998**, *19*, 529–535. [PubMed]
5. Stohs, S.J.; Bagchi, D. Oxidative mechanisms in the toxicity of metal ions. *Free Radic. Biol. Med.* **1995**, *18*, 321–336. [CrossRef]
6. Blom, A.; Harder, W.; Matin, A. Unique and overlapping pollutant stress proteins of *Escherichia coli*. *Appl. Environ. Microbiol.* **1992**, *58*, 331–334. [PubMed]
7. Feriance, P.A.; Farewell, A.; Nystrom, T. The cadmium-stress stimulon of *Escherichia coli* K-12. *Microbiology* **1998**, *144*, 1045–1050. [CrossRef] [PubMed]
8. Coogan, T.P.; Bare, R.M.; Waalkes, M.P. Cadmium-induced DNA strand damage in cultured liver cells: Reduction in cadmium genotoxicity following zinc pretreatment. *Toxicol. Appl. Pharmacol.* **1992**, *113*, 227–233. [CrossRef]
9. Rossman, T.G.; Roy, N.K.; Lin, W.C. Is cadmium genotoxic? *IARC Sci. Publ.* **1992**, *118*, 367–375. [PubMed]
10. Smith, J.B.; Dwyer, S.C.; Smith, L. Lowering extracellular pH evokes inositol polyphosphate formation and calcium mobilization. *J. Biol. Chem.* **1989**, *264*, 8723–8728. [PubMed]
11. Th'evenod, F.; Jones, S.W. Cadmium block of calcium current in frog sympathetic neurons. *Biophys. J.* **1992**, *63*, 162–168. [CrossRef]
12. Suszkiw, J.; Toth, G.; Murawsky, M.; Cooper, G.P. Effects of $Pb^{2+}$ and $Cd^{2+}$ on acetylcholine release and $Ca^{2+}$ movements in synaptosomes and subcellular fractions from rat brain and Torpedo electric organ. *Brain Res.* **1984**, *323*, 31–46. [CrossRef]
13. Dally, H.; Hartwig, A. Induction and repair inhibition of oxidative DNA damage by nickel (II) and cadmium (II) in mammalian cells. *Carcinogenesis* **1997**, *18*, 1021–1026. [CrossRef] [PubMed]
14. Oh, S.H.; Lim, S.C. A rapid and transient ROS generation by cadmium triggers apoptosis via caspase-dependent pathway in HepG2 cells and this is inhibited through *N*-acetylcysteine-mediated catalase upregulation. *Toxicol. Appl. Pharmacol.* **2006**, *212*, 212–223. [CrossRef] [PubMed]
15. Pathak, N.; Khandelwal, S. Oxidative stress and apoptotic changes in murine splenocytes exposed to cadmium. *Toxicology* **2006**, *220*, 26–36. [CrossRef] [PubMed]
16. Valko, M.; Rhodes, C.J.; Moncol, J.; Izakovic, M.; Mazur, M. Free radicals, metals and antioxidants in oxidative stress-induced cancer. *Chem. Biol. Interact.* **2006**, *160*, 1–40. [CrossRef] [PubMed]
17. Mosmann, T. Rapid colorimetric assay for cellular growth and survival: Applications to proliferation and cytotoxicity assays. *J. Immunol. Method* **1983**, *65*, 55–63. [CrossRef]
18. Tchounwou, P.B.; Wilson, B.; Schneider, J.; Ishaque, A.; Morrow, J.D.; Roberts, L.J. Cytogenic assessment of arsenic trioxide toxicity in the Mutatox, Ames II and CAT-Tox assays. *Metal Ions Biol. Med.* **2000**, *6*, 89–91.
19. Halliwell, B.; Gutteridge, J.M. The antioxidants of human extracellular fluids. *Arch. Biochem. Biophys.* **1990**, *280*, 1–8. [CrossRef]
20. Morrow, J.M.; Roberts, L.J. The isoprostanes: Unique bioactive products of lipid peroxidation. *Prog. Lipid Res.* **1997**, *36*, 1–21. [CrossRef]

21. Kumar, S.; Yedjou, C.G.; Tchounwou, P.B. Arsenic trioxide induces oxidative stress, DNA damage, and mitochondrial pathway of apoptosis in human leukemia (HL-60) cells. *J. Exp. Clin. Cancer Res.* **2014**. [CrossRef] [PubMed]
22. Velma, V.; Tchounwou, P.B. Oxidative stress and DNA damage induced by chromium in liver and kidney of goldfish, *Carassius auratus*. *Biomark Insights* **2013**, *8*, 43–51. [PubMed]
23. Patlolla, A.K.; Hackett, D.; Tchounwou, P.B. Silver nanoparticle-induced oxidative stress-dependent toxicity in Sprague-Dawley rats. *Mol. Cell Biochem.* **2015**. in press. [CrossRef] [PubMed]
24. Collins, A.R. Comet Assay for DNA damage and repair: Principles, applications and limitations. *Mol. Biotechnol.* **2001**, *26*, 249–261. [CrossRef]
25. Collins, A.R.; Dusinská, M.; Horská, A. Detection of alkylation damage in human lymphocyte DNA with the comet assay. *Acta Biochim. Pol.* **2001**, *48*, 611–614. [PubMed]
26. Yedjou, C.G.; Tchounwou, P.B. *In-vitro* genotoxic effect of arsenic trioxide to human leukemia (HL-60) cells using the comet assay. *Mol. Cell Biochem.* **2007**, *301*, 123–130. [CrossRef] [PubMed]
27. Tchounwou, C.K.; Yedjou, C.G.; Farah, I.; Tchounwou, P.B. D-Glucose-induced cytotoxic, genotoxic, and apoptotic effects on human breast adenocarcinoma (MCF-7) cells. *J. Cancer Sci. Ther.* **2014**, *6*, 156–160. [CrossRef] [PubMed]
28. Yedjou, C.G.; Tchounwou, P.B.; Jerkins, J.; McMurray, R. Basic mechanisms of arsenic trioxide (ATO)-induced apoptosis in human leukemia (HL-60) cells. *J. Hematol. Oncol.* **2010**. [CrossRef] [PubMed]
29. Morselt, A.F. Environmental pollutants and diseases. A cell biological approach using chronic cadmium exposure in the animal model as a paradigm case. *Toxicology* **1991**, *70*, 1–132. [PubMed]
30. World Health Organization. *Environmental Health Criteria 134: Cadmium*, 1st ed.; World Health Organization: Geneva, Switzerland, 1992.
31. Song, N.H.; Koh, J.W. Effects of cadmium chloride on the cultured human lens epithelial cells. *Mol. Vis.* **2012**, *18*, 983–988. [PubMed]
32. Tsuzuki, K.; Sugiyama, M.; Haramaki, N. DNA single-strand breaks and cytotoxicity induced by chromate (VI), cadmium (II), and mercury (II) in hydrogen peroxide-resistant cell lines. *Environ. Health Perspect.* **1994**, *102*, 341–342. [CrossRef] [PubMed]
33. Mukherjee, S.; Das, S.K.; Kabiru, W.; Russell, K.R.; Greaves, K.; Ademoyero, A.A.; Rhaney, F.; Hills, E.R.; Archibong, A.E. Acute cadmium toxicity and male reproduction. *Adv. Reprod.* **2002**, *6*, 143–155.
34. Mitra, R.S. Protein synthesis in *Escherichia coli* during recovery from exposure to low levels of $Cd^{2+}$. *Appl. Environ. Microbiol.* **1984**, *47*, 1012–1016. [PubMed]
35. Baselt, R.C.; Cravey, R. *Disposition of Toxic Drugs and Chemicals in Man*; Chemical Technology Institute: Foster City, CA, USA, 1995; Volume 4, pp. 105–107.
36. Baselt, R.C. *Disposition of Toxic Drugs and Chemicals in Man*; Chemical Toxicology Institute: Forster City, CA, USA, 2000; pp. 3–4.
37. International Agency for Research on Cancer (IARC). Berylium, cadmium, mercury and exposures in the glass manufacturing industry. In *International Agency for Research on Cancer Monographs on the Evaluation of Carcinogenic Risks to Humans*; IARC Scientific Publications: Lyon, France, 1993; Volume 58, pp. 119–237.
38. Paschal, D.C.; Burt, V.; Caudill, S.P.; Gunter, E.W.; Pirkle, J.L.; Sampson, E.J.; Miller, D.T.; Jackson, R.J. Exposure of the U.S. population aged 6 years and older to cadmium: 1988–1994. *Arch. Environ. Contam. Toxicol.* **2000**, *38*, 377–383. [CrossRef] [PubMed]
39. Agency for Toxic Substances and Disease Registry (ATSDR). *Draft Toxicological Profile for Cadmium*; ATDSR: Atlanta, GA, USA, 2008.
40. Satarug, S.; Baker, J.R.; Urbenjapol, S.M.; Haswell-Elkins, M.; Reilly, P.E.; Williams, D.J.; Moore, M.R. A global perspective on cadmium pollution and toxicity in non-occupationally exposed population. *Toxicol. Lett.* **2003**, *137*, 65–83. [CrossRef]
41. Bashandy, S.A.; Alhazza, I.M.; Mubark, M. Role of zinc in the protection against cadmium induced hepatotoxicity. *Int. J. Pharmacol.* **2006**, *2*, 79–88.
42. Szuster-Ciesielska, A.; Stachura, A.; Slotwinska, M.; Kaminska, T.; Sniezko, R.; Paduch, R.; Abramczyk, D.; Filar, J.; Kandefer-Szerszen, M. The inhibitory effect of zinc on cadmium-induced cell apoptosis and reactive oxygen species (ROS) production in cell cultures. *Toxicology* **2006**, *145*, 159–171. [CrossRef]
43. Patra, R.C.; Swarup, D.; Dwivedi, S.K. Antioxidant effects of α tocopherol, ascorbic acid and L-methionine on lead induced oxidative stress to the liver, kidney and brain in rats. *Toxicology* **2001**, *162*, 81–88. [CrossRef]

44. Patra, R.C.; Swarup, D.; Senapat, S.K. Effects of cadmium on lipid peroxides and superoxide dismutasein hepatic, renal and testicular tissue of rats. *Vet. Hum. Toxicol.* **1999**, *41*, 65–67. [PubMed]
45. Almazan, G.; Liu, H.N.; Khorchid, A. Exposure of developing oligodendrocytes to cadmium causes HSP72 induction, free radical generation, reduction in glutathione levels, and cell death. *Free Radic. Biol. Med.* **2000**, *29*, 858–869. [CrossRef]
46. Ikediobi, C.O.; Badisa, V.L.; Ayuk-Takem, L.T.; Latinwo, L.M.; West, J. Response of antioxidant enzymes and redox metabolites to cadmium-induced oxidative stress in CRL-1439 normal rat liver cells. *Int. J. Mol. Med.* **2004**, *14*, 87–92. [CrossRef] [PubMed]
47. Casalino, E.; Casalino, G.; Calzaretti, C.S. Molecular inhibitory mechanisms of antioxidant enzymes in rat liver and kidney by cadmium. *Toxicology* **2002**, *179*, 37–50. [CrossRef]
48. Pourahmad, J.; O'Brien, P.J.; Jokar, F.; Daraei, B. Carcinogenimetal induced sites of reactive oxygen species formation in hepatocytes. *Toxicol. In vitro* **2003**, *17*, 803–810. [CrossRef]
49. Traore, A.; Ruiz, S.; Baudrimont, I.; Sanni, A.; Dano, S.D.; Guarigues, P.; Narbonne, J.F.; Creppy, E.E. Combined effects of okadaic acid and cadmium on lipid peroxidation and DNA bases modifications (m5dC and 8-(OH)-dG) in Caco-2 cells. *Arch. Toxicol.* **2000**, *74*, 79–84. [PubMed]
50. Yiin, S.J.; Chern, C.L.; Sheu, J.Y.; Lin, L.H. Cadmium-induced liver, heart, and spleen lipid peroxidation in rats and protection by selenium. *Biol. Trace Elem. Res.* **2000**, *78*, 219–230. [CrossRef]
51. Abshire, M.K.; Devor, D.E.; Diwan, B.A.; Shaughnessy, J.D., Jr.; Waalkes, M.P. *In vitro* exposure to cadmium in rat L6 myoblasts can result in both enhancement and suppression of malignant progression *in vivo*. *Carcinogenesis* **1996**, *17*, 1349–1356. [CrossRef] [PubMed]
52. Durnam, D.M.; Palmiter, R.D. Transcriptional regulation of the mouse metallothionein-I gene by heavy metals. *J. Biol. Chem.* **1981**, *256*, 5712–5716. [PubMed]
53. Hwua, Y.; Yang, J. Effect of 3-aminotriazole on anchorage independence and mutagenicity in cadmium- and lead-treated diploid human fibroblasts. *Carcinogenesis* **1998**, *19*, 881–888. [CrossRef] [PubMed]
54. Landolph, J. Molecular mechanisms of transformation of CH3/10T1/2 C1 8 mouse embryo cells and diploid human fibroblasts by carcinogenic metal compounds. *Environ. Health Perspect.* **1994**, *102*, 119–125. [PubMed]
55. Bertin, G.; Averbeck, D. Cadmium: cellular effects, modifications of biomolecules, modulation of DNA repair and genotoxic consequences (a review). *Biochimie* **2006**, *88*, 1549–1559. [CrossRef] [PubMed]
56. Beyersmann, D.; Hechtenberg, S. Cadmium, Gene Regulation, and Cellular Signaling in Mammalian Cells. *Toxicol. Appl. Pharmacol.* **1997**, *144*, 247–261. [CrossRef] [PubMed]
57. Hart, B.A.; Lee, C.H.; Shukla, G.S.; Shukla, A.; Osier, M.; Eneman, J.D.; Chiu, J.F. Characterization of cadmium-induced apoptosis in rat lung epithelial cells: Evidence for the participation of oxidant stress. *Toxicology* **1999**, *133*, 43–58. [CrossRef]
58. Zhou, T.; Zhou, G.; Song, W.; Eguchi, N.; Lu, W.; Lundin, E; Jin, T.; Nordberg, G. Cadmium-induced apoptosis and changes in expression of p53, c-jun and MT-I genes in testes and ventral prostate of rats. *Toxicology* **1999**, *142*, 1–13. [CrossRef]
59. Somji, S.; Sens, D.A.; Garrett, S.H.; Sens, M.A.; Todd, J.H. Heat shock protein 27 expression in human proximal tubule cells exposed to lethal and sublethal concentrations of CdCl2. *Environ. Health Perspect.* **1999**, *107*, 545–552. [CrossRef] [PubMed]
60. Habeebu, S.S.; Liu, J.; Klaassen, C.D. Cadmium-induced apoptosis in mouse liver. *Toxicol. Appl. Pharmacol.* **1998**, *149*, 203–209. [CrossRef] [PubMed]
61. Tsangaris, G.T.; Tzortzatou-Stathopoulou, F. Cadmium induces apoptosis differentially on immune system cell lines. *Toxicology* **1998**, *128*, 143–150. [CrossRef]
62. Lawal, A.O.; Ellis, E.M. Differential sensitivity and responsiveness of three human cell lines HepG2, 1321N1 and HEK 293 to cadmium. *J. Toxicol. Sci.* **2010**, *35*, 465–478. [CrossRef] [PubMed]
63. Oh, S.H.; Lim, S.C. A rapid and transient ROS generation by cadmium trigger apoptosis via caspase dependent pathway in HepG2 cells and this is inhibited through *N* acetyl cysteine mediated catalase regulation. *Toxicol. Appl. Pharmacol.* **2006**, *212*, 212–223. [CrossRef] [PubMed]

3

# Heavy Metal Pollution in a Soil-Rice System in the Yangtze River Region of China

**Zhouping Liu, Qiaofen Zhang, Tiqian Han, Yanfei Ding, Junwei Sun, Feijuan Wang * and Cheng Zhu ***

Academic Editor: Paul Tchounwou

College of Life Sciences, China Jiliang University, Hangzhou 310018, China; jslzp89@163.com (Z.L.); zhangqf025@163.com (Q.Z.); hantq96@163.com (T.H.); dingyanfei1984@126.com (Y.D.); Juville@cjlu.edu.cn (J.S.)

* Correspondence: wfj0311@cjlu.edu.cn (F.W.); pzhch@cjlu.edu.cn (C.Z.);

**Abstract:** Heavy metals are regarded as toxic trace elements in the environment. Heavy metal pollution in soil or rice grains is of increasing concern. In this study, 101 pairs of soil and rice samples were collected from the major rice-producing areas along the Yangtze River in China. The soil properties and heavy metal (*i.e.*, Cd, Hg, Pb and Cr) concentrations in the soil and rice grains were analyzed to evaluate the heavy metal accumulation characteristics of the soil-rice systems. The results showed that the Cd, Hg, Pb and Cr concentrations in the soil ranged from 0.10 to 4.64, 0.01 to 1.46, 7.64 to 127.56, and 13.52 to 231.02 $mg \cdot kg^{-1}$, respectively. Approximately 37%, 16%, 60% and 70% of the rice grain samples were polluted by Cd, Hg, Pb, and Cr, respectively. The degree of heavy metal contamination in the soil-rice systems exhibited a regional variation. The interactions among the heavy metal elements may also influence the migration and accumulation of heavy metals in soil or paddy rice. The accumulation of heavy metals in soil and rice grains is related to a certain extent to the pH and soil organic matter (SOM). This study provides useful information regarding heavy metal accumulation in soil to support the safe production of rice in China. The findings from this study also provide a robust scientific basis for risk assessments regarding ecological protection and food safety.

**Keywords:** soil-rice system; pollution characteristics; heavy metal accumulation; Yangtze River region

---

## 1. Introduction

Heavy metals are regarded as nonessential and toxic trace elements in the environment [1]. Due to their potential toxicity, as well as their persistent and irreversible accumulation characteristics, heavy metals, such as Cd, Pb, Hg, and Cr, are listed as key monitoring pollutants by the Chinese Ministry of Environmental Protection [2]. Cr is classified as a priority pollutant in Category A (human carcinogen), while Cd and Pb are classified in Category B (probable human carcinogen) by the United States Environmental Protection Agency [3]. Soil is the most important reservoir of heavy metals in the terrestrial ecosystem, and the content of heavy metals in soils is an important indicator of environmental quality [4,5]. Soils are also the main source of heavy metals in plants, where they become the main source of heavy metals in plant-derived foods [6]. Heavy metals that accumulate and migrate in soil not only influence plant growth and yield, but also have the potential to accumulate in the human body through the food chain, thus posing a serious threat to human health [7–9]. Recently, the issue of heavy metal pollution in agricultural soils and plants has become increasingly serious [10–12]. Paddy rice is one of the main global food sources, and China is the largest producer of rice in the world; thus, the contamination and accumulation of heavy metals, such as Cd, Pb, Hg, and Cr, in rice grains have become a topic of great public concern [12,13]. Numerous studies have been performed on Cd, Pb, Cr, and Hg in soils or rice [13–15]. In China, approximately 10% of planting land is contaminated by heavy metals (Cd, Pb, Hg, Cr, *etc.*), and approximately ten million tons of grains are contaminated every year [14]. In a random survey, researchers discovered that 10% of the commercially available rice grain in China had excessive levels of Cd [15].

As one of the fastest developing regions in China, the Yangtze River region is also famous for being the "Home of Rice and Fish". Due to its vast and fertile farmlands, the Yangtze River region, including Hubei, Hunan and Jiangxi Provinces, is China's main rice-growing area. In recent years, the heavy metal pollution in rice grains and rice products has become increasingly serious in some parts of the Yangtze River region [12,13,16,17]. Cd, Pb, and Cr contamination in paddy soils from Hubei Province [16] and Cd-polluted rice in Hunan Province [17] have been reported.

Researchers have performed numerous surveys and experiments regarding the sources of heavy metal pollution, the bioavailability of metals, and other issues in agricultural soil or plants [16–19]. However, most previous studies on the accumulation of heavy metals in rice were performed using pot or field experiments or focused on special use areas (e.g., mining areas). Due to the complexity of soil-plant relationships and the heterogeneity of a natural agricultural system, regional field investigations should be encouraged due to their greater practical significance. Moreover, there is a lack of information related to the migration and accumulation characteristics of heavy metal trace elements in soil-rice systems of a natural agricultural system and a lack of information about the factors (soil pH and soil organic matter) that directly influence the migration and accumulation of heavy metals in natural soil-rice systems.

Therefore, a regional-scale survey was performed to study Cd, Pb, Hg, and Cr contamination and their influence factors in a soil-rice system under actual field conditions. The aims of this study were to: (1) study the accumulation status of heavy metals (Cd, Pb, Cr, and Hg) in soils and the corresponding rice grain and establish a corresponding traceable database; (2) study the relationships between the heavy metals (Cd, Pb, Cr, and Hg) in rice grain and the corresponding soils; and (3) clarify the factors (soil pH and soil organic matter) that control heavy metal accumulation in soil and rice grain.

## 2. Experimental Section

### 2.1. Sample Collection and Study Area

Pairs of agricultural soil and rice grain samples (101) of were collected from large, thriving, representative paddy fields at the mature stage of rice growth along the midstream and downstream reaches of the Yangtze River in Hubei, Hunan, and Jiangxi Provinces (Figure 1). All the samples were collected in October 2012.

We designed a suitable route, chose the sampling areas, and went to the rice planting areas that were generally located between two towns. We collected three to six samples of soil and rice grain at each sampling point. Five to ten mature rice panicles were harvested from each paddy field. Soil samples (approximately 300 g) were collected from 10 to 15 cm above the surface from the same paddy fields.

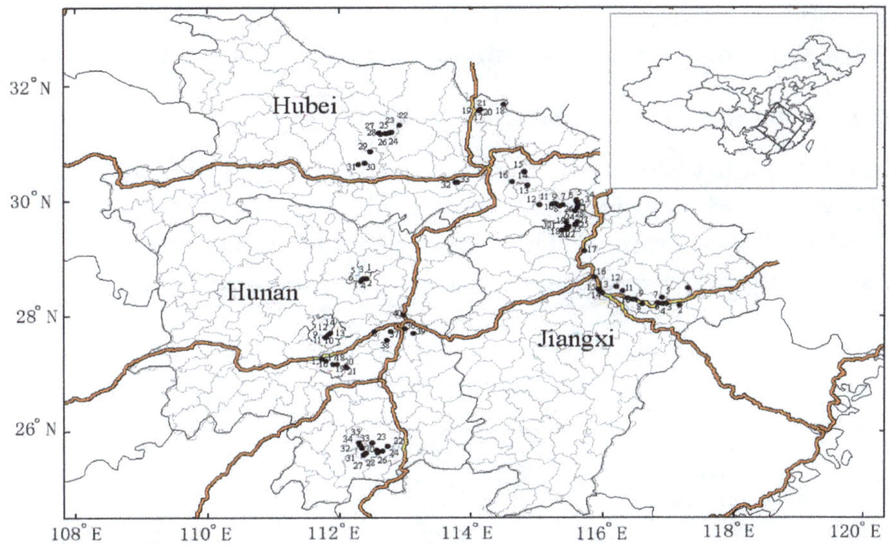


**Figure 1.** Distribution of sampling sites in the Yangtze River region (Hubei, Hunan and Jiangxi). The sampling areas in Hubei.: 1–6 Wuxue City, 7–12 Yangxin City, 13–16 Ezhou City, 121 Dawu City, 228 Zhongxiang City, 29–31 Shayang City, 32–33 Wuhan City; the sampling areas in Hunan: 1–7 Yiyang City, 8–14 Liangyu City, 15–18 Shaodong City, 121 Yiyang City, 22–26 Guiyang City, 27–30 Jiahe City, 31–35 Gongzhou City, 36 Zhuzhou City, 37–38 Xiangtan City, 39–40 Changsha City; the sampling areas in Jiangxi: 1–2 Guixi City, 3–6 Yujiang City, 7–8 Dongxiang City, 9–12 Jinxian City, 13–16 Nanchang City, 17–18 Yongxiu City, 19–28 Ruichang City.

## 2.2. *Quality of the Applied Methodology*

Soil samples were air-dried indoors at room temperature. They were crushed with a wooden stick, then passed through a 2.5 mm nylon sieve to remove grit and plant residues and then were finely ground with an agate mortar. They were screened through a 100-mesh nylon sieve, and a subsample (100 g) was bagged and stored in a dry location for heavy metal concentration and soil property analyses. Simultaneously, the samples of rice grain were rinsed with tap water, washed with distilled water and dried at 105 °C for 1 h and then at 70 °C to a constant weight in the oven. They were then shelled with a rice husking machine and ground to a powder for further analysis.

Soil samples were digested in a $HNO_3$/HF mixture (3:1) in a microwave digester (CEM-MARS: Boston, MA, USA). Approximately 0.2 g of each of the mentioned soil samples was weighed and placed into a digestion vessel. High purity $HNO_3$ (6 mL) and high purity HF (2 mL) were added to the vessels, which were then placed in a ventilation cabinet for 2–3 h. The samples were then digested in the microwave digester. Simultaneously, rice grain samples were digested using $HNO_3$ and $H_2O_2$. Rice grain samples (0.3 g) were weighed and placed into digestion vessels. The vessels were stored in a ventilation cabinet for 2–3 h after 6 mL high purity $HNO_3$ and 0.2 mL $H_2O_2$ were added, and then digested in the microwave digester. After digestion, all the mixtures were reduced to 1 mL at high temperature. The solutions were then diluted into 25 mL volumetric flasks with ultra-pure water, and the clarified samples thus obtained were stored in the refrigerator at 4 °C for further analysis.

The concentrations of the four heavy metals in soil and rice grain were measured using two different methods. Cd, Pb, and Cr concentrations were measured by graphite furnace atomic absorption spectrometry (AA7000, SHIMADZU: Kyoto, Japan); Hg concentrations were determined with hydride generation atomic absorption spectrometry (AA7000, SHIMADZU: Kyoto, Japan). All samples were measured in triplicate. Reagent blanks, a standard reference soil sample (GBW07443) and standard plant samples (GBW10010), obtained from the Centre of Standard Materials of China, were used to monitor the determination quality of the soil and rice samples, respectively. The recovery of samples spiked with standards ranged from 89% to 101.4%. The detection limits for Cd, Pb, Cr, and Hg were 0.0025, 0.027, 0.005, and 0.0002 $\mu g \cdot L^{-1}$, respectively. Soil pH was measured using the glass electrode method of Chaturvedi and Sankar [20], with a water to soil ratio of 2.5:1. Soil organic matter (SOM) content was determined by the chromic acid titration method [21].

### *2.3. Evaluation Standards for Soil and Rice Pollution*

The Environmental Quality Standards for Soils in China [22] and heavy metal standards for food safety [23] (Table 1) were used as guideline values for the assessment of heavy metal accumulation and pollution in soil and rice grains. The Environmental Quality Standards for Soils includes three levels, and the second one was adopted as the guideline value for agricultural production.

**Table 1.** Environmental quality standard for soils in China (GB 15618-1995) [22], heavy metal standards for food safety (GB2715–2005) [23] (mg· kg$^{-1}$).

| Heavy Metals | Grade 1 | Grade 2 | | | Grade 3 | Food Safety Standards (In Rice) |
|---|---|---|---|---|---|---|
| | National Backgrounds | pH < 6.5 | pH 6.5–pH 7.5 | pH > 7.5 | pH > 6.5 | |
| Cd ⩽ | 0.20 | 0.30 | 0.30 | 0.6 | 1.0 | 0.20 |
| Hg ⩽ | 0.15 | 0.30 | 0.50 | 1.0 | 1.5 | 0.02 |
| Pb ⩽ | 35 | 250 | 300 | 350 | 500 | 0.20 |
| Cr ⩽ | 90 | 250 | 300 | 350 | 400 | 1.0 |

### *2.4. Statistical Analysis*

A standard statistical analysis (mean, standard deviation, *etc.*) was performed to describe soil properties (SOM and pH) in three study areas. Significant differences among areas required evaluation by ANOVA and Kruskal–Wallis tests ($\alpha = 0.01$). To study the relationship between soil element pollution and rice element contents, canonical correlation analyses (CCorA) were used. The location and sample map was produced using the ArcGIS 10.0 software (ESRI: Redlands, CA, USA). All statistical analysis data were statistically analyzed by the XLSTAT (Addinsoft: Brooklyn, NY, USA) and SPSS 18.0 programs (IBM: Chicago, IL, USA).

## 3. Results and Discussion

### *3.1. Heavy Metal Contamination in Soil and Soil Properties*

The concentrations of heavy metals in soils from Hubei, Hunan and Jiangxi Provinces are presented in Table 2. Compared with the Environmental Quality Standards for Soils in China [22] (Table 1), 78.78%, 6.06%, 3.03% and 21.21% of the 33 soil samples from Hubei Province exceeded the local soil national backgrounds for Cd, Hg, Pb and Cr, respectively, which indicated that some heavy metals had accumulated in the surveyed soil. Combined with the corresponding pH values shown in Table 2, among the 33 sampling points from Hubei Province, 45.45% and 75.76% of the soil samples exceeded the Grade 2 thresholds for Cd and Hg, respectively, whereas the Pb and Cr concentrations of soil samples from Hubei Province were under the Grade 2 thresholds.

In Hunan Province, 92.5%, 27.5%, 15% and 10% of the 40 surveyed soil samples exceeded the local soil backgrounds for Cd, Hg, Pb and Cr, respectively, indicating that Cd had accumulated considerably in the surveyed soil. Combined with the corresponding pH values, except for the samples at sites 6, 16 and 17, the remaining soil samples exceeded the Grade 2 thresholds for Cd, and some samples exceeded the Grade 3 thresholds. Only 7.5% of the tested soil samples were not contaminated by Cd. Three soil samples (7.5%) exceeded the Grade 2 thresholds for Hg. The Pb and Cr concentrations in soil samples from Hunan Province were under the Grade 2 thresholds.

Similarly, in the 28 soil samples from Jiangxi Province, 78.57%, 25%, 7.14% and 32.14% of the samples had Cd, Hg, Pb and Cr concentrations exceeding the local soil backgrounds, respectively. As shown in Table 2, the Cd concentration of 19 soil samples (67.8%) exceeded the Grade 2 thresholds. Two soil samples showed Hg concentrations exceeding the Grade 2 thresholds, whereas all samples had Pb and Cr concentrations below the Grade 2 thresholds.

**Table 2.** Soil properties and heavy metal concentrations (Cd, Hg, Pb, Cr) in the soil and rice of Hubei, Hunan and Jiangxi.

| | Location | Soil | | | | | | Rice Grain | | | |
|---|---|---|---|---|---|---|---|---|---|---|---|
| | | Cd (mg·kg$^{-1}$) | Hg (mg·kg$^{-1}$) | Pb (mg·kg$^{-1}$) | Cr (mg·kg$^{-1}$) | pH | SOM (mg·kg$^{-1}$) | Cd (mg·kg$^{-1}$) | Hg (mg·kg$^{-1}$) | Pb (mg·kg$^{-1}$) | Cr (mg·kg$^{-1}$) |
| | 1 | 0.19 ± 0.05 | 1.46 ± 0.13 | 27.95 ± 1.11 | 212.19 ± 20.68 | 5.83 ± 0.05 | 19.26 ± 1.38 | 0.03 ± 0.02 | 0.015 ± 0.003 | 0.18 ± 0.05 | 1.09 ± 0.22 |
| | 2 | 0.66 ± 0.09 | 0.68 ± 0.12 | 16.8 ± 1.12 | 42.87 ± 2.56 | 6.41 ± 0.03 | 27.52 ± 1.06 | 0.05 ± 0.03 | 0.017 ± 0.002 | 0.18 ± 0.06 | 0.65 ± 0.10 |
| | 3 | 0.21 ± 0.04 | 0.49 ± 0.06 | 18.27 ± 1.34 | 49.49 ± 9.08 | 6.09 ± 0.25 | 20.64 ± 2.10 | 0.02 ± 0.00 | 0.018 ± 0.002 | 0.23 ± 0.06 | 0.64 ± 0.16 |
| | 4 | 0.31 ± 0.01 | 0.37 ± 0.05 | 18.27 ± 1.56 | 47.82 ± 10.22 | 4.46 ± 0.44 | 20.64 ± 1.78 | 0.17 ± 0.03 | 0.002 ± 0.001 | 0.15 ± 0.03 | 0.62 ± 0.20 |
| | 5 | 0.35 ± 0.04 | 0.49 ± 0.04 | 32.61 ± 2.46 | 53.79 ± 8.87 | 6.14 ± 0.23 | 13.76 ± 0.67 | 0.06 ± 0.02 | 0.001 ± 0.000 | 0.30 ± 0.09 | 0.8 ± 0.15 |
| | 6 | 0.31 ± 0.03 | 0.6 ± 0.08 | 20.15 ± 1.23 | 34.43 ± 3.22 | 4.75 ± 0.55 | 27.52 ± 1.99 | 0.14 ± 0.05 | 0.001 ± 0.000 | 0.21 ± 0.06 | 0.68 ± 0.11 |
| | 7 | 0.95 ± 0.10 | 0.77 ± 0.10 | 29.37 ± 2,78 | 224.99 ± 45.38 | 6.13 ± 0.04 | 27.52 ± 0.34 | 0.09 ± 0.02 | 0.001 ± 0.000 | 0.20 ± 0.04 | 33.82 ± 6.29 |
| | 8 | 0.43 ± 0.09 | 0.49 ± 0.08 | 27.91 ± 1.34 | 114.54 ± 26.11 | 5.98 ± 0.02 | 28.89 ± 0.98 | 0.04 ± 0.01 | 0.000 ± 0.000 | 0.26 ± 0.03 | 0.52 ± 0.12 |
| | 9 | 0.51 ± 0.04 | 0.72 ± 0.29 | 36.48 ± 3.12 | 231.02 ± 11.98 | 6.86 ± 0.06 | 41.96 ± 7.10 | 0.04 ± 0.02 | 0.002 ± 0.000 | 0.20 ± 0.03 | 29.89 ± 6.98 |
| | 10 | 0.82 ± 0.11 | 0.69 ± 0.18 | 34.35 ± 2.45 | 220.04 ± 34.39 | 6.62 ± 0.12 | 39.21 ± 6.02 | 0.03 ± 0.00 | 0.007 ± 0.001 | 0.19 ± 0.08 | 1.32 ± 0.18 |
| | 11 | 0.88 ± 0.12 | 0.7 ± 0.21 | 24.52 ± 2.45 | 38.73 ± 4.56 | 6.8 ± 0.22 | 41.27 ± 9.12 | 0.04 ± 0.01 | 0.01 ± 0.003 | 0.19 ± 0.07 | 0.87 ± 0.15 |
| | 12 | 0.41 ± 0.10 | 0.51 ± 0.18 | 32.88 ± 5.45 | 38.49 ± 1.89 | 5.98 ± 0.68 | 20.64 ± 2.89 | 0.13 ± 0.06 | 0.011 ± 0.002 | 0.21 ± 0.10 | 0.81 ± 0.07 |
| | 13 | 0.82 ± 0.11 | 0.90 ± 0.34 | 34.38 ± 6.86 | 24.05 ± 2.43 | 6.76 ± 0.24 | 30.95 ± 3.56 | 0.03 ± 0.00 | 0.011 ± 0.002 | 0.30 ± 0.01 | 0.59 ± 0.15 |
| | 14 | 0.44 ± 0.07 | 0.44 ± 0.11 | 20.76 ± 1.28 | 45.2 ± 6.56 | 6.91 ± 0.41 | 16.51 ± 1.64 | 0.04 ± 0.01 | 0.011 ± 0.002 | 0.19 ± 0.05 | 7.01 ± 3.22 |
| | 15 | 0.17 ± 0.01 | 0.21 ± 0.09 | 15.47 ± 1.50 | 47.15 ± 2.44 | 7.02 ± 0.06 | 12.31 ± 2.37 | 0.02 ± 0.00 | 0.012 ± 0.002 | 0.35 ± 0.08 | 1.10 ± 0.37 |
| | 16 | 0.3 ± 0.10 | 0.37 ± 0.05 | 32.73 ± 3.88 | 41.76 ± 3.51 | 6.7 ± 0.34 | 18.57 ± 1.88 | 0.04 ± 0.02 | 0.015 ± 0.004 | 0.28 ± 0.03 | 1.22 ± 0.34 |
| Hubei | 17 | 0.25 ± 0.09 | 0.22 ± 0.01 | 14.54 ± 3.11 | 34.93 ± 1.45 | 4.94 ± 0.08 | 14.45 ± 1.28 | 0.20 ± 0.09 | 0.011 ± 0.001 | 0.27 ± 0.09 | 0.90 ± 0.16 |
| | 18 | 0.41 ± 0.13 | 0.46 ± 0.17 | 16.82 ± 4.80 | 35.77 ± 2.22 | 6.09 ± 0.27 | 19.95 ± 3.56 | 0.10 ± 0.04 | 0.016 ± 0.003 | 0.19 ± 0.05 | 0.74 ± 0.21 |
| | 19 | 0.36 ± 0.08 | 0.35 ± 0.16 | 15.61 ± 4.41 | 42.95 ± 5.10 | 6.00 ± 0.06 | 12.38 ± 0.61 | 0.08 ± 0.02 | 0.014 ± 0.003 | 0.13 ± 0.01 | 5.91 ± 1.11 |
| | 20 | 0.49 ± 0.07 | 0.39 ± 0.10 | 13.18 ± 1.28 | 39.84 ± 1.00 | 6.21 ± 0.06 | 41.27 ± 4.14 | 0.16 ± 0.01 | 0.022 ± 0.01 | 0.13 ± 0.02 | 9.84 ± 0.60 |
| | 21 | 1.24 ± 0.10 | 0.29 ± 0.01 | 21.03 ± 2.12 | 213.29 ± 11.38 | 6.19 ± 0.04 | 32.33 ± 2.09 | 0.02 ± 0.00 | 0.027 ± 0.004 | 0.23 ± 0.03 | 38.96 ± 0.98 |
| | 22 | 0.22 ± 0.07 | 0.43 ± 0.08 | 21.55 ± 0.98 | 64.11 ± 4.50 | 6.06 ± 0.05 | 21.32 ± 1.32 | 0.10 ± 0.00 | 0.009 ± 0.002 | 0.19 ± 0.02 | 4.07 ± 0.22 |
| | 23 | 0.27 ± 0.09 | 0.63 ± 0.09 | 17.78 ± 0.26 | 48.17 ± 4.92 | 6.43 ± 0.31 | 16.51 ± 0.98 | 0.04 ± 0.01 | 0.001 ± 0.001 | 0.22 ± 0.02 | 2.42 ± 0.52 |
| | 24 | 0.29 ± 0.07 | 0.46 ± 0.08 | 24.02 ± 0.67 | 46.58 ± 3.30 | 6.26 ± 0.38 | 27.52 ± 1.58 | 0.06 ± 0.01 | 0.009 ± 0.001 | 0.18 ± 0.01 | 1.89 ± 0.34 |
| | 25 | 0.18 ± 0.10 | 0.38 ± 0.06 | 30.02 ± 2.33 | 62.83 ± 4.42 | 6.30 ± 0.28 | 12.38 ± 1.04 | 0.05 ± 0.00 | 0.019 ± 0.006 | 0.27 ± 0.02 | 6.60 ± 0.51 |
| | 26 | 0.10 ± 0.01 | 0.57 ± 0.11 | 21.3 ± 1.04 | 83.18 ± 3.67 | 6.20 ± 0.40 | 7.57 ± 0.88 | 0.09 ± 0.01 | 0.016 ± 0.004 | 0.25 ± 0.01 | 9.90 ± 1.02 |
| | 27 | 0.20 ± 0.01 | 0.33 ± 0.08 | 23.41 ± 1.48 | 46.17 ± 3.08 | 6.41 ± 0.08 | 22.01 ± 3.13 | 0.04 ± 0.01 | 0.015 ± 0.00 | 0.23 ± 0.01 | 6.01 ± 0.80 |
| | 28 | 0.26 ± 0.01 | 0.5 ± 0.03 | 17.17 ± 0.34 | 43.18 ± 5.10 | 6.31 ± 0.08 | 15.82 ± 1.34 | 0.07 ± 0.02 | 0.018 ± 0.007 | 0.29 ± 0.03 | 2.15 ± 0.06 |
| | 29 | 0.76 ± 0.04 | 0.32 ± 0.03 | 17.82 ± 0.67 | 42.4 ± 2.88 | 5.37 ± 0.06 | 19.26 ± 1.02 | 0.15 ± 0.01 | 0.017 ± 0.003 | 0.36 ± 0.04 | 3.06 ± 0.50 |
| | 30 | 0.32 ± 0.02 | 0.42 ± 0.07 | 18.42 ± 1.45 | 220.63 ± 13.09 | 4.24 ± 0.04 | 20.64 ± 1.34 | 0.40 ± 0.08 | 0.027 ± 0.004 | 0.19 ± 0.01 | 2.85 ± 0.08 |
| | 31 | 0.25 ± 0.01 | 0.32 ± 0.10 | 17.28 ± 0.72 | 40.89 ± 1.06 | 4.13 ± 0.07 | 15.13 ± 0.90 | 0.49 ± 0.03 | 0.006 ± 0.002 | 0.24 ± 0.01 | 1.92 ± 0.22 |
| | 32 | 0.47 ± 0.09 | 0.22 ± 0.05 | 23.96 ± 1.06 | 68.93 ± 4.26 | 6.69 ± 0.42 | 21.32 ± 1.38 | 0.08 ± 0.04 | 0.011 ± 0.002 | 0.17 ± 0.01 | 1.80 ± 0.19 |
| | 33 | 0.49 ± 0.06 | 0.19 ± 0.02 | 27.78 ± 0.45 | 80.48 ± 5.72 | 6.64 ± 0.02 | 33.02 ± 2.64 | 0.08 ± 0.01 | 0.01 ± 0.005 | 0.19 ± 0.01 | 1.45 ± 0.06 |

**Table 2.** *Cont.*

| | Location | Soil | | | | | | Rice Grain | | | |
|---|---|---|---|---|---|---|---|---|---|---|---|
| | | Cd (mg·kg$^{-1}$) | Hg (mg·kg$^{-1}$) | Pb (mg·kg$^{-1}$) | Cr (mg·kg$^{-1}$) | pH | SOM (mg·kg$^{-1}$) | Cd (mg·kg$^{-1}$) | Hg (mg·kg$^{-1}$) | Pb (mg·kg$^{-1}$) | Cr (mg·kg$^{-1}$) |
| Hunan | 1 | 0.55 ± 0.10 | 0.01 ± 0.01 | 25.08 ± 2.34 | 140.7 ± 8.38 | 4.11 ± 0.12 | 30.27 ± 2.81 | 0.68 ± 0.12 | 0.007 ± 0.001 | 0.21 ± 0.03 | 0.53 ± 0.10 |
| | 2 | 3.44 ± 0.22 | 0.03 ± 0.00 | 22.2 ± 1.02 | 53.74 ± 3.46 | 6.82 ± 0.06 | 49.53 ± 3.54 | 0.25 ± 0.01 | 0.011 ± 0.01 | 0.20 ± 0.02 | 0.77 ± 0.08 |
| | 3 | 0.50 ± 0.04 | 0.05 ± 0.00 | 20.08 ± 0.49 | 61.82 ± 2.85 | 3.90 ± 0.10 | 23.39 ± 4.05 | 1.36 ± 0.13 | 0.013 ± 0.005 | 0.21 ± 0.01 | 0.41 ± 0.09 |
| | 4 | 0.47 ± 0.04 | 0.18 ± 0.02 | 19.21 ± 1.21 | 65.71 ± 5.24 | 5.91 ± 0.20 | 30.27 ± 3.18 | 0.10 ± 0.01 | 0.011 ± 0.004 | 0.21 ± 0.01 | 0.49 ± 0.01 |
| | 5 | 0.79 ± 0.03 | 0.06 ± 0.02 | 22.89 ± 3.44 | 64.62 ± 6.06 | 4.85 ± 0.15 | 46.78 ± 6.02 | 0.44 ± 0.01 | 0.019 ± 0.005 | 0.18 ± 0.02 | 0.35 ± 0.05 |
| | 6 | 0.29 ± 0.00 | 0.05 ± 0.01 | 25.65 ± 2.05 | 40.75 ± 3.56 | 6.51 ± 0.07 | 12.38 ± 3.11 | 0.24 ± 0.03 | 0.018 ± 0.005 | 0.17 ± 0.01 | 6.32 ± 0.90 |
| | 7 | 1.64 ± 0.10 | 0.03 ± 0.01 | 19.99 ± 1.05 | 68.54 ± 2.69 | 6.08 ± 0.12 | 39.9 ± 1.89 | 0.03 ± 0.01 | 0.019 ± 0.01 | 0.29 ± 0.01 | 0.47 ± 0.05 |
| | 8 | 0.97 ± 0.06 | 0.26 ± 0.05 | 21.56 ± 2.18 | 61.15 ± 3.06 | 6.30 ± 0.08 | 33.02 ± 2.34 | 1.02 ± 0.06 | 0.015 ± 0.002 | 0.49 ± 0.08 | 7.64 ± 0.03 |
| | 9 | 0.59 ± 0.04 | 0.07 ± 0.03 | 29.81 ± 0.88 | 82.65 ± 3.12 | 5.00 ± 0.06 | 30.27 ± 4.20 | 0.02 ± 0.00 | 0.017 ± 0.001 | 0.39 ± 0.04 | 3.30 ± 0.08 |
| | 10 | 1.46 ± 0.14 | 0.31 ± 0.06 | 15.15 ± 0.45 | 89.35 ± 5.15 | 6.95 ± 0.07 | 57.78 ± 7.44 | 0.09 ± 0.01 | 0.007 ± 0.001 | 0.36 ± 0.02 | 3.21 ± 0.20 |
| | 11 | 1.94 ± 0.09 | 0.20 ± 0.03 | 38.26 ± 2.65 | 33.43 ± 2.83 | 6.40 ± 0.03 | 45.4 ± 2.08 | 0.03 ± 0.00 | 0.011 ± 0.006 | 0.68 ± 0.06 | 3.67 ± 0.11 |
| | 12 | 3.44 ± 0.67 | 0.10 ± 0.02 | 23.67 ± 1.67 | 43.25 ± 2.04 | 6.19 ± 0.44 | 50.9 ± 2.42 | 0.115 ± 0.01 | 0.005 ± 0.001 | 0.36 ± 0.05 | 3.49 ± 0.09 |
| | 13 | 0.32 ± 0.04 | 0.04 ± 0.01 | 23.39 ± 3.81 | 83.41 ± 6.72 | 6.13 ± 0.50 | 19.26 ± 1.09 | 0.065 ± 0.01 | 0.006 ± 0.001 | 0.28 ± 0.02 | 3.42 ± 0.31 |
| | 14 | 0.70 ± 0.11 | 0.14 ± 0.03 | 11.44 ± 0.56 | 64.56 ± 4.38 | 6.86 ± 0.20 | 22.01 ± 0.84 | 0.02 ± 0.00 | 0.007 ± 0.002 | 0.48 ± 0.02 | 2.84 ± 0.42 |
| | 15 | 1.08 ± 0.42 | 0.19 ± 0.02 | 25.56 ± 1.89 | 75.97 ± 5.47 | 6.61 ± 0.04 | 38.52 ± 2.55 | 0.03 ± 0.01 | 0.006 ± 0.002 | 0.38 ± 0.02 | 2.97 ± 0.36 |
| | 16 | 0.17 ± 0.02 | 0.11 ± 0.01 | 23.53 ± 0.85 | 41.53 ± 2.70 | 6.7 ± 0.08 | 8.25 ± 1.00 | 0.01 ± 0.00 | 0.009 ± 0.003 | 0.44 ± 0.04 | 2.67 ± 0.04 |
| | 17 | 0.30 ± 0.05 | 0.04 ± 0.01 | 19.61 ± 2.31 | 73.7 ± 4.36 | 6.73 ± 0.06 | 8.25 ± 0.48 | 0.07 ± 0.01 | 0.013 ± 0.005 | 0.21 ± 0.01 | 2.51 ± 0.06 |
| | 18 | 0.71 ± 0.0 | 0.01 ± 0.00 | 32.39 ± 3.58 | 77.21 ± 2.99 | 6.28 ± 0.12 | 39.90 ± 2.56 | 0.03 ± 0.00 | 0.015 ± 0.005 | 0.30 ± 0.04 | 2.56 ± 0.05 |
| | 19 | 0.45 ± 0.10 | 0.06 ± 0.01 | 24.15 ± 4.08 | 56.11 ± 4.00 | 6.20 ± 0.44 | 24.76 ± 1.08 | 0.19 ± 0.05 | 0.013 ± 0.006 | 0.33 ± 0.08 | 2.54 ± 0.07 |
| | 20 | 1.46 ± 0.62 | 0.1 ± 0.05 | 26.89 ± 1.67 | 78.4 ± 3.98 | 6.11 ± 0.54 | 30.27 ± 2.22 | 0.62 ± 0.14 | 0.008 ± 0.003 | 0.39 ± 0.03 | 2.82 ± 0.16 |
| | 21 | 2.38 ± 0.56 | 0.03 ± 0.01 | 32.72 ± 3.28 | 69.02 ± 17.45 | 5.39 ± 0.48 | 17.88 ± 0.67 | 0.77 ± 0.21 | 0.019 ± 0.006 | 0.56 ± 0.05 | 3.41 ± 0.67 |
| | 22 | 3.34 ± 0.33 | 0.11 ± 0.05 | 127.56 ± 6.82 | 62.14 ± 6.12 | 6.14 ± 0.14 | 39.9 ± 2.28 | 0.18 ± 0.01 | 0.016 ± 0.005 | 0.38 ± 0.03 | 4.72 ± 1.16 |
| | 23 | 1.27 ± 0.25 | 0.13 ± 0.0 | 21.28 ± 1.62 | 70.41 ± 3.88 | 6.10 ± 0.12 | 45.4 ± 5.67 | 1.46 ± 0.03 | 0.014 ± 0.007 | 0.43 ± 0.02 | 4.22 ± 0.86 |
| | 24 | 0.57 ± 0.04 | 0.79 ± 0.0 | 89.37 ± 4.50 | 59.41 ± 4.21 | 5.86 ± 0.08 | 26.14 ± 1.32 | 0.73 ± 0.04 | 0.006 ± 0.001 | 0.44 ± 0.04 | 4.59 ± 0.42 |
| | 25 | 1.14 ± 0.08 | 0.01 ± 0.00 | 21.82 ± 2.00 | 32.86 ± 5.38 | 5.77 ± 0.14 | 44.02 ± 3.52 | 0.014 ± 0.01 | 0.021 ± 0.007 | 0.27 ± 0.01 | 4.80 ± 0.61 |
| | 26 | 1.06 ± 0.06 | 0.11 ± 0.03 | 13.39 ± 0.56 | 60.45 ± 3.45 | 6.11 ± 0.16 | 38.52 ± 2.65 | 0.03 ± 0.01 | 0.023 ± 0.008 | 0.31 ± 0.05 | 4.32 ± 0.35 |
| | 27 | 0.38 ± 0.02 | 0.01 ± 0.00 | 10.47 ± 0.99 | 77.19 ± 6.11 | 5.93 ± 0.48 | 24.76 ± 1.28 | 0.04 ± 0.01 | 0.021 ± 0.001 | 0.23 ± 0.01 | 0.33 ± 0.01 |
| | 28 | 0.47 ± 0.08 | 0.07 ± 0.01 | 21.21 ± 3.45 | 49.93 ± 5.45 | 4.83 ± 0.16 | 41.27 ± 4.45 | 0.05 ± 0.01 | 0.013 ± 0.001 | 0.31 ± 0.01 | 4.47 ± 0.60 |
| | 29 | 0.37 ± 0.09 | 0.29 ± 0.07 | 30.02 ± 6.98 | 37.37 ± 3.22 | 5.91 ± 0.18 | 17.88 ± 1.65 | 0.05 ± 0.00 | 0.022 ± 0.002 | 0.18 ± 0.01 | 0.23 ± 0.02 |
| | 30 | 0.35 ± 0.08 | 0.08 ± 0.02 | 12.8 ± 1.78 | 67.58 ± 5.45 | 6.18 ± 0.40 | 34.39 ± 3.55 | 0.044 ± 0.01 | 0.024 ± 0.003 | 0.29 ± 0.00 | 4.20 ± 0.29 |
| | 31 | 0.49 ± 0.16 | 0.14 ± 0.0 | 16.29 ± 0.99 | 80.35 ± 6.68 | 6.34 ± 0.12 | 30.27 ± 2.86 | 0.02 ± 0.01 | 0.024 ± 0.003 | 0.40 ± 0.08 | 4.09 ± 0.82 |
| | 32 | 1.12 ± 0.09 | 0.17 ± 0.00 | 19.17 ± 1.34 | 39.17 ± 1.67 | 6.19 ± 0.18 | 26.14 ± 1.64 | 0.02 ± 0.00 | 0.009 ± 0.002 | 0.23 ± 0.02 | 0.64 ± 0.08 |
| | 33 | 0.62 ± 0.14 | 0.63 ± 0.08 | 36.57 ± 7.22 | 78.45 ± 9.02 | 5.88 ± 0.20 | 16.51 ± 1.12 | 0.09 ± 0.01 | 0.009 ± 0.002 | 0.31 ± 0.02 | 3.95 ± 1.23 |
| | 34 | 1.38 ± 0.32 | 0.15 ± 0.02 | 52.69 ± 6.12 | 84.26 ± 6.12 | 6.25 ± 0.14 | 44.02 ± 4.34 | 0.42 ± 0.08 | 0.01 ± 0.002 | 0.42 ± 0.04 | 71.04 ± 9.08 |
| | 35 | 2.83 ± 0.23 | 0.06 ± 0.00 | 29.49 ± 4.56 | 77.46 ± 2.55 | 5.98 ± 0.08 | 48.15 ± 2.85 | 1.81 ± 0.31 | 0.028 ± 0.002 | 0.22 ± 0.01 | 5.92 ± 3.82 |
| | 36 | 0.64 ± 0.09 | 0.26 ± 0.04 | 59.8 ± 7.45 | 85.38 ± 4.33 | 5.18 ± 0.12 | 34.39 ± 2.66 | 1.03 ± 0.10 | 0.032 ± 0.003 | 0.32 ± 0.02 | 5.60 ± 1.08 |
| | 37 | 0.54 ± 0.12 | 0.3 ± 0.08 | 32.93 ± 2.34 | 75.52 ± 1.32 | 5.19 ± 0.08 | 26.14 ± 1.05 | 1.34 ± 0.09 | 0.017 ± 0.003 | 0.31 ± 0.05 | 4.25 ± 0.49 |

**Table 2.** *Cont.*

| | Location | Soil | | | | | | Rice Grain | | | |
|---|---|---|---|---|---|---|---|---|---|---|---|
| | | Cd (mg·kg$^{-1}$) | Hg (mg·kg$^{-1}$) | Pb (mg·kg$^{-1}$) | Cr (mg·kg$^{-1}$) | pH | SOM (mg·kg$^{-1}$) | Cd (mg·kg$^{-1}$) | Hg (mg·kg$^{-1}$) | Pb (mg·kg$^{-1}$) | Cr (mg·kg$^{-1}$) |
| | 38 | 0.71 ± 0.21 | 0.1 ± 0.03 | 17.18 ± 0.78 | 52.76 ± 4.34 | 5.91 ± 0.01 | 22.01 ± 1.87 | 1.64 ± 0.09 | 0.012 ± 0.01 | 0.49 ± 0.03 | 4.25 ± 1.02 |
| | 39 | 2.76 ± 0.30 | 0.6 ± 0.09 | 17.59 ± 1.56 | 140.45 ± 10.20 | 5.34 ± 0.22 | 34.39 ± 2.35 | 1.40 ± 0.06 | 0.014 ± 0.00 | 0.29 ± 0.02 | 4.12 ± 0.87 |
| | 40 | 1.19 ± 0.17 | 0.14 ± 0.05 | 27.21 ± 4.23 | 103.22 ± 6.72 | 5.66 ± 0.08 | 37.15 ± 2.44 | 0.03 ± 0.01 | 0.001 ± 0.000 | 0.19 ± 0.00 | 4.12 ± 0.65 |
| Jiangxi | 1 | 0.26 ± 0.04 | 0.19 ± 0.05 | 15.36 ± 1.22 | 47.05 ± 3.56 | 4.34 ± 0.04 | 13.76 ± 1.22 | 0.40 ± 0.08 | 0.005 ± 0.002 | 0.31 ± 0.02 | 2.34 ± 0.10 |
| | 2 | 1.22 ± 0.14 | 0.06 ± 0.02 | 13.47 ± 0.89 | 34.5 ± 10.28 | 4.36 ± 0.12 | 16.51 ± 1.56 | 0.15 ± 0.03 | 0.013 ± 0.002 | 0.23 ± 0.01 | 1.44 ± 0.08 |
| | 3 | 0.59 ± 0.08 | 0.22 ± 0.07 | 14.46 ± 0.67 | 62.86 ± 12.43 | 4.09 ± 0.06 | 7.57 ± 2.11 | 0.05 ± 0.01 | 0.015 ± 0.007 | 0.35 ± 0.02 | 1.05 ± 0.09 |
| | 4 | 0.60 ± 0.12 | 0.04 ± 0.01 | 7.64 ± 1.11 | 13.52 ± 5.02 | 4.16 ± 0.08 | 25.45 ± 3.65 | 0.21 ± 0.06 | 0.022 ± 0.004 | 0.25 ± 0.02 | 3.25 ± 0.11 |
| | 5 | 0.17 ± 0.04 | 0.09 ± 0.02 | 9.86 ± 2.34 | 100.88 ± 10.53 | 4.25 ± 0.12 | 15.13 ± 1.20 | 0.28 ± 0.02 | 0.021 ± 0.007 | 0.50 ± 0.08 | 59.93 ± 6.08 |
| | 6 | 0.36 ± 0.08 | 0.12 ± 0.01 | 12.00 ± 4.12 | 37.43 ± 1.67 | 3.79 ± 0.08 | 52.97 ± 7.02 | 0.10 ± 0.09 | 0.024 ± 0.004 | 0.36 ± 0.02 | 77.54 ± 11.01 |
| | 7 | 0.15 ± 0.04 | 0.16 ± 0.03 | 15.64 ± 3.34 | 138.54 ± 8.88 | 3.99 ± 0.20 | 9.63 ± 2.44 | 0.26 ± 0.04 | 0.011 ± 0.004 | 0.38 ± 0.03 | 6.45 ± 0.91 |
| | 8 | 1.24 ± 0.24 | 0.10 ± 0.02 | 11.14 ± 1.08 | 73.95 ± 4.94 | 4.09 ± 0.09 | 20.64 ± 1.06 | 0.19 ± 0.03 | 0.009 ± 0.002 | 0.25 ± 0.02 | 6.95 ± 1.08 |
| | 9 | 0.14 ± 0.04 | 0.15 ± 0.04 | 14.08 ± 2.78 | 114.07 ± 10.02 | 4.85 ± 0.11 | 28.2 ± 0.86 | 0.39 ± 0.06 | 0.014 ± 0.002 | 0.44 ± 0.04 | 3.16 ± 0.20 |
| | 10 | 0.14 ± 0.02 | 0.48 ± 0.13 | 18.54 ± 3.56 | 86.06 ± 3.98 | 4.84 ± 0.14 | 13.76 ± 0.29 | 0.89 ± 0.13 | 0.012 ± 0.004 | 0.28 ± 0.01 | 2.66 ± 0.48 |
| | 11 | 0.36 ± 0.10 | 0.17 ± 0.02 | 18.78 ± 3.17 | 110.06 ± 16.67 | 5.09 ± 0.22 | 35.77 ± 3.18 | 0.34 ± 0.08 | 0.006 ± 0.002 | 0.25 ± 0.03 | 1.58 ± 0.52 |
| | 12 | 0.15 ± 0.05 | 0.15 ± 0.01 | 16.91 ± 1.42 | 78.13 ± 4.24 | 4.90 ± 0.12 | 41.96 ± 1.67 | 0.09 ± 0.01 | 0.009 ± 0.001 | 0.61 ± 0.09 | 10.28 ± 2.00 |
| | 13 | 0.24 ± 0.06 | 0.14 ± 0.02 | 10.33 ± 1.29 | 93.52 ± 10.08 | 5.00 ± 0.04 | 26.83 ± 1.35 | 0.15 ± 0.03 | 0.03 ± 0.003 | 0.29 ± 0.02 | 93.74 ± 3.91 |
| | 14 | 0.24 ± 0.04 | 0.35 ± 0.07 | 19.69 ± 2.61 | 90.27 ± 4.67 | 6.2 ± 0.06 | 27.52 ± 0.79 | 0.10 ± 0.04 | 0.013 ± 0.003 | 0.22 ± 0.01 | 10.32 ± 1.18 |
| | 15 | 0.17 ± 0.07 | 0.15 ± 0.01 | 8.89 ± 1.52 | 43.53 ± 5.12 | 5.46 ± 0.09 | 29.58 ± 1.86 | 0.07 ± 0.03 | 0.008 ± 0.004 | 0.25 ± 0.02 | 2.23 ± 0.25 |
| | 16 | 0.71 ± 0.12 | 0.09 ± 0.02 | 12.81 ± 0.89 | 63.21 ± 12.44 | 5.54 ± 0.02 | 31.64 ± 3.22 | 0.06 ± 0.02 | 0.008 ± 0.001 | 0.32 ± 0.04 | 3.93 ± 0.72 |
| | 17 | 4.64 ± 0.51 | 0.06 ± 0.00 | 26.75 ± 1.28 | 57.53 ± 1.98 | 5.99 ± 0.04 | 34.39 ± 1.67 | 0.35 ± 0.06 | 0.004 ± 0.001 | 0.22 ± 0.02 | 1.23 ± 0.22 |
| | 18 | 2.66 ± 0.26 | 0.04 ± 0.00 | 25.74 ± 2.31 | 97.76 ± 7.51 | 5.78 ± 0.12 | 39.91 ± 6.08 | 0.22 ± 0.07 | 0.003 ± 0.000 | 0.12 ± 0.01 | 0.81 ± 0.13 |
| | 19 | 3.58 ± 1.28 | 0.13 ± 0.01 | 36.18 ± 4.72 | 90.95 ± 2.32 | 6.29 ± 0.17 | 38.52 ± 8.02 | 0.44 ± 0.09 | 0.09 ± 0.004 | 0.23 ± 0.01 | 0.78 ± 0.04 |
| | 20 | 2.40 ± 0.67 | 0.11 ± 0.00 | 27.72 ± 1.78 | 90.33 ± 7.15 | 5.84 ± 0.03 | 44.02 ± 4.25 | 0.64 ± 0.17 | 0.002 ± 0.001 | 0.15 ± 0.03 | 1.25 ± 0.32 |
| | 21 | 3.42 ± 0.22 | 0.06 ± 0.01 | 27.82 ± 3.45 | 86.6 ± 14.12 | 5.81 ± 0.15 | 52.28 ± 7.31 | 0.96 ± 0.21 | 0.005 ± 0.001 | 0.34 ± 0.02 | 1.35 ± 0.19 |
| | 22 | 1.43 ± 0.36 | 0.09 ± 0.00 | 28.49 ± 4.02 | 89.24 ± 10.58 | 5.88 ± 0.19 | 53.65 ± 7.04 | 0.67 ± 0.14 | 0.001 ± 0.000 | 0.17 ± 0.01 | 0.93 ± 0.06 |
| | 23 | 2.12 ± 0.48 | 0.04 ± 0.01 | 31.21 ± 3.33 | 68.05 ± 3.98 | 6.24 ± 0.03 | 35.77 ± 2.43 | 0.29 ± 0.05 | 0.003 ± 0.001 | 0.18 ± 0.04 | 1.05 ± 0.11 |
| | 24 | 0.64 ± 0.14 | 0.03 ± 0.00 | 21.7 ± 1.21 | 78.3 ± 5.82 | 5.51 ± 0.06 | 24.76 ± 2.89 | 0.42 ± 0.04 | 0.005 ± 0.001 | 0.22 ± 0.03 | 0.90 ± 0.06 |
| | 25 | 0.74 ± 0.12 | 0.06 ± 0.01 | 19.65 ± 0.99 | 78.32 ± 3.77 | 5.89 ± 0.14 | 46.78 ± 11.45 | 0.61 ± 0.06 | 0.004 ± 0.001 | 0.25 ± 0.05 | 0.76 ± 0.05 |
| | 26 | 1.42 ± 0.36 | 0.07 ± 0.01 | 24.12 ± 4.18 | 63.61 ± 9.22 | 6.04 ± 0.08 | 42.65 ± 4.88 | 0.37 ± 0.04 | 0.008 ± 0.003 | 0.23 ± 0.01 | 0.86 ± 0.06 |
| | 27 | 0.91 ± 0.21 | 0.05 ± 0.01 | 18.01 ± 3.09 | 55.37 ± 1.62 | 5.79 ± 1.02 | 30.27 ± 3.67 | 0.28 ± 0.07 | 0.001 ± 0.000 | 0.28 ± .002 | 0.77 ± 0.11 |
| | 28 | 2.86 ± 0.34 | 0.12 ± 0.02 | 54.64 ± 6.29 | 65.86 ± 6.08 | 5.39 ± 0.28 | 44.02 ± 5.23 | 0.52 ± 0.14 | 0.005 ± 0.001 | 0.28 ± 0.03 | 0.90 ± 0.09 |

Note: Results are represented as the mean ± standard deviation ($n$ = 3 biological replicates); The sampling areas in **Hubei**: 1–6 Wuxue City, 7–12 Yangxin City, 13–16 Ezhou City, 17–21Dawu City, 22–28 Zhongxiang City, 29–31 Shayang City, 32–33 Wuhan City; the sampling areas in **Hunan**: 1–7 Yiyang City, 8–14 Liangyu City, 15–18 Shaodong City, 19–21 Yiyang City, 22–26 Guiyang City, 27–30 Jiahe City, 31–35 Gongzhou City 36 Zhuzhou City 37–38 Xiangtan City 39–40 Changsha City; the sampling areas in **Jiangxi**: 1–2 Guixi City, 3–6 Yujiang City, 7–8 Dongxiang City, 9–12 Jinxian City, 13–16 Nanchang City, 17–18 Yongxiu City, 19–28 Ruichang City.

The average concentrations of heavy metals in soil are presented in Table 3. The average concentrations of Cd, Hg, Pb, and Cr in soils from Hubei Province were 0.43, 0.50, 23.16, and 81.24 mg$\cdot$ kg$^{-1}$, respectively. The average concentrations of Cd, Hg, Pb, and Cr in soils from Hunan Province were 1.12, 0.15, 29.00, and 69.00 mg$\cdot$ kg$^{-1}$, respectively. The average concentrations of Cd, Hg, Pb, and Cr in soils from Jiangxi Province were 1.20, 0.13, 20.06, and 75.34 mg$\cdot$ kg$^{-1}$, respectively. The Cd accumulation in soils among the test samples was in the order of Hunan = Jiangxi > Hubei; the Hg contamination of the soils was in order the of Hubei > Hunan = Jiangxi; the Pb concentration of the soils was in the order of Hunan > Hubei > Jiangxi; and the Cr contamination of the soils among the test samples from three provinces showed no significant difference. The results show that some soil samples from this region were polluted by Cd and Hg, and heavy metal contamination is serious at some sample sites. Cd and Hg contamination in agricultural soils may originate from: (1) atmospheric fallout from industrial and urban activities; (2) the application of sewage sludge and manure as fertilizer; and (3) the widespread use of organic and complex fertilizers [11,24–26]. Previous studies have shown that the application of K fertilizer significantly affects the Cd concentration in soils [27].

**Table 3.** The average concentrations of heavy metals in soil and rice.

| Experimental Province | Soil (mg$\cdot$ kg$^{-1}$) | | | | Rice Grain (mg$\cdot$ kg$^{-1}$) | | | |
|---|---|---|---|---|---|---|---|---|
| | Cd | Hg | Pb | Cr | Cd | Hg | Pb | Cr |
| Hubei | 0.43 ± 0.26 $^{a}$ | 0.50 ± 0.24 $^{b}$ | 23.16 ± 6.74 $^{a,b}$ | 81.24 ± 65.75 $^{a}$ | 0.09 ± 0.10 $^{a}$ | 0.012 ± 0.01 $^{a}$ | 0.22 ± 0.06 $^{a}$ | 5.52 ± 9.65 $^{a}$ |
| Hunan | 1.12 ± 0.92 $^{b}$ | 0.15 ± 0.17 $^{a}$ | 29.00 ± 21.33 $^{b}$ | 69.00 ± 23.44 $^{a}$ | 0.41 ± 0.54 $^{b}$ | 0.014 ± 0.01 $^{a}$ | 0.33 ± 0.12 $^{b}$ | 4.91 ± 10.88 $^{a}$ |
| Jiangxi | 1.20 ± 1.24 $^{b}$ | 0.13 ± 0.09 $^{a}$ | 20.06 ± 10.02 $^{a}$ | 75.34 ± 26.79 $^{a}$ | 0.34 ± 0.24 $^{b}$ | 0.013 ± 0.02 $^{a}$ | 0.29 ± 0.11 $^{b}$ | 10.66 ± 24.02 $^{a}$ |

Note: Results are represented as mean ± standard deviation; different letters $^{a, b, c}$ in the same column denote that the differences are statistically significant (LSD, $p < 0.05$).

Soil properties have a considerable influence on the environmental chemistry properties and distribution of heavy metals [28,29]. The soil organic matter (SOM) concentration and soil pH were determined for all soil samples (Table 2). The soil pH values in the samples from Hubei, Hunan, and Jiangxi Provinces were in the range of 4.13–7.02, 3.90–6.95, and 3.79–6.29, respectively, with average values of 6.05, 5.92, and 5.20, respectively. The average SOM content in the soil samples from Hubei Province was 23.2 mg$\cdot$ kg$^{-1}$, ranging from 7.57 to 41.96 mg$\cdot$ kg$^{-1}$; these values are similar to those obtained by Wang *et al.* [30]. The average SOM content for the soil samples from Hunan Province was 32.26 mg$\cdot$ kg$^{-1}$, ranging from 8.25 to 57.78 mg$\cdot$ kg$^{-1}$. The SOM content in the soil samples from Jiangxi Province ranged from 7.57 to 53.65 mg$\cdot$ kg$^{-1}$, with an average value of 31.57 mg$\cdot$ kg$^{-1}$. The soil pH and SOM contents varied between soil samples from the different provinces.

### 3.2. Heavy Metal Content in Rice Grains

Table 2 presents the heavy metal concentrations in the rice grain samples. The concentrations of four heavy metals in the rice grain from Hubei Province exceeded the National Food Hygiene Standard values [23]. Specifically, the Cd concentrations in rice grain from sites 30 and 31 exceeded the National Food Health Standards of China [23] (Cd $\leqslant$ 0.2 mg$\cdot$ kg$^{-1}$), whereas the Hg concentrations in rice grain samples from sites 20, 21, and 29 exceeded the National Food Health Standards (Hg $\leqslant$ 0.02 mg$\cdot$ kg$^{-1}$). The Pb concentrations in 17 samples exceeded the National Food Health Standards (Pb $\leqslant$ 0.2 mg$\cdot$ kg$^{-1}$), and the Pb and Cr standards were exceeded in some samples. The Cr concentrations in 19 samples were 1 to 12 times greater than the National Food Health Standards (Cr $\leqslant$ 1 mg$\cdot$ kg$^{-1}$). The limits for Cd, Hg, Pb, and Cr were exceeded in 6.06%, 9.09%, 51.52%, and 57.58% of the samples, respectively.

The concentrations of Cd, Hg, Pb, and Cr in rice samples from Hunan Province exceeded the thresholds of the National Food Hygiene Standard for rice grain (Table 1). The Cd threshold value was exceeded in 17 tested rice grain samples, whereas the threshold for Hg was exceeded in six samples. The Pb concentrations in all except two samples exceeded the National Food Hygiene threshold values. In the samples that exceeded the threshold value, the Pb concentrations ranged from 0.17 to 0.68 mg· $kg^{-1}$. The Cr concentrations were below the threshold value in 10 samples; the concentrations in the remaining samples were 2–9 times greater than the standard value. The threshold values for Cd, Hg, Pb, and Cr were exceeded in 42.50%, 15%, 95%, and 75% of all rice grain samples, respectively.

The concentrations of four heavy metals in the rice grain samples from Jiangxi Province were not below the National Food Sanitation Standards. As shown in Table 2, the National Food Sanitation Standards for Cd, Hg, Pb, and Cr were exceeded in 18, five, seven, and 18 samples from Jiangxi Province, respectively. Among the tested samples, the maximum Cd concentration was 0.96 mg· $kg^{-1}$, which was nearly five times greater than the national limit. The maximum concentration of Cr was 93.74 mg· $kg^{-1}$, whereas the national health standard is 1.0 mg· $kg^{-1}$. Of the samples collected from Jiangxi Province, 42.50%, 15%, 95%, and 75% exceeded the thresholds for Cd, Hg, Pb, and Cr, respectively.

There is a clear regional pattern in the pollution: some regions are specifically polluted by certain metals. Compared with other regions, the Cr detection rate in rice grain samples was higher in Zhongxiang, Shayang and Wuhan of Hubei Province, whereas the detection rate of Pb was higher for Zhongxiang and Wuxue. Cd pollution of rice grains was more serious in Yiyang, Hengyang, and Zhuzhou of Hunan Province than in other locations. The results of this study are similar to those found by Ming *et al.* [31]. The Cd pollution of rice grain samples was more serious in some locations of Jiangxi Province than others; this trend was also confirmed by Zhang *et al.* [32].

The average concentrations of heavy metals in the rice grain samples are presented in Table 3. The average concentrations of Cd, Hg, Pb, and Cr in the rice grain from Hubei Province were 0.09, 0.012, 0.22, and 5.52 mg· $kg^{-1}$, respectively. The average concentrations of Cd, Hg, Pb, and Cr in the rice grain from Hunan Province were 0.41, 0.014, 0.33, and 4.91 mg· $kg^{-1}$, respectively. The average concentrations of Cd, Hg, Pb, and Cr in the rice grain from Jiangxi Province were 0.34, 0.013, 0.29, and 10.66 mg· $kg^{-1}$, respectively. Comparing the samples from the three provinces (Tables 2 and 3), The Cd and Pb accumulation of the rice grains among the test samples was in the order of Hunan = Jiangxi > Hubei. The Cr contamination of the soils among the test samples from the three provinces showed no significant difference. The accumulation of heavy metals in rice grains is likely influenced by more than the heavy metal accumulation in soil. Studies have shown that heavy metal contamination in rice plants may also be due to air pollution, sewage water irrigation, and pesticide spraying [33–35].

### *3.3. Relationship between Heavy Metal Concentrations in Soil and Rice Grains*

The concentrations of Cd, Hg, Pb, and Cr in soils and rice grain from the Yangtze River region are shown in Table 4. The results show that 70.90%, 29.7%, 0%, and 0% of the 101 soil samples collected in the study area were polluted by Cd, Hg, Pb, and Cr, respectively. There was considerable accumulation of Cd in the soil. The concentrations of Cd, Hg, Pb, and Cr in rice grain from the Yangtze River region ranged from 0.01 to 1.81, 0.00 to 0.09, 0.12 to 0.68, and 0.23 to 93.74 mg· $kg^{-1}$, respectively. The average Cd, Hg, Pb, and Cr concentrations in rice grain were 0.29, 0.01, 0.25, and 6.70 mg· $kg^{-1}$, respectively. Of the samples that had excessive heavy metal concentrations, the number of samples in which the Pb concentrations exceeded the standards was greatest, followed by Cr and Cd; these results indicate that rice grain was seriously polluted by Cd and Cr. Compared with the trace metal contents in agricultural soils in other areas in China [36] and in other countries [37], the mean concentrations of Cd and Cr in the Yangtze River region were generally higher, whereas Pb was generally lower in our study.

**Table 4.** The concentrations of four heavy metals (Cd, Hg, Pb, Cr) in soil and rice grain from the Yangtze River of China.

| Items | Soil | | | | Rice Grain | | | |
|---|---|---|---|---|---|---|---|---|
| | Cd | Hg | Pb | Cr | Cd | Hg | Pb | Cr |
| Number | 101 | 101 | 101 | 101 | 101 | 101 | 101 | 101 |
| Max (mg· $kg^{-1}$) | 4.64 | 1.46 | 127.56 | 231.02 | 1.81 | 0.09 | 0.68 | 93.74 |
| Min (mg· $kg^{-1}$) | 0.10 | 0.01 | 7.64 | 13.52 | 0.01 | 0 | 0.12 | 0.23 |
| Overall mean (mg· $kg^{-1}$) | 0.79 | 0.18 | 24.62 | 74.76 | 0.29 | 0.01 | 0.25 | 6.70 |
| Standard deviation (mg· $kg^{-1}$) | 0.94 | 0.25 | 15.27 | 44.13 | 0.39 | 0.01 | 0.11 | 15.42 |
| Median (mg· $kg^{-1}$) | 0.55 | 0.16 | 21.55 | 64.62 | 0.10 | 0.01 | 0.25 | 2.66 |
| Percentage of excessive amount | 70.90% | 29.7% | 0% | 0% | 37% | 16% | 60% | 70% |

Our results showed that the soil was not polluted by Pb and Cr, but the concentrations of Pb and Cr exceeded the threshold values in rice grain; this may be related to sewage water irrigation on the soil surface, heavy metal deposition from the air, fertilizers and the variety of rice being cultivated [25,38–40]. Many studies have shown that transport emissions can influence the concentrations of heavy metals in plants [41–44], particularly Pb concentrations [45]. Furthermore, the rice grain grown in Cd-contaminated soil was not polluted by Cd; this may be related to the varieties of rice being cultivated [40].

CCorA was applied to study the relationship between the heavy metal concentrations in rice grain and the heavy metal concentrations in soil. The CCorA results are plotted in Figure 2. The first two factors (F1 and F2) in the multivariate analysis represented 90.95% of the total variance and showed that the Cd and Pb concentrations in rice grain are positively related with the Cd and Pb concentrations in the corresponding soil, as expected. However, the results also showed that the Hg concentrations in soil are positively related to the Cr concentrations in the soil and that the concentrations of Cd and Pb in soil are positively correlated. The concentration of Hg in soils was found to be influenced by the accumulation of Pb and Cd in the rice grain. These results illustrate that the interactions among the heavy metal elements may also directly influence the migration and accumulation of heavy metals in soil or paddy rice. This conclusion was also obtained by Adams *et al.* [46].

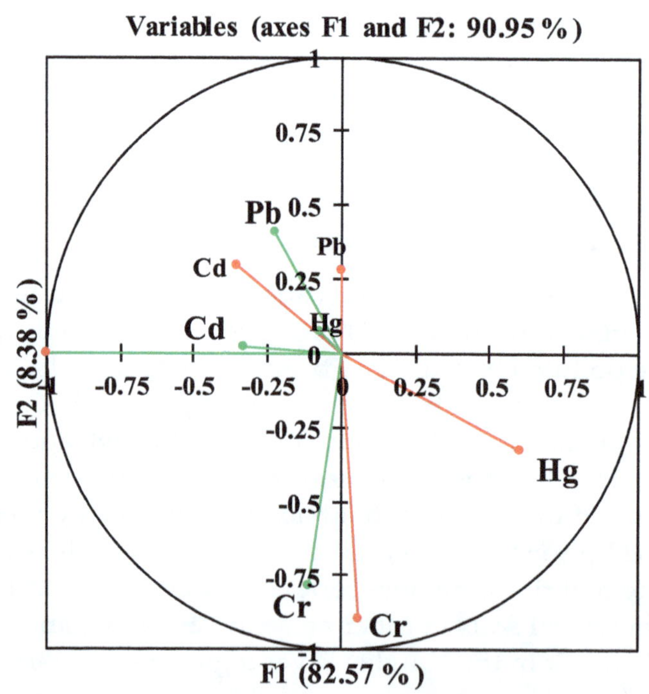


**Figure 2.** Ordination diagram based on the CCA of the heavy metal (Cd, Hg, Pb, and Cr) in soil (in black) and heavy metal (Cd, Hg, Pb, and Cr) in the rice grain (in red).

The heavy metal concentrations in rice grain and soil do not exhibit a one-to-one correlation. Rogival *et al.* [47] showed that the concentrations of heavy metals in soil were not the sole determinant of heavy metal accumulation in crops. Thus, the concentrations of heavy metals in soils were not the sole determinant of the heavy metal accumulation in the corresponding rice grain, which suggests that other factors may influence heavy metal accumulation in rice grain.

### 3.4. Relationship between Soil Properties and Heavy Metal Concentrations in the Soil-Rice System

The pH value and SOM content are important physicochemical properties of soil. Numerous studies have shown that soil pH and SOM control the accumulation of heavy metals in soil and plants [48,49]. CCorA was used to study the relationships between the heavy metal concentrations in soil and the soil properties (soil pH and SOM), as well as the relationships between the heavy metal concentrations in rice grain and soil properties. The results are shown in Figures 3 and 4. As shown in Figure 3, the first two factors (F1 and F2) in the multivariate analysis represented 96.97% of the total variance. The relationships between the soil properties and heavy metal concentrations in soil are not straightforward. There was a positive relationship between the soil pH values and Cr concentrations in soil. The SOM content was positively correlated with the Cd concentrations in soil samples. However, the relationships between Pb/Hg in soil and SOM content were not significant.

As shown in Figure 4, the first two factors (F1 and F2) in the multivariate analysis represented 99.94% of the total variance. The concentrations of Cd, Hg, Pb, and Cr in rice grain were negatively correlated with the soil pH value. The concentrations of Cd, Cr and Hg in rice grain were positively correlated with SOM content. The Cd and Cr concentrations in rice grain tended to be positively correlated with SOM content, which may favor the retention of these metals in the soil.

Tables 2 and 3 illustrate that rice grain samples that had high Cd concentrations generally had soil pH values less than 5.0. In general, as the pH value decreased, the absorption of Cd in rice grain increased. When the soil pH was higher than 6.0, the corresponding Cd concentration in the rice grain was within the national limits for rice grain, even when the Cd concentration in the soil was relatively high.

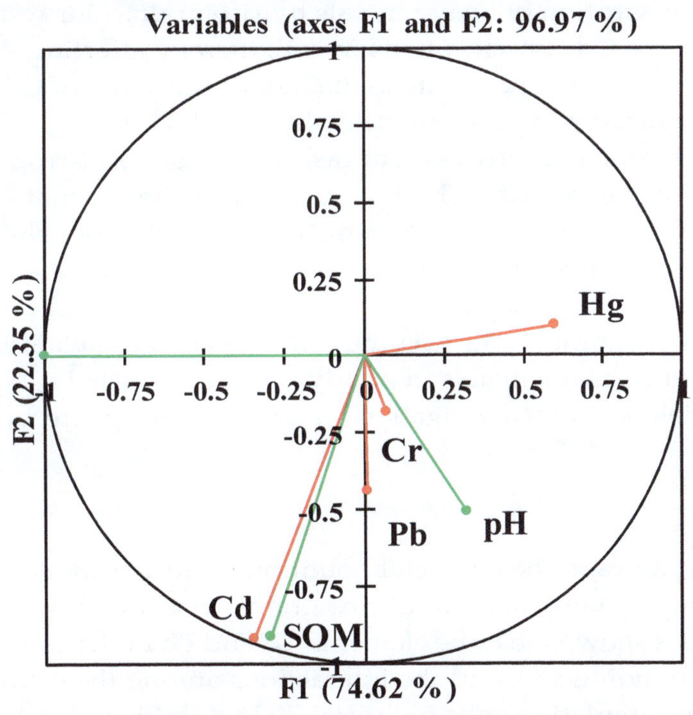


**Figure 3.** Ordination diagram based on the CCA of soil properties (in black) and heavy metals (Cd, Hg, Pb, and Cr) in the soil (in red).

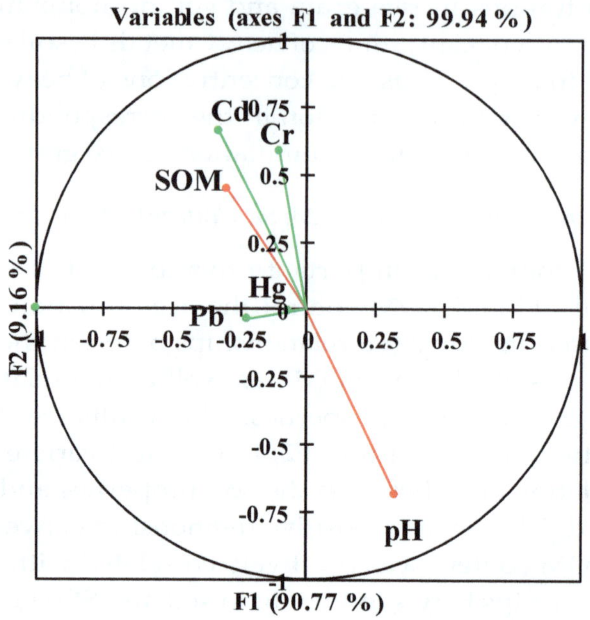


**Figure 4.** Ordination diagram based on the CCA of soil properties (in black) and heavy metals (Cd, Hg, Pb, and Cr) in rice grain (in red).

Soil pH has the ability to significantly alter the chemical speciation of the heavy metals in soil and influences the availability of heavy metals in soil and the ability of plants to absorb heavy metals [50–52]. Our study demonstrated that soil pH has an influence on heavy metal (Cd, Cr and Hg) concentrations in rice grain; as the soil pH value increased, the heavy metal concentrations in the rice grain decreased to a certain extent. This finding is similar to the findings of other studies [49,51,53] and illustrates that low pH may result in increased solubility and a high availability of heavy metals in soil [53], influencing the absorption of heavy metals by rice grain. Our results showed that SOM could significantly influence the Cd concentrations in soil, thereby affecting the absorption of Cd in rice grain. The effect of SOM on Cd could be due to the higher solubility of Cd in soils with increases in SOM. This result was in contrast to a previous study by Ghosh [54].

The migration and transformation processes of heavy metals in soil-crop systems are extremely complex, and many factors are involved. The total amount of heavy metals in soil, soil pH and SOM are major controlling factors in the migration of heavy metals from soil to plants. Numerous studies have indicated that differences in soil type, interactions between the elements, soil electrical conductivity, clay content, nutrients, enzyme activity, and other physical and chemical properties could directly influence the migration and accumulation of heavy metals in soil, as well as the absorption and accumulation of heavy metals in paddy rice [46,49,54–56]. Other possible influences include the type of paddy field examined, irrigation water, the atmosphere, and soil microbiological activity [57,58].

## 4. Conclusions

Based on the guideline values for heavy metal pollution, some paddy fields of the Yangtze River of China showed Cd and Hg contamination, and rice grains were polluted by Cd, Hg, Pb, and Cr. Canonical correlation analyses showed that the elements Cd and Pb in rice are related to the elemental concentrations in the corresponding soil, and the interactions among the elements Cd, Pb and Cr in soil directly influences the accumulation of heavy metals (Hg, Cd, Pb and Cr) in paddy soil and rice. Furthermore, soil pH and SOM were the important factors controlling the accumulation of Cd, Pb and Cr in soil and rice. These findings have important implications for the development of pollution prevention and reduction strategies to reduce heavy metal pollution for regions. This study can serve as a reference for the regional management and environment quality assessment of the study area. In further work, some experiment, which combined with soil properties, could be used for rational site-specific management in paddy fields.

**Acknowledgments:** This work was supported by the "Five-twelfth" National Science and Technology Support Program (2012BAK17B03), the National Nature Science Foundation of China (31401299, 31170251, 31401356), and Zhejiang Jinhua Science and Technology project (2015-2-012).

**Author Contributions:** This study was designed by Cheng Zhu, Zhouping Liu and Feijuan Wang. Field sampling was conducted by Junwei Sun and Zhouping Liu. Chemical analysis in the laboratory was performed by Zhouping Liu, Qiaofen Zhang, and Tiqian Han. All data was analysised by Zhouping Liu, Cheng Zhu, Feijuan Wang and Yanfei Ding. The manuscript was wrote by Zhouping Liu.

## References

1. Xiao, R.; Bai, J.H.; Gao, H.F. Distribution and contamination assessment of heavy metals in water and soils from the college town in the Pearl River Delta, China. *Clean (Weinh)* **2013**, *40*, 1167–1173. [CrossRef]
2. Ministry of Environmental protection of the People's Republic of China. "Twelfth Five Year Plan" for the Comprehensive Prevention and Control of Heavy Metal Pollution. 2011. Available online: http://www.mep.gov.cn/zwgk/hjgh/shierwu/ (accessed on 20 December 2015).
3. US EPA. *Integrated Risk Information System (IRIS)*; National Center for Environmental Assessment: Washington, DC, USA, 1999.
4. Franco-Uria, A.; Lopez-Mateo, C.; Roca, E.; Fernandez-Marcos, M.L. Source identification of heavy metals in pastureland by multivariate analysis in NW Spain. *J. Hazard. Mater.* **2009**, *165*, 1008–1015. [CrossRef] [PubMed]
5. Li, X.Y. Heavy metal contamination of urban soil in an old industrial city (Shenyang) in Northeast China. *Geoderma* **2013**, *192*, 50–58. [CrossRef]
6. Liu, J.G.; Qian, M.; Cai, G.L.; Yang, J.C.; Zhu, Q.S. Uptake and translocation of Cd in different rice cultivars and the relation with Cd accumulation in rice grain. *J. Hazard. Mater.* **2007**, *143*, 443–447. [CrossRef] [PubMed]
7. Abrahams, P.W. Soils: Their implications to human health. *Sci. Total Environ.* **2002**, *291*, 1–32. [CrossRef]
8. Chabukdhara, M.; Nema, A.K. Heavy metals assessment in urban soil around industrial clusters in Ghaziabad, India: Probabilistic health risk approach. *Ecotoxicol. Environ. Saf.* **2012**, *87*, 57–64. [CrossRef] [PubMed]
9. Fu, S.; Wei, C.Y. Multivariate and spatial analysis of heavy metal sources and variations in a large old antimony mine, China. *J. Soils Sediments* **2013**, *13*, 106–116. [CrossRef]
10. Hu, Y.N.; Liu, X.P.; Bai, J.M.; Kaimin, S.; Eddy, Y.Z.; Hefa, C. Assessing heavy metal pollution in the surface soils of a region that had undergone three decades of intense industrialization and urbanization. *Environ. Sci. Pollut. Res.* **2013**, *20*, 6150–6159. [CrossRef] [PubMed]
11. Bai, L.Y.; Zeng, X.B.; Su, S.M.; Duan, R.; Wang, Y.N.; Gao, X. Heavy metal accumulation and source analysis in greenhouse soils of Wuwei District, Gansu Province, China. *Environ. Sci. Pollut. Res.* **2015**, *22*, 5359–5369. [CrossRef] [PubMed]
12. Khan, A.; Khan, S.; Khan, M.A.; Qamar, Z.; Waqas, M. The uptake and bioaccumulation of heavy metals by food plants, their effects on plants nutrients, and associated health risk: A review. *Environ. Sci. Pollut. Res. Int.* **2015**, *22*, 72–99. [CrossRef] [PubMed]
13. Fang, Y.; Sun, X.Y.; Yang, W.J.; Ma, N.; Xin, Z.H.; Fu, J.; Liu, X.C.; Liu, M.; Mariga, A.M.; Zhu, X.F. Concentrations and health risks of lead, cadmium, arsenic, and mercury in rice and edible mushrooms in China. *Food Chem.* **2014**, *147*, 147–151. [CrossRef] [PubMed]
14. Teng, Y.G.; Ni, S.J.; Wang, J.S.; Zuo, R.; Yang, J. A geochemical survey of trace elements in agricultural and non-agricultural topsoil in Dexing area, China. *J. Geochem. Explor.* **2010**, *104*, 118–127. [CrossRef]
15. Zhen, Y.H.; Cheng, V.J.; Pan, G.X.; Li, L.Q. Cd, Zn and Se content of the polished rice samples from some Chinese open markets and their relevance to food safety. *J. Saf. Environ.* **2008**, *8*, 119–122.

16. Yin, C.Q.; Sun, Q.B.; Zhao, X.Q. Assessment of heavy metal contamination in paddy soils from daye mining area of Hubei province, China. *Adv. Mater. Res. Vol.* **2012**, *599*, 434–440. [CrossRef]
17. Hang, X.S.; Wang, H.Y.; Zhou, J.M.; Ma, C.L.; Du, W.C.; Chen, X.Q. Risk assessment of potentially toxic element pollution in soils and rice (*Oryza sativa*) in a typical area of the Yangtze River Delta. *Environ. Pollut.* **2009**, *157*, 2542–2549. [CrossRef] [PubMed]
18. Fu, J.J.; Zhou, Q.F.; Liu, J.M.; Liu, W.; Wang, T.; Zhang, Q.H.; Jiang, G.B. High levels of heavy metals in rice (*Oryza sativa* L.) from a typical E-waste recycling area in southeast China and its potential risk to human health. *Chemosphere* **2008**, *71*, 1269–1275. [CrossRef] [PubMed]
19. Zhao, Y.C.; Wang, Z.G.; Sun, W.X.; Huang, B.; Shi, X.Z.; Ji, J.F. Spatial interrelations and multi-scale sources of soil heavy metal variability in a typical urban-rural transition area in Yangtze River Delta region of China. *Geoderma* **2010**, *156*, 216–227. [CrossRef]
20. Chaturvedi, R.K.; Sankar, K. *Laboratory Manual for the Physicochemical Analysis of Soil, Water and Plant*; Wildlife Institute of India: Dehradun, India, 2006; p. 9.
21. Ryan, J.; Estefan, G.; Rashid, A. *Soil and Plant Analysis Laboratory Manual*, 2nd ed.; International Center for agricultural Research in the dry Areas (ICARDA): Aleppo, Syria; National agricultural Research Center (NARC): Aleppo, Syria, 2001; pp. 46–48.
22. GB2715–2005. Environmental Quality Standard for Food Stuff. Available online: http://datacenter.mep.gov.cn/trs/query.action (accessed on 20 December 2015).
23. GB15618–1995. National Quality Standard for Soils. Available online: http://www.moh.gov.cn/moh/main?fid=open&fun=show_dbquery_pdf&nid=104 (accessed on 20 December 2015).
24. Arao, T.; Ishikawa, S.; Murakami, M.; Abe, K.; Maejima, Y.; Makino, T. Heavy metal contamination of agricultural soil and countermeasures in Japan. *Paddy Water Environ.* **2010**, *8*, 247–257. [CrossRef]
25. Mar, S.S.; Okazaki, M.; Motobayashi, T. The influence of phosphate fertilizer application levels and cultivars on cadmium uptake by Komatsuna (*Brassica rapa* L. var. *perviridis*). *Soil Sci. Plant Nutr.* **2012**, *58*, 492–502. [CrossRef]
26. Fekri, M.; Kaveh, S. Heavy metal accumulation in soil after application of organic wastes. *Arab. J. Geosci.* **2013**, *6*, 463–467. [CrossRef]
27. Nouri, J.; Mahvi, A.H.; Jahed, G.R.; Babaei, A.A. Regional distribution pattern of groundwater heavy metals resulting from agricultural activities. *Environ. Geol.* **2008**, *55*, 1337–1343. [CrossRef]
28. Ramos-Miras, J.J.; Roca-Perez, L.; Guzmán-Palomino, M.; Boluda, R.; Gil, C. Background levels and baseline values of available heavy metals in Mediterranean greenhouse soils (Spain). *J. Geochem. Explor.* **2011**, *110*, 186–192. [CrossRef]
29. Zhou, L.; Yang, B.; Xue, N.; Li, F.; Seip, H.M.; Cong, X.; Yan, Y.; Liu, B.; Han, B.; Li, H. Ecological risks and potential sources of heavy metals in agricultural soils from Huanghuai Plain, China. *Environ. Sci. Pollut. Res.* **2014**, *21*, 1360–1369. [CrossRef] [PubMed]
30. Wang, W.N.; Lu, J.W.; Lu, M.X.; Dai, Z.G.; Li, X.K. Status quo and variation of soil fertility in paddy field—A case study of Hubei province. *Acta Pedol. Sin.* **2012**, *49*, 319–330.
31. Lei, M.; Zeng, M.; Wang, L.H. Arsenic, lead, and cadmium pollution in rice from Hunan markets and contaminated areas and their health risk assessment. *Acta Scient. Circumstant.* **2010**, *30*, 2314–2320.
32. Zhang, L.Y.; LI, L.Q.; Pan, G.X. Variation of Cd, Zn and Se Contents of Polished Rice and the Potential Health Risk for Subsistence-Diet Farmers from Typical Areas of South China. *Environ. Sci.* **2009**, *30*, 2792–2797.
33. Cui, Y.J.; Zhu, Y.G.; Zhai, R.H.; Huang, Y.Z.; Qiu, Y.; Liang, J.Z. Exposure to metal mixtures and human health impacts in a contaminated area in Nanning, China. *Environ. Int.* **2005**, *l31*, 784–790. [CrossRef] [PubMed]
34. Chary, N.S.; Kamala, C.T.; Raj, D.S.S. Assessing risk of heavy metals from consuming food grown on sewage irrigated soils and food chain transfer. *Ecotoxicol. Environ. Saf.* **2008**, *69*, 513–524. [CrossRef] [PubMed]
35. Huang, S.S.; Liao, Q.L.; Hua, M.; Wu, X.; Bi, K.; Yan, C.; Chen, B.; Zhang, X. Survey of heavy metal pollution and assessment of agricultural soil in Yangzhong district, Jiangsu Province, China. *Chemosphere* **2007**, *67*, 2148–2155. [CrossRef] [PubMed]

36. Lv, J.; Liu, Y.; Zhang, Z.; Zhou, R.; Zhu, Y. Distinguishing anthropogenic and natural sources of trace elements in soils undergoing recent 10-year rapid urbanization: A case of Donggang, Eastern China. *Environ. Sci. Pollut. Res.* **2015**, *1*, 1–12. [CrossRef] [PubMed]
37. Kelepertzis, E. Accumulation of heavy metals in agricultural soils of Mediterranean: Insights from Argolida basin, Peloponnese, Greece. *Geoderma* **2014**, *221*, 82–90. [CrossRef]
38. Chung, B.; Song, C.; Park, B.; Cho, J. Heavy Metals in Brown Rice and Soil After Long-Term Irrigation of Wastewater Discharged from Domestic Sewage Treatment Plants. *Pedosphere* **2011**, *21*, 621–627. [CrossRef]
39. Feng, J.F.; Wang, Y.X.; Zhao, J.; Zhu, L.Q.; Bian, X.M.; Zhang, W.J. Source attributions of heavy metals in rice plant along highway in Eastern China. *J. Environ. Sci.* **2011**, *23*, 1158–1164. [CrossRef]
40. Luo, C.; Liu, C.; Wang, Y.; Liu, X.; Li, F.; Zhang, G.; Li, X. Heavy metal contamination in soils and vegetables near an e-waste processing site, South China. *J. Hazard. Mater.* **2011**, *186*, 481–490. [CrossRef] [PubMed]
41. Bakirdere, S.; Yaman, M. Determination of lead, cadmium and copper in roadside soil and plants in Elazig, Turkey. *Environ. Monit. Assess.* **2008**, *136*, 401–410. [CrossRef] [PubMed]
42. Chen, X.; Xia, X.; Zhao, Y.; Zhang, P. Heavy metal concentrations in roadside soils and correlation with urban traffic in Beijing, China. *J. Hazard. Mater.* **2010**, *181*, 640–646. [CrossRef] [PubMed]
43. Nabulo, G.; Oryem-Origa, H.; Diamond, M. Assessment of lead, cadmium, and zinc contamination of roadside soils, surface films, and vegetables in Kampala City, Uganda. *Environ. Res.* **2006**, *101*, 42–52. [CrossRef] [PubMed]
44. Saeedi, M.; Hosseinzadeh, M.; Jamshidi, A.; Pajooheshfar, S.P. Assessment of heavy metals contamination and leaching characteristics in highway side soils, Iran. *Environ. Monit. Assess.* **2009**, *151*, 231–241. [CrossRef] [PubMed]
45. Feng, J.F.; Zhao, J.; Bian, X.M.; Zhang, W. Spatial distribution and controlling factors of heavy metals contents in paddy soil and crop grains of rice-wheat cropping system along highway in East China. *Environ. Geochem. Health* **2012**, *34*, 605–614. [CrossRef] [PubMed]
46. Adams, M.; Zhao, F.J.; McGrath, S.P.; Nicholson, F.A.; Chambers, B.J. Predicting cadmium concentrations in wheat and barley grain using soil properties. *J. Environ. Qual.* **2004**, *33*, 532–541. [CrossRef] [PubMed]
47. Rogival, D.; Scheirs, J.; Blust, R. Transfer and accumulation of metals in a soil-diet-wood mouse food chain along a metal pollution gradient. *Environ. Pollut.* **2007**, *145*, 516–528. [CrossRef] [PubMed]
48. Zalidis, G.; Barbayiarinis, N.; Matsi, T. Forms and distribution of heavy metals in soils of the Axios Delta of Northern Greece. *Commun. Soil Sci. Plant Anal.* **1999**, *30*, 817–827. [CrossRef]
49. Zhao, K.; Liu, X.M.; Xu, J.M.; Selim, H.M. Heavy metal contaminations in a soil-rice system: Identification of spatial dependence in relation to soil properties of paddy fields. *J. Hazard. Mater.* **2010**, *181*, 778–787. [CrossRef] [PubMed]
50. Zeng, F.R.; Ali, S.; Zhang, H.T.; Ouyang, Y.; Qiu, B.; Wu, F.; Zhang, G. The influence of pH and organic matter content in paddy soil on heavy metal availability and their uptake by rice plants. *Environ. Pollut.* **2011**, *159*, 84–91. [CrossRef] [PubMed]
51. Bang, J.; Hesterberg, D. Dissolution of trace element contaminants from two coastal plain soils as affected by pH. *J. Environ. Qual.* **2004**, *33*, 891–901. [CrossRef] [PubMed]
52. Sukreeyapongse, O.; Holm, P.E.; Strobel, B.W.; Panichsakpatana, S.; Magid, J.; Hansen, H.C.B. pH-dependent release of cadmium, copper, and lead from natural and sludge-amended soils. *J. Environ. Qual.* **2002**, *31*, 1901–1909. [CrossRef] [PubMed]
53. Zhao, K.L.; Fu, W.J.; Ye, Z.Q.; Zhang, C.S. Contamination and Spatial Variation of Heavy Metals in the Soil-Rice System in Nanxun County, Southeastern China. *Int. J. Environ. Res. Public Health* **2015**, *12*, 1577–1594. [CrossRef] [PubMed]
54. Ghosh, A.K.; Bhatt, M.A.; Agrawal, H.P. Effect of long-term application of treated sewage water on heavy metal accumulation in vegetables grown in Northern India. *Environ. Monitor. Assess.* **2012**, *184*, 1025–1036. [CrossRef] [PubMed]
55. McCauley, A.; Jones, C.; Jacobsen, J. Soil pH and organic matter. *Nutr. Manag. Module* **2009**, *8*, 1–11.
56. Li, W.L.; Xu, B.B.; Song, Q.J.; Liu, X.M.; Xu, J.M.; Brookes, P.C. The identification of "hotspots" of heavy metal pollution in soil–rice systems at a regional scale in Eastern China. *Sci. Total Environ.* **2014**, *472*, 407–420. [CrossRef] [PubMed]
57. Li, P.J.; Wang, X.; Allinson, G.; Li, X.; Xiong, X. Risk assessment of heavy metals in soil previously irrigated with industrial wastewater in Shenyang, China. *J. Hazard. Mater.* **2009**, *161*, 516–521. [CrossRef] [PubMed]

# 4

# Spatiotemporal Pattern Analysis of Scarlet Fever Incidence in Beijing, China, 2005–2014

**Gehendra Mahara [1,2,†], Chao Wang [1,2,†], Da Huo [3], Qin Xu [1,2], Fangfang Huang [1,2], Lixin Tao [1,2], Jin Guo [1,2], Kai Cao [1,2], Liu Long [1,2], Jagadish K. Chhetri [4], Qi Gao [1,2], Wei Wang [1,2,5], Quanyi Wang [3,*] and Xiuhua Guo [1,2,*]**

Academic Editor: Paul B. Tchounwou

1 Department of Epidemiology and Biostatistics, School of Public Health, Capital Medical University, Beijing 100069, China; gbmahara@gmail.com (G.M.); 13810147054@139.com (C.W.); xuqinhuachong@126.com (Q.X.); hff87hff87@163.com (F.H.); taolixin.2008@163.com (L.T.); guojin5827501@163.com (J.G.); anzhen602@163.com (K.C.); llstatistics@163.com (L.L.); gaoqi@ccmu.edu.cn (Q.G.); wei.wang@ecu.edu.au (W.W.)

2 Beijing Municipal Key Laboratory of Clinical Epidemiology, Beijing 100069, China

3 Institute for Infectious Disease & Endemic Disease Control, Beijing Center for Disease Prevention & Control (CDC), Beijing Center for Preventive Medical Research, Beijing 100069, China; howardhuo@139.com

4 Department of Geriatrics, XuanWu Hospital of Capital Medical University, Beijing 100069, China; chhetri_jk@hotmail.com

5 Systems and Intervention Research Centre for Health, School of Medical Sciences, Edith Cowan University, Perth 6027, Australia

* Correspondence: bjcdcxm@126.com (Q.W.); guoxiuh@ccmu.edu.cn (X.G.);

† These authors contributed equally to this work.

**Abstract:** *Objective*: To probe the spatiotemporal patterns of the incidence of scarlet fever in Beijing, China, from 2005 to 2014. *Methods*: A spatiotemporal analysis was conducted at the district/county level in the Beijing region based on the reported cases of scarlet fever during the study period. Moran's autocorrelation coefficient was used to examine the spatial autocorrelation of scarlet fever, whereas the Getis-Ord Gi* statistic was used to determine the hotspot incidence of scarlet fever. Likewise, the space-time scan statistic was used to detect the space-time clusters, including the relative risk of scarlet fever incidence across all settings. *Results*: A total of 26,860 scarlet fever cases were reported in Beijing during the study period (2005–2014). The average annual incidence of scarlet fever was 14.25 per 100,000 population (range, 6.76 to 32.03 per 100,000). The incidence among males was higher than that among females, and more than two-thirds of scarlet fever cases (83.8%) were among children 3–8 years old. The seasonal incidence peaks occurred from March to July. A higher relative risk area was mainly in the city and urban districts of Beijing. The most likely space-time clusters and secondary clusters were detected to be diversely distributed in every study year. *Conclusions*: The spatiotemporal patterns of scarlet fever were relatively unsteady in Beijing from 2005 to 2014. The at-risk population was mainly scattered in urban settings and dense districts with high population, indicating a positive relationship between population density and increased risk of scarlet fever exposure. Children under 15 years of age were the most susceptible to scarlet fever.

**Keywords:** scarlet fever; spatiotemporal patterns; children; Beijing

---

## 1. Introduction

Scarlet fever, also known as "scarlatina", is a contagious disease caused by erythrogenic exotoxin, a substance produced by *Streptococcus pyogenes* (Group A Beta Haemolytic Streptococcus, GABHS), which usually occurs in winter and spring and commonly affects children [1–5]. The incubation period ranges from 12 h to 7 days following contact with an infected person. Scarlet fever usually transfers through respiratory droplets (bacteria are spread through the nose) or direct contact with the mucus, saliva or skin of an infected person in the acute illness phase or sub-clinical stages [2]. However, it can also be transferred through food or contaminated milk [3,6]. The incidence of scarlet fever is high in overcrowded places, such as schools and other institutional settings. Although common among children 5–15 years of age, adults are also infected by the disease [7]. However, scarlet fever rarely occurs in children younger than 2 years of age because they are protected by maternal antibodies [8]. Notably, to date, there is no vaccine that prevents scarlet fever [9]. After treatment with antibiotics, scarlet fever can be completely cured and further complications can be prevented. However, if not treated in time, the disease can result in complications, such as septicaemia, vasculitis, hepatitis, glomerulonephritis and rheumatic fever, *etc.* [10–12].

Cases of scarlet fever have been documented since the early 18th century [13]. It became a common childhood illness in the 20th century. Today, scarlet fever rarely occurs in developed countries. However, it is a notifiable disease in developing countries, and outbreaks in teaching institutions have been recorded worldwide, including in Australia [14], Hong Kong [15], Taiwan [16], the United Kingdom and Mainland China [3,4,17,18]. In England, more than 1200 scarlet fever cases were reported from January to March in 2015 [19]. Thus far, scarlet fever has been continuously reported in China, the United Kingdom and Poland [19–21].

Several epidemiological studies of scarlet fever have been previously conducted in different parts of the world, for example, in Poland in 2012 [20], Hong Kong in 2011 [15,22], Northern Taiwan from 2001 to 2002 [23], Granada, Spain in 2012 [24], and in Beijing, Shanghai, and Hefei city in China in different years [3,25–28]. These studies have mainly focused on the molecular and epidemiological characteristics of the disease, as well as its antibiotic sensitivity [17,23,25,28,29]. However, a spatiotemporal model of scarlet fever seems to be lacking, and such a model might be helpful in determining high-risk areas during specific time periods. Therefore, a more location-specific public health intervention to control and prevent scarlet fever could be initiated [30,31]. The objective of this study was to identify the spatiotemporal patterns of scarlet fever at the district/county level based on the surveillance data gathered during our study period. Spatial, temporal, and spatiotemporal analyses were conducted to understand the trends and dynamics of transmission and to identify the high-risk areas and time frames of scarlet fever infection in Beijing, thereby providing reliable information concerning where and when prevention and control measures should be focused.

## 2. Methodology

### 2.1. Study Area

Beijing is the capital city of the People's Republic of China. It is situated at the northern tip of the North China Plain, at 39°56′ N and 116°20′ E region, and it is surrounded by mountains to the north, northwest and west. The topography of the Beijing area slopes from the northwest to the southeast, and there is a continental climate characterized by hot, humid summers and cold, windy, and dry winters. Beijing measures 16,410.54 Sq km, with a population of 21.15 million. It consists of 16 districts (there were 18 districts before 2010) and 223 towns [32]. The districts are named as follows: (1) Dongcheng, (2) Xicheng, (3) Chaoyang, (4) Fengtai, (5) Shijingshan, (6) Haidian, (7) Fangshan, (8) Tongzhou, (9) Shunyi, (10) Changping, (11) Daxing, (12) Mentougou, (13) Huairou, (14) Pinggu, (15) Miyun, and (16) Yanqing [32].

### 2.2. *Data Source*

All reported cases (including the patient's age, sex, occupation, and address) of scarlet fever from 1 January 2005 to 31 November 2014 were extracted from The Centre for Disease Prevention and Control China (China CDC) [5]. Additionally, we retrieved population data from the Scientific Data Sharing Center of Public Health, which was developed by the China CDC. During the study period, socio-demographic data from each district were obtained from the Beijing Bureau of Statistics of China. Maps of Beijing were downloaded from Data Sharing Infrasturucture of Earth System Science [33].

### 2.3. *Case Definition*

The diagnosis of scarlet fever is based on the clinical criteria established by the Law of Communicable Diseases Prevention and Control of The People's Republic China and Guidance offered by the Chinese Ministry of Health. Scarlet fever is a notifiable Group B infectious disease, according to the China National Notifiable Infectious Disease Surveillance System (NNIDSS) [34]. The clinical manifestations of scarlet fever (ICD A38.01) are acute onset of fever, pharyngitis with strawberry tongue, red rash with a sandpaper feel, and itching, as well as a throat swab culture and stain and a skin smear stain to confirm *Streptococcus bacteria* (Group A) infection. By law, clinicians in China are required to report clinical cases of scarlet fever to the NNIDSS within 6 h after a diagnosis, and then the case will be added to the NNIDSS database [34].

#### 2.3.1. Ethical Approval

We obtained ethical approval from the ethical review committee of the School of Public Health, Capital Medical University (SPHCMU) Approval No. IRB00009511, Beijing, China. Consent from each individual subject was not required because we used only aggregated data from the CDC, Beijing, China.

#### 2.3.2. Human Participant Protection

This study was approved under the Institutional Review Committee for Healthcare Research and Quality protocol at Capital Medical University, Beijing, China, with secondary analysis of confidential data from the CDC, Beijing, China.

### 2.4. *Statistical Analysis*

All frequencies of scarlet fever were summarized annually, corresponding to each geographic area (e.g., district), and the incidence of scarlet fever per 100,000 population in each year was calculated by scarlet fever case counts, divided by that year's population. Socio-demographic characteristics of scarlet fever patients in Beijing from 2005 to 2014 was calculated by using SPSS V-20 (SPSS Statistics 20, IBM: Corporation 1 New Orchard Road Armonk, New York, NY, 10504–1722, USA). Likewise, the address information and other data from each case were inputted into a Microsoft Excel 2010 spreadsheet (Microsoft: Redmond, WA, USA) and then linked to the corresponding location in ArcGIS software V 10.1 (ESRI, Inc., Redlands, CA, USA) [35]. For analysis, all collected data were managed in text format for support and recognition by the designed software.

#### 2.4.1. Spatial Autocorrelation Analysis

The spatial autocorrelation (Global Moran's *I*) statistic measures were simultaneously based on the featured areas and values to analyse the degree of the dependency among the different observations in geographical space. Similarly, it also evaluates whether the disease patterns are clustered, dispersed or randomly distributed in the area [35]. In fact, the Global spatial autocorrelation statistic assesses the overall pattern and trend of the data, and it is used more extensively than other methods [36]. Thus, the autocorrelation statistic (Global Moran's *I)* was used to detect the spatial autocorrelation of scarlet fever in the study locations by year. We used ArcGIS 10.1 software to compute the Moran's *I* statistic.

In general, Moran's $I$ index values close to +1.0 indicate disease clustering, whereas $I$ index values close to −1.0 indicate the dispersion of disease. Similarly, Moran's $I$ index value zero suggests that there is no disease clustering, indicating that the disease is randomly distributed in the study area [37,38].

We used local autocorrelation Moran's $I$ statistic to investigate the local level cluster (hotspot) locations of scarlet fever incidents. Anselin Local Moran's $I$ was used to determine whether there were positive spatial correlations (high-high clusters and low-low clusters) or negative spatial correlations (high-low clusters and low-high clusters). In fact, this statistic was used to detect the core clusters/outliers of districts with high scarlet fever incidences. Anselin spatial weights matrix is based on the inverse distance and weight between polygon centroids, which were constructed to identify the spatial relationships among the districts [39]. ArcGIS 10.1 was used to perform the analysis [40].

#### 2.4.2. Hot Spot Analysis

The local Getis-Ord statistic (Gi*), also known as hotspot analysis, was used to provide additional information that indicated the intensity and stability of core hotspot or cold spot clusters of scarlet fever in the entire study location [41–45]. The Getis-Ord Gi* statistic serves as an indicator for local autocorrelation, which measures how spatial autocorrelation varies locally over the study location and then calculates a statistic for each point data [46]. This method provides more intuitive results with a better visual exploration and also has the advantage of distinguishing high value clusters (hot spot) or low value clusters (cold spot). The degree of clustering and its statistical significance is estimated based on a confidence level, according to the Z-scores. If the Z (Gi*) score is positive and significant, it shows that one district and its neighbouring regions have a relatively high frequency of scarlet fever incidents, which is a hotspot (a spatial cluster of high data values); on the contrary, if the Z (Gi*) score is negative and significant, it indicates a cold-spot (a spatial cluster of low data values) [35]. Districts with Z-scores > 2.58 were considered to be significant at a 99% confidence level ($p < 0.01$), and they remained in the hot spot category. Districts with a Z-score between 1.65–1.96 and 1.96–2.58 were considered to be significant at 90% and 95% confidence levels ($p < 0.10$, and 0.05), and they were categorized as high risk districts. Whereas, a Z-score < −2.58 indicates the clustering of low values, indicating a cold spot [47]. The Gi* index was calculated as described in the previous study [41]. The output of the Gi* statistic presented on a map indicated the location of the spatial clusters in the study area, with the help of geographic information system (GIS) (ArcGIS, V-10.1).

#### 2.4.3. Retrospective SaTScan Analysis

SaTScan™ Software version-9.1 using the Kulldorff method of retrospective space-time scan statistic, was used to detect the geographical clusters of scarlet fever at the district/county level based on a discrete Poisson model. The space-time scan statistic is defined by a cylindrical window, with a circular geographic base and with height corresponding to the time period of possible clusters [48,49]. The base is centred around one of the several possible centroids placed in the study area, with the radius of the base varying continuously, according to the population range of the area, from zero to the maximum cluster size of the total population who might be at risk [50,51]. The window is then moved in space and time; therefore, it is possible to check all geographic locations and sizes, as well as each possible time period [52]. As a result, we obtained an infinite number of overlapping cylinders of windows of various sizes and shapes, jointly covering the entire study region, in which each cylinder reflects a possible cluster [49]. Thus, we performed the cluster analysis with the following default setting: maximum spatial cluster size, 50% of the population. The statistical significance of each cluster in the study area was based on comparing the likelihood ratio (*LLR*) against a null distribution achieved from the Monte Carlo simulation, with the number of replications set to 999 to ensure the sufficient statistical power. Additionally, the window with the maximum *LLR* is assumed to be the most likely cluster, that is, the cluster least likely to be caused by chance, and other windows with a statistically significant *LLR* were measured as secondary clusters [53].

In the likelihood ratio test, for each location and size of the scanning window, the alternative hypothesis is that there is an elevated rise within the window compared to outside the window. Under the Poisson assumption, the likelihood function for a specific window is proportional to Equation (1):

$$\left(\frac{c}{E[c]}\right)^{c}\left(\frac{C-c}{C-E[c]}\right)^{C-c} I() \tag{1}$$

where "C" is the total number of cases, "c" is the observed number of cases within the window and "E[c]" is the covariate adjusted expected number of cases within the window under the null hypothesis. Because the analysis is conditioned on the total number of cases observed, "C−E[c]" is the expected number of cases outside the window. "I()" is an indicator function. When SaTScan is set to scan only for clusters with high rates, "I()" is equal to 1 when the window has more cases than expected under the null-hypothesis, and it is equal to zero otherwise. The opposite is true when the SaTScan is set to scan only for clusters with low rates. When the program scans for clusters with either high or low rates, then $I() = 1$ for all windows [53,54]. The scan window with the larger *LLR* value is defined as the most likely cluster; other scan windows with the *LLR* values with statistical significance are defined as secondary clusters. The relative risk (RR) of the incidence inside and outside the window is considered to be statistically significant if $p < 0.05$, which is evaluated by a Monte Carlo simulation [54,55].

In this study, the spatial units of space-time scan analysis were the 16 districts of Beijing, and the time units were 10 years (from 2005 to 2014). The time frame of the scan analysis was set to be one month to control for the time trends and to observe the cluster changes in the entire study period [54,55].

## 3. Results

### 3.1. Demographic Characteristics

A total of 27,688 scarlet fever cases were reported in Beijing from January 2005 to November 2014. We excluded 788 of these cases because the patients did not reside in Beijing, and we did not include 40 additional cases due to unavailable resident information; a total of 26,860 cases were included in the final analyses. The annual average incidence rate of scarlet fever was 14.25 per 100,000 populations (range, 6.76/100,000 in 2009 to 32.03/100,000 in 2011) (Table 1).

**Table 1.** Incidence of scarlet fever in Beijing from 2005 to 2014.

| Year | No. of Cases | Incidence Rate (/$10^5$) |
|---|---|---|
| 2005 | 2164 | 14.7 |
| 2006 | 2268 | 14.16 |
| 2007 | 2392 | 14.27 |
| 2008 | 1798 | 10.15 |
| 2009 | 1258 | 6.76 |
| 2010 | 1665 | 8.50 |
| 2011 | 6466 | 32.03 |
| 2012 | 3367 | 16.27 |
| 2013 | 2170 | 10.26 |
| 2014 | 3321 | 15.43 |
| Total | 26,860 | 14.25 (Average) |

Table 2 summarizes the sociodemographic characteristics of scarlet fever cases from 2005 to 2014. The reported cases of scarlet fever varied during the study period. The number was high in 2011 (6466 cases), followed by 2012 (3367 cases). The lowest number of cases was observed in 2009 (1258 cases). Among the total 26,860 cases, 63.2% (16,972) were male and 36.8% (9888) were female, with an average male-female sex ratio of 1.71:1. Most scarlet fever patients were 3–8 years old (83.8%, 22,511), and 9–15-year-old patients accounted for 10.3% (2757) of the cases over the study period. Children, who were spending their daytime at kindergarten or school, accounted for 40% (10,893), with primary level and secondary level students in the majority (48.7%, 13,074). Likewise, the morbidity rate of scarlet fever cases was higher in urban areas (62.2%, 16,716) than in rural areas (37.8%, 10,144) (Table 2).

### 3.2. Spatial Pattern

The distribution of scarlet fever varied at the district/county level in the Beijing region during the study period (Figure 1). A relatively high incidence appeared in the districts of *Chaoyang* and *Xicheng*, and a slightly lower incidence was observed in the Haiden and Fengtai districts. However, a lower incidence of scarlet fever was observed in Changping, Fangshan, Tongzhou and Daxing districts across the entire study period (Figure 1).

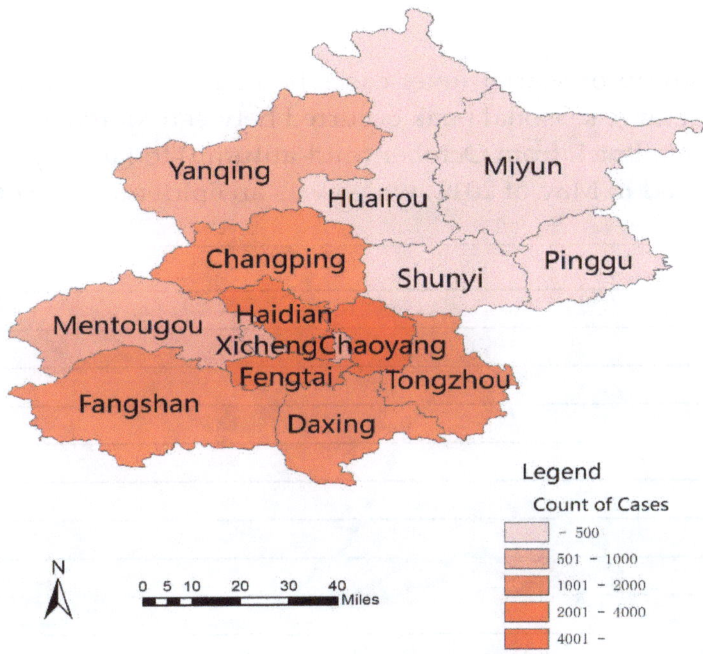


**Figure 1.** The distribution of scarlet fever cases in different districts in Beijing, 2005–2014.

**Table 2.** Socio-demographic characteristics of scarlet fever patients in Beijing, 2005–2014.

| Group | 2005 | 2006 | 2007 | 2008 | 2009 | 2010 | 2011 | 2012 | 2013 | 2014 | Total |
|---|---|---|---|---|---|---|---|---|---|---|---|
| Gender | | | | | | – | | | | | |
| Male | 1358 (62.8) | 1419 (62.6) | 1523 (63.7) | 1111 (62.1) | 766 (60.9) | 1050 (63.1) | 4158 (64.3) | 2173 (64.5) | 1350 (62.2) | 2064 (62.1) | 16,972 (63.2) |
| Female | 806 (37.2) | 849 (37.4) | 869 (36.3) | 678 (37.9) | 492 (39.1) | 615 (36.9) | 2308 (35.7) | 1194 (35.5) | 820 (37.8) | 1257 (37.9) | 9888 (36.8) |
| Sex Ratio | 1.68 | 1.67 | 1.75 | 1.64 | 1.55 | 1.71 | 1.76 | 1.82 | 1.64 | 1.64 | 1.71 |
| Age group | | | | | | – | | | | | |
| Younger than 2 years | 48 (2.2) | 100 (4.4) | 115 (4.8) | 58 (3.2) | 40 (3.2) | 81 (4.9) | 314 (4.9) | 163 (4.8) | 82 (3.8) | 152 (4.6) | 1153 (4.3) |
| 3–8 years | 1981 (91.5) | 1987 (87.6) | 2041 (85.3) | 1545 (86.3) | 1101 (87.5) | 1448 (87.0) | 5357 (82.8) | 2637 (78.3) | 1675 (77.2) | 2739 (82.5) | 22,511 (83.8) |
| 9–15 years | 123 (5.7) | 167 (7.4) | 215 (9.0) | 163 (9.1) | 112 (8.9) | 129 (7.7) | 668 (10.3) | 463 (13.8) | 350 (16.1) | 367 (11.1) | 2757 (10.3) |
| Older than 16 years | 12 (0.6) | 14 (0.6) | 21 (0.9) | 23(1.3) | 5 (0.4) | 7 (0.4) | 127 (2.0) | 104 (3.1) | 63 (2.9) | 63 (1.9) | 439 (1.6) |
| Districts | | | | | | – | | | | | |
| Urban | 1803 (83.3) | 1950 (86.0) | 1956 (85.7) | 1533 (85.7) | 1076 (85.7) | 958 (57.5) | 3460 (53.5) | 1896 (56.3) | 1117 (51.5) | 965 (29.1) | 16,716 (62.2) |
| Rural | 361 (16.7) | 318 (14.0) | 436 (18.2) | 256 (14.3) | 180 (14.3) | 707 (42.5) | 3006 (46.5) | 1471 (43.7) | 1053 (48.5) | 2356 (70.9) | 10,144 (37.8) |

**Table 2.** *Cont.*

| Group | 2005 | 2006 | 2007 | 2008 | 2009 | 2010 | 2011 | 2012 | 2013 | 2014 | Total |
|---|---|---|---|---|---|---|---|---|---|---|---|
| Occupation | | | | | | – | | | | | |
| Kindergarten Children | 975 (45.1) | 975 (43.0) | 1024 (42.8) | 919 (51.1) | 518 (41.2) | 840 (50.5) | 2595 (40.1) | 1175 (34.9) | 669 (30.8) | 1173 (35.3) | 10863 (40.4) |
| Scattered children | 150 (6.9) | 274 (12.1) | 267 (11.2) | 164 (9.2) | 107 (8.5) | 218 (13.1) | 673 (10.4) | 310 (9.2) | 190 (8.8) | 397 (12.0) | 2750 (10.2) |
| Student | 1036 (47.9) | 1015 (44.8) | 1097 (45.9) | 696 (38.9) | 632 (50.2) | 604 (36.3) | 3142 (48.6) | 1846 (54.8) | 1280 (59.0) | 1726 (52.0) | 13,074 (48.7) |
| Others | 3 (0.1) | 4 (0.2) | 4 (0.2) | 10 (0.6) | 1 (0.1) | 3 (0.2) | 56 (0.9) | 36 (1.1) | 31 (1.4) | 25 (0.8) | 173 (0.6) |

Student (above than pre-primary level) and others (teachers, staffs of school). The bracketed figures indicate the percentage of cases in the corresponding year and group, and the urban districts included Haidian, Shijingshan, Fengtai, Chaoyang, Dongcheng, and Xicheng; the rural districts included Changping, Mentougou, Fanshan, Daxing, Tongzhou, Shunyi, Yanqing, Huairou, Miyun, and Pinggu for further analysis.

### 3.3. Temporal Pattern

The monthly distribution of scarlet fever cases had significant seasonality and periodicity in Beijing (Figure 2). The first large seasonal peak occurred between March (early spring) and July (early summer), followed by a small pick from October (mid-autumn) to January (mid-winter). The highest incidence peak was observed in May of 2011, which was an epidemic year of scarlet fever in Beijing.

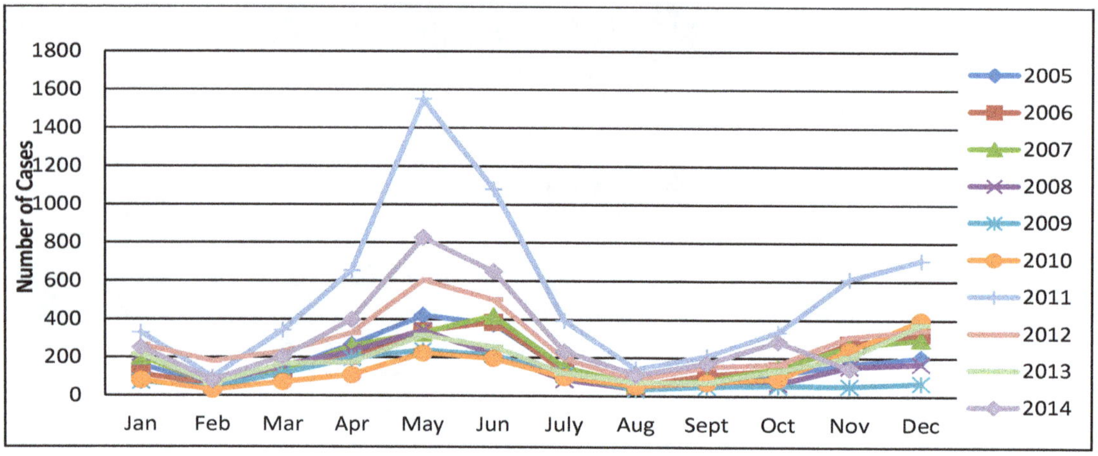


**Figure 2.** Monthly incidence of scarlet fever in Beijing, China, 2005–2014.

### 3.4. Spatial Autocorrelation Analysis

Table 3 describes the result of spatial autocorrelation (Global Moran's *I*) analysis in Beijing. The value of Global Moran's *I* increased from −0.004 to 0.141, and the Z score peaked (0.629 to 2.063) between 2005 and 2014; consequently, the distribution of scarlet fever was not spatially auto-correlated, indicating that the scarlet fever incidence was distributed randomly over that time period in Beijing.

**Table 3.** The results of the global spatial autocorrelation test of scarlet fever cases in Beijing, 2005–2014.

| Years | Moran *I* | Z-Score | *p*-Value |
|---|---|---|---|
| 2005 | 0.059 | 1.218 | 0.223 |
| 2006 | −0.004 | 0.629 | 0.529 |
| 2007 | 0.053 | 1.228 | 0.220 |
| 2008 | 0.100 | 1.703 | 0.088 |
| 2009 | 0.141 | 2.063 | 0.039 |
| 2010 | 0.095 | 1.641 | 0.101 |
| 2011 | 0.064 | 1.412 | 0.158 |
| 2012 | 0.040 | 1.142 | 0.253 |
| 2013 | 0.040 | 1.172 | 0.241 |
| 2014 | 0.019 | 0.946 | 0.344 |

Anselin Local Moran's *I* spatial analysis of scarlet fever incidence detected diverse patterns of spatial association from 2005 to 2014. This analysis shows that the strengthened high-high positive spatial association of scarlet fever incidence was in Fengtai district in 2005, which is an urban district in Beijing, and it also identified a high-high positive association in Chaoyang, Xcheng, Dongcheng and Fengtai districts in 2009; these areas are also urban districts, as well as the most densely populated districts of Beijing (Figure S1).

Figure 3 demonstrates the hotspots of scarlet fever incidence at the district level in Beijing, during the study period. The score of Giz >2.58 reflects the presence of disease aggregation, with a probability of 99%, using Getis-Ord Gi* statistic. The locations and size of hotspots varied in each year. The main hotspots of scarlet fever were identified in urban districts, such as Xicheng, Haidian, Fengtai, and Shijingshan, and two rural districts (Doxing and Changping). Notably, no hotspot areas were detected in the rural districts of Beijing, except for Doxing and Changping districts (Figure 3).

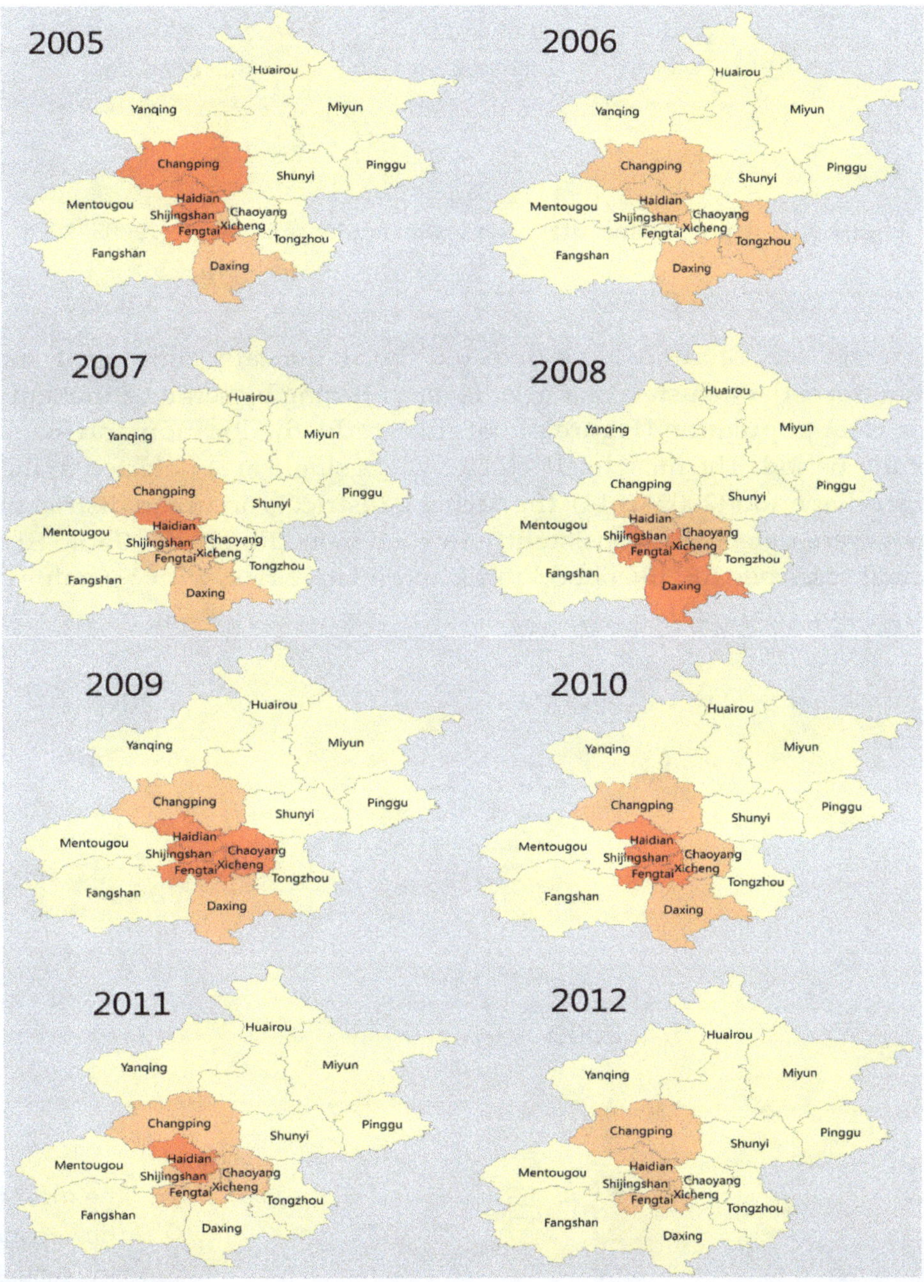


**Figure 3.** *Cont.*

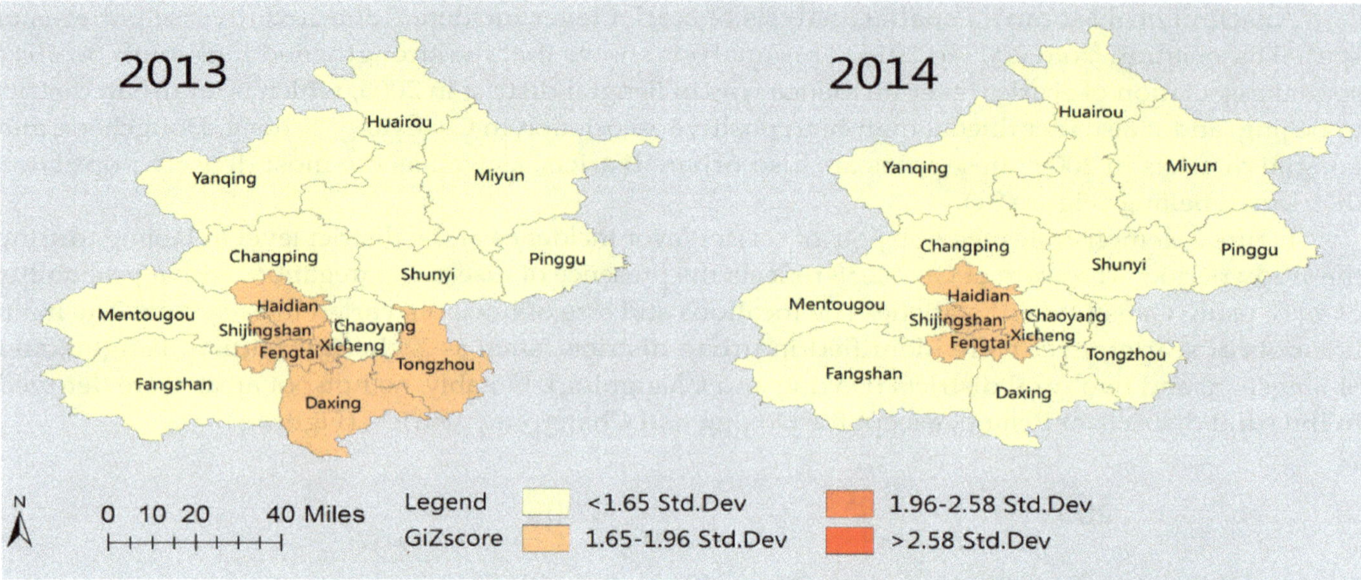


**Figure 3.** Hotspot clusters of scarlet fever incidence in Beijing, 2005–2014.

*3.5. Spatial Clusters of Scarlet Fever Disease*

Using the discrete Poisson model, we detected the statistically significant, most likely spatial clusters, and several secondary clusters of scarlet fever in Beijing by setting a maximum spatial cluster size of 50% of the total population (Figure 4). In this study, the location and size of these clusters varied over the study period. The most likely cluster was a single area detected in the Xicheng district during the entire study period (2005–2014); the Xicheng district is an urban district of Beijing. Several secondary clusters were detected inconsistently in each year (Figure 4). Table S1 summarizes the location, size, annual cases per $10^5$, likelihood ratio, and relative risk, with a $p$-value of the most likely spatial clusters in Beijing (2005–2014).

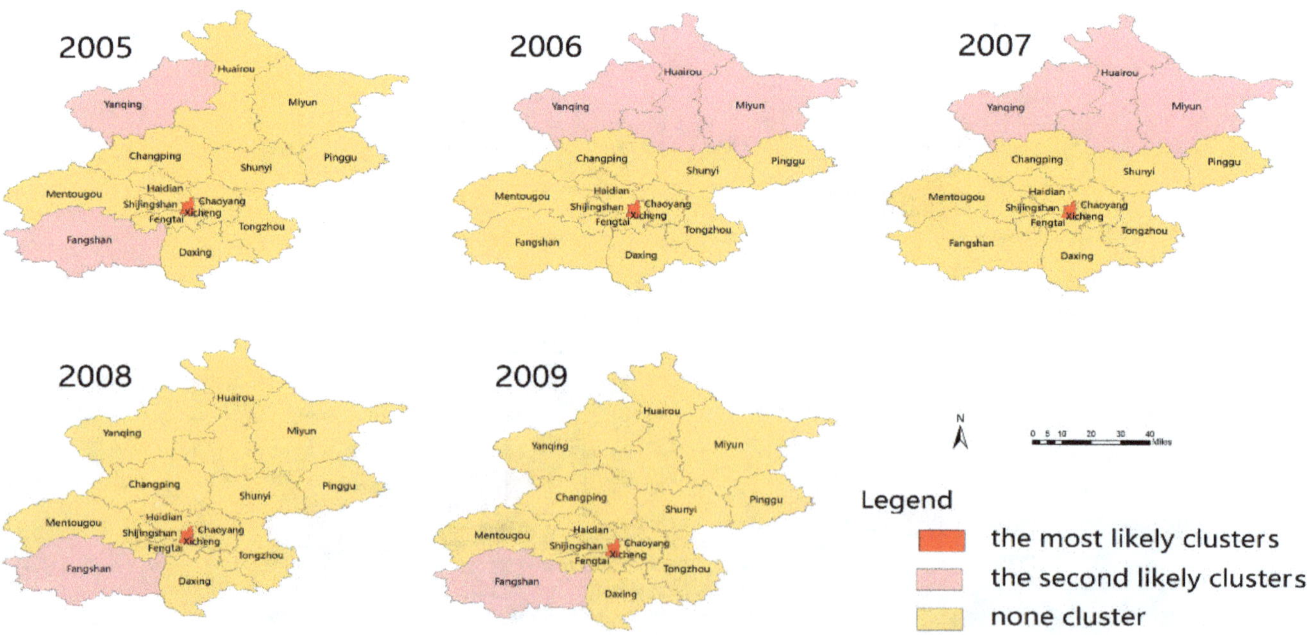


**Figure 4.** *Cont.*

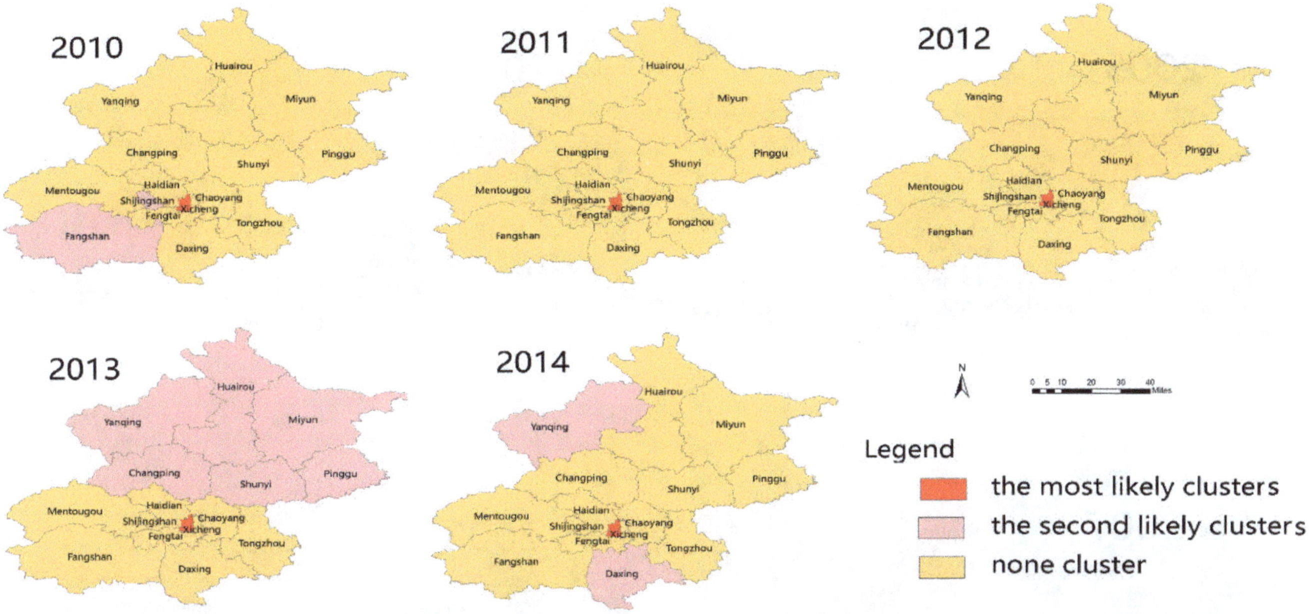


**Figure 4.** The most likely clusters and the secondary clusters of scarlet fever incidence in Beijing were detected using the Purely Spatial Poisson model, 2005–2014. Red indicates the most likely cluster, slightly red indicates secondary clusters and light red indicates no clusters in the study areas.

### *3.6. Space-Time Clusters*

Using a space-time permutation model, we identified the statistically significant monthly spatiotemporal clusters with a high incidence of scarlet fever from 2005 to 2014 (Table 4, Figure 5). The incident of scarlet fever had space-time heterogeneity in the Beijing region during the study period (Figure 5).

**Table 4.** The most likely high-risk clusters of scarlet fever incidence detected using a space-time permutation model (setting 50% as the maximum cluster size) in Beijing, 2005–2011.

| Scan Year | Cluster Time Frame | Center(Latitude, Longitude)/Radius (km) | Cluster Districts | Relative Risk | *p*-Value |
|---|---|---|---|---|---|
| 2005 | 1 January to 30 April | 116.156 E, 40.537 N/0 | 1 | 1.90 | 0.014 |
| 2006 | 1 May to 31 August | 117.135 E, 40.205 N/0 | 1 | 1.97 | 0.000049 |
| 2007 | 1–31 January | 116.728 E, 39.801 N/35 | 4 | 1.55 | 0.012 |
| 2008 | 1 January to 29 February | 116.507 E, 39.949 N/0 | 1 | 1.82 | 0.00029 |
| 2009 | 1 January to 28 February | 39.99 N, 115.78 E/30.98 | 2 | 2.22 | 0.109 |
| 2010 | 1–31 December | 116.410 E, 39.911 N/4.34 | 2 | 1.28 | 0.081 |
| 2011 | 1 September to 31 December | 116.156 E, 40.537 N/0 | 1 | 2.97 | <0.001 |
| 2012 | 1 April to 30 June | 117.135 E, 40.205 N/56.68 | 4 | 1.49 | 0.0000014 |
| 2013 | 1 July to 30 September | 116.579 E, 40.628 N/55.86 | 5 | 1.87 | 0.017 |
| 2014 | 1 September to 30 November | 39.991 N, 115.787 E/33.31 | 3 | 1.43 | 0.107 |

Note: "Scan year" indicates the boundary of time points input into the scanning analysis, and the "Cluster time frame" indicates the boundary of time points identified by the scanning analysis.

The most likely clusters detected varied in each study year in the study area. Five clusters were detected in 2013 in the following rural districts: Huairou, Miyun, Yanqing, Shunyi, and Changping), and the cluster time ranged from 1 July to 30 September; the cluster centre was (116.57° E, 40.62° N), and the radius was 55.86 Km, with a relative risk (RR) value of 1.87 ($p > 0.017$), followed by four clusters in 2007 (Tongzhou, Chaoying, Dongcheng, Shunyi districts) and 2012 (Pinggu, Shunyi, Miyun and Tongzhou districts). In addition, we identified three most likely clusters in 2014 and two clusters in 2009 and 2010; the rest of the year, we detected a single cluster (Figure 5, Table 4). There were several secondary clusters identified apart from the most likely clusters (Figure 5).

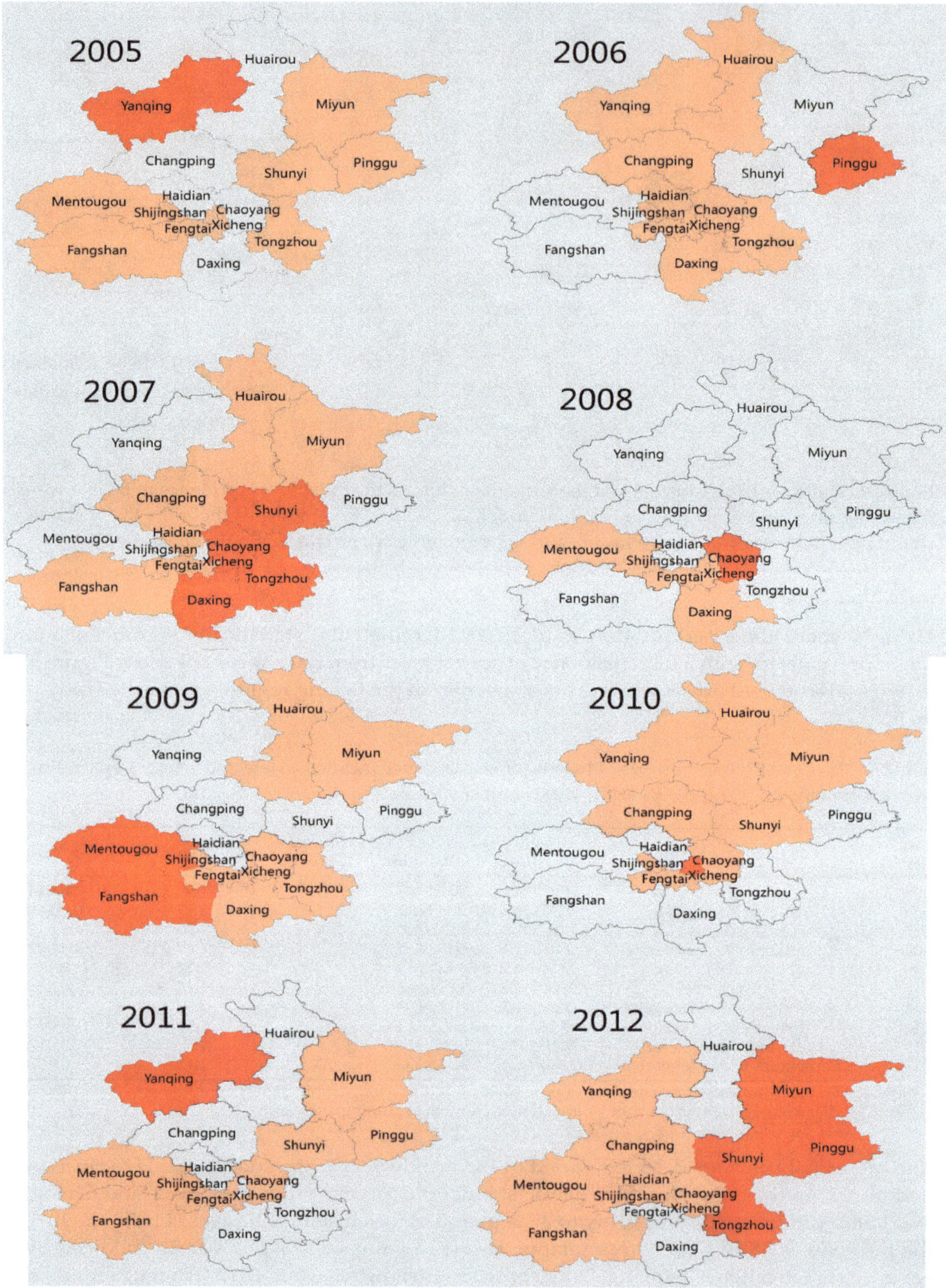


**Figure 5.** *Cont.*

**Figure 5.** The most likely clusters and the secondary clusters of scarlet fever incidence in Beijing using a space-time permutation model, 2005–2014.

## 4. Discussion

We have provided geographical clusters of scarlet fever incidence in Beijing through exploratory disease mapping, spatial autocorrelation analysis and hotspot analysis. We detected that the incidence of scarlet fever varied greatly over time, and the Beijing region experienced the highest number of reported cases (6566) in 2011. The annual incidence of scarlet fever was persistently greater among males than among females, particularly among nursery, kindergarten and pre-primary school children, who were found to be more vulnerable to scarlet fever during 2005–2014. The incidence of scarlet fever was high among children 3–8 years of age (83.8%) during epidemic and non-epidemic years in Beijing (Table 2). A study in Hefei, China, also reported similar results from 2004 to 2008 [28]. Likewise, a previous study by Chen, *et al.* reported that the maximum incidence (85.9%) was among children 4–9 years of age. Peng *et al.* [17] reported similar results in Beijing from 2006 to 2011, with a higher incidence among males than among females. Additionally, several studies in Hong Kong [22] and Poland [20] found a higher incidence among children 4–7 years of age. The WHO and Public Health, UK, stated that the scarlet fever infection is most common among children 5–15 years old [2,56].

Our study revealed that the incidence of scarlet fever seemed to be slightly increased from 2005 to 2008; however, it was notably lower in 2009 and 2010 (6.76 cases/100,000 population and 8.50 cases/100,000 population, respectively). Unexpectedly, scarlet fever broke out in 2011 with 32.03 cases/100,000 population in Beijing, which is consistent with the findings of Peng *et al.* [17], who reported 7.0 to 14.3 cases/100,000 populations from 2006 to 2010 and 31.4 cases/100,000 populations in 2011 in the same study area. Interestingly, the Luk, *et al.* study [22] reported that the incidence of scarlet fever showed a decreasing trend in Hong Kong from 2008 to 2010, whereas another study suggested that the incidence of scarlet fever had decreased prior to the 2008 Olympic Games in Beijing [26]. However, our finding showed that the scarlet fever incidence demonstrated an increasing trend from 2005 (14.7) until 2007 (14.27), and then dropped in 2008 (10.15 cases/100,000 population). Furthermore, the study of You, *et al.* [25] stated that the incidence of scarlet fever was at a lower level in the eight years prior to the scarlet fever incidence in 2011. Indeed, 2011 was an epidemic year of scarlet fever in Beijing, Shanghai, Hong Kong, Macau and Taiwan [2].

The autocorrelation analysis of Global Moran's $I$ value demonstrated that the spatial distribution of scarlet fever became more uneven (Table 3). However, Anselin Local Moran's $I$ analysis revealed that the circular area with a high-high positive spatial association with the incidence of scarlet fever was located around the urban districts (Xicheng, Chaoyang, and Fengtai) in Beijing (Figure S1). In addition, the Getis-Ord Gi* statistic indicated several hotspot clusters in urban districts (Haidian, Shijingshan, Fengtai, Xicheng and Chaoyang) and few in the rural Changping and Daxing districts) in densely populated areas, supporting a positive relationship between population density and an increased risk of scarlet fever exposure.

The temporal analysis showed that during the epidemic year 2011 and throughout 2005 to 2014, scarlet fever cases peaked twice yearly. The first peak occurred in May (at the end of spring), and a smaller, second peak was observed in December (early winter) in Beijing. More clearly, the peak began in March and reached its highest level in May, and a second smaller peak began in October and reached its highest level in December in each year. A consistent finding was reported by Peng *et al.* [17] and Chen *et al.* [29] in Beijing and Shanghai, respectively. Some studies have reported that the incidence peak was observed in March in Hefei, China, [28] and in the United Kingdom [21], and in June in Hong Kong [22]. It has been widely known that scarlet fever primarily occurs in the winter/spring seasons [2,56,57]. Notably, the seasonal peaks varied in different regions. These differences might be partially attributed to climate, geographic distribution, and social factors. This persuasive evidence indicates that scarlet fever transmits more easily in spring and winter compared to the other season, which is consistent with previous studies, as well with WHO data [2].

Furthermore, spatial cluster scanning results showed that the centre of the most likely cluster was detected in Xicheng district in the urban area during our entire study period and that the secondary cluster detected was diverse in each district in the different time periods. The highest relative risk (10.12), with annual cases of 228.2/100,000 population, a likelihood ratio of 2850.73, and a $p$-value < 0.001, was observed in 2011 in Beijing, which was an epidemic year for scarlet fever [2]. Additionally, the most likely clusters, along with secondary cluster locations, were found in densely populated districts, with major transmission spots, such as markets, railways and highways, which suggest that scarlet fever might be transmitted through railways and highways from one place to another. This result also reminds us that prevention and control measures for scarlet fever should focus more on the most likely and secondary cluster areas, including neighbouring districts, to maximize the cost-effectiveness of prevention efforts.

Moreover, our study found the most likely space-time clusters in diverse patterns in Beijing, including time frame, cluster centre, relative risk with and $p$ value. The most likely clusters were observed in rural districts (Huairou, Yanqing, Miyun, Shunyi, Pinggu, Changping) and in urban districts (Xicheng, Chaoyang and Shijingshan), and several secondary clusters were found in urban and rural districts (Figure 5, Table 4). Information about high risk clustered and secondary clustered districts, including neighbouring districts, might be helpful in preventing disease infection and to manage public health planning and resource allocation in advance, to minimize costs and reduce complications.

There are several limitations in this study. First, we used data from reported cases of scarlet fever (*i.e.*, confirmed cases using diagnosis criteria), however, we were unable to explore the characteristics of group A streptococcus GAS *emm typing* stain (e.g., *emm1, emm3 emm4, emm6, emm12, or emm22)*, according to the protocol described by the US CDC due to unavailable data for such information. Second, scarlet fever has been listed as a notifiable infectious disease in China since 2004, and we could not assure the accuracy of the relatively new surveillance monitoring system because some mild cases might be missed or might not be reported (they might utilize home remedies). In this context, the patient information might be lost. Third, this study used circular spatial scanned statistic, which does not allow for irregular geographic shapes. Likewise, the small clusters in the early years could not be identified when data from all 119 months were used. We treated this limitation by scanning each year independently, which has the disadvantage of losing information from each year in the further clusters. Fourth, environmental factors, such as temperature, humidity, rainfall, precipitation and wind pressure, could influence scarlet fever infection, but such data have been available since only 2012 in Beijing. Therefore, we were unable to include these data in our study.

## 5. Conclusions

In conclusion, our study provides a good understanding of the spatiotemporal patterns of scarlet fever in Beijing from 2005 to 2014. The at-risk population was mostly distributed in urban areas and highly and densely populated districts, which indicates a positive relationship between population density and increased risk of scarlet fever exposure. Children under 15 years of age were more vulnerable to scarlet fever infection. With the help of spatiotemporal framework analyses, disinfection should be strengthened and management should focus on the scarlet fever cluster areas and cluster times (*i.e.*, in the urban districts of Beijing). This focus may help control the scarlet fever epidemic more effectively and reduce its harm to children.

**Supplementary Materials:** Table S1: The most likely clusters of scarlet fever disease in Beijing, 2005–2014, using purely spatial analysis of the discrete Poisson model (setting 50% as the maximum cluster size). Figure S1. Local Moran's *I* analysis of scarlet fever in Beijing, 2005–2014.

**Acknowledgments:** The authors thank the Ministry of Science and Technology of Beijing, China, National Centre for Public Health Surveillance and Information Services, China Centre for Disease Control and Prevention, Beijing, China, for providing the scarlet fever case data from Beijing. We also thank the Beijing Municipal Bureau of Statistics and National Geometric Centre of Beijing for the baseline data. Moreover, we thank our colleagues at the Department of Epidemiology and Health Statistics, School of Public Health, Capital Medical University, for their valuable suggestions and software support. Funding: National Science and Technology, Major Project (2012ZX10005009-003 and 2014ZX10004005-001), National 973 Project (serial No.: 2011CB505404), National "12th Five-Year" Plan for Science and Technology Support, China (2012BAI37B03), and National Natural Science Foundation of China (serial No.: 81302516), Edith Cowan University-SRF 2015–2016, and Edith Cowan University Industry Collaboration Scheme-2013 (G1001368). Wei Wang is supported by the Importation and Development of High-Caliber Talents Project of Beijing Municipal Institutions.

**Author Contributions:** Gehendra Mahara and Xiuhua Guo originated the study concept and design, Chao Wang, Fangfang Huang, Lixin Tao, Qi Gao and Kai Cao performed the initial analysis. Gehendra Mahara, Jagadish K. Chhetri and Jin Guo drafted the initial manuscript. Wei Wang and Quanyi Wang cross supervised the study conceptualization and design and critically reviewed the manuscript. Kai Cao, Qin Xu, Liu Long, Fangfang Huang, Lixin Tao, Da Huo and Chao Wang assisted with the data analysis and interpretation. Wei Wang and Xiuhua Guo supervised the study design conceptualization and critically reviewed the manuscript, and Gehendra Mahara and Jagadish. K. Chhetri prepared the final draft of the manuscript.

## References

1. Edward, Z., Jr. Scarlet Fever. Medscape.com. 2014. Available online: http://emedicine.medscape.com/article/1053253-overview (accessed on 7 July 2015).
2. World Health Organization. Western Pacific Region. Scarlet Fever: Factsheet of Scarlet Fever in China. Available online: http://www.wpro.who.int/china/mediacentre/factsheets/scarlet_fever/en/ (accessed on 5 June 2015).
3. Yang, S.G.; Dong, H.J.; Li, F.R.; Xie, S.Y.; Cao, H.C.; Xia, S.C.; Yu, Z.; Li, L.J. Report and analysis of a scarlet fever outbreak among adults through food-borne transmission in China. *J. Infect.* **2007**, *55*, 419–424. [CrossRef]
4. Dong, H.; Xu, G.; Li, S.; Song, Q.; Liu, S.; Lin, H.; Chai, Y.; Zhou, A.; Fang, T.; Zhang, H.; *et al.* Beta-haemolytic group A streptococci emm75 carrying altered pyrogenic exotoxin A linked to scarlet fever in adults. *J. Infect.* **2008**, *56*, 261–267. [CrossRef] [PubMed]
5. CDC, Chinese Center for Disease Control and Prevention Beijing. Available online: http://www.chinacdc.cn/en/aboutus/200506/t20050604_22209.html (accessed on 1 June 2015).
6. Eyler, J.M. Epidemiology of milk-born scarlet fever: The case of Edwardian Brighton. *Am. J. Public Health* **1986**, *76*, 573–584. [CrossRef] [PubMed]
7. Cunningham, M.W. Pathogenesis of group A streptococcal infections. *Clin. Microbiol. Rev.* **2000**, *13*, 470–511. [CrossRef] [PubMed]

8. Leslie, D.L.; Kozma, L.; Martin, A.; Landeros, A.; Katsovich, L.; King, R.A.; Leckman, J.F. Neuropsychiatric disorders associated with streptococcal infection: A case-control study among privately insured children. *J. Am. Acad. Child Adolesc. Psychiatry* **2008**, *47*, 1166–1172. [CrossRef] [PubMed]
9. Centers for Disease Control and Prevention (CDC). Scarlet Fever: A Group A Streptococcal Infection. Available online: http://www.cdc.gov/features/scarletfever/ (accessed on 9 October 2015).
10. Gómez-Carrasco, J.A.; Lassaletta, A.; Ruano, D. Acute hepatitis may form part of scarlet fever. *An. Pediatr.* **2004**, *60*, 382–383. [CrossRef]
11. Güven, A. Hepatitis and hematuria in scarlet fever. *Indian J. Pediatr.* **2002**, *69*, 985–986. [CrossRef] [PubMed]
12. Lau, S.K.; Woo, P.C.; Yuen, K.Y. Toxic scarlet fever complicating cellulites: Early clinical diagnosis is crucial to prevent a fatal outcome. *New Microbiol.* **2004**, *27*, 203–206. [PubMed]
13. Rolleston, J.D. The history of scarlet fever. *BMJ* **1928**, *2*, 926–929. [CrossRef] [PubMed]
14. Feeney, K.T.; Dowse, G.K.; Keil, A.D.; Mackaay, C.; McLellan, D. Epidemiological features and control of an outbreak of scarlet fever in a Perth primary school. *Commun. Dis. Intell.* **2005**, *29*, 386–390.
15. Hsieh, Y.C.; Huang, Y.C. Scarlet fever outbreak in Hong Kong, 2011. *J. Microbiol. Immunol. Infect.* **2011**, *44*, 409–411. [CrossRef] [PubMed]
16. Yan, J.J.; Liu, C.C.; Ko, W.C.; Hsu, S.Y.; Wu, H.M.; Lin, Y.S.; Lin, M.T.; Chuang, W.J.; Wu, J.J. Molecular analysis of group A streptococcal isolates associated with scarlet fever in Southern Taiwan between 1993 and 2002. *J. Clin. Microbiol.* **2003**, *41*, 4858–4861. [CrossRef] [PubMed]
17. Yang, P.; Peng, X.; Zhang, D.; Wu, S.; Liu, Y.; Cui, S.; Lu, G.; Duan, W.; Shi, W.; Liu, S.; *et al.* Group A Streptococcus strains circulating during scarlet fever epidemic, Beijing, China-2011. *Emerg. Infect. Dis.* **2013**, *19*, 909–915. [CrossRef] [PubMed]
18. Zhang, T.; Yang, M.; Xiao, X.; Feng, Z.; Li, C.; Zhou, Z.; Ren, Q.; Li, X. Spectral analysis based on fast Fourier transformation (FFT) of surveillance data: The case of scarlet fever in China. *Epidemiol. Infect.* **2014**, *142*, 520–529. [CrossRef] [PubMed]
19. Daily News Paper of England. Available online: http://www.dailymail.co.uk/health/article-2994236/Scarlet-fever-England-doubles-one-year-doctors-saying-300-cases-reported-week-alone.html (accessed on 10 August 2015).
20. Staszewska, E.; Kondej, B.; Czarkowski, M.P. Scarlet fever in Poland in 2012. *Prz. Epidemiol.* **2014**, *68*, 209–212.
21. Guy, R.; Williams, C.; Irvine, N.; Reynolds, A.; Coelho, J.; Saliba, V.; Thomas, D.; Doherty, L.; Chalker, V.; von Wissmann, B.; *et al.* Increase in scarlet fever notifications in the United Kingdom, 2013/2014. *Euro Surveill.* **2014**, *19*. [CrossRef]
22. Luk, E.Y.Y.; Lo, J.Y.C.; Li, A.Z.L.; Lau, M.C.K.; Cheung, T.K.M.; Wong, A.Y.M.; Wong, M.M.H.; Wong, C.W.; Chuang, S.; Tsang, T. Scarlet fever epidemic, Hong Kong, 2011. *Emerg. Infect. Dis.* **2012**, *18*, 1658–1661.
23. Chen., Y.Y.; Huang, C.T.; Yao, S.M.; Chang, Y.C.; Shen, P.W.; Chou, C.Y.; Li, S.Y. Molecular epidemiology of group A Streptococcus causing scarlet fever in northern Taiwan, 2001–2002. *Diagn. Microbiol. Infect. Dis.* **2007**, *58*, 289–295. [CrossRef] [PubMed]
24. Fernández-Prada, M.; Martínez-Diz, S.; López, A.C.; Nievas., D.A.; Romero, B.M.; Martínez, J.H. Scarlet fever outbreak in a Public school in Granada in 2012. *An. Pediatr.* **2014**, *80*, 249–253. [CrossRef]
25. You, Y.H.; Song, Y.Y.; Yan, X.M.; Wang, H.B.; Zhang, M.H.; Tao, X.X.; Li, L.L.; Zhang, Y.X.; Jiang, X.H.; Zhang, B.H.; *et al.* Molucular epidemiology characteristics of *Streptococcus pyogenes* strains involved in an outbreak of scarlet fever in China, 2011. *Biomed. Environ. Sci.* **2013**, *26*, 877–885. [PubMed]
26. Quian, H.K.; Yang, P.; Zhang, Y.; Wang, X.L.; Duan, W.; Wang, Q.Y. Spatial-Temporal scan statistics on scarlet fever cases in Beijing from 2005–2010. *Dis. Surveill.* **2011**, *26*, 435–238.
27. Liu, Z.; Wang, B.X.; Wang, S.C. The analysis of dynamics of scarlet fever cases in China, 2005–2008. *J. Public Health Prev. Med.* **2009**, *20*, 21–22.
28. Wang, J.; Zhang, J.Q.; Pan, H.F.; Zhu, Y.; He, Q. Epidemiological investigation of scarlet fever in Hefei City, China, from 2004 to 2008. *Trop. Dr.* **2010**, *40*, 4225–4226. [CrossRef] [PubMed]
29. Chen, M.; Yao, W.; Wang, X.; Li, Y.; Chen, M.; Wang, G.; Zhang, X.; Pan, H.; Hu, J.; Zeng, M. Outbreak of scarlet fever associated with *emm*12 type group A Streptococcus in 2011 in Shanghai, China. *Pediatr. Infect. Dis. J.* **2012**, *31*, 158–162. [CrossRef] [PubMed]
30. Lawson, A.B. *Statistical Methods in Spatial Epidemiology*, 2nd ed.; John Wiley & Sons Ltd.: Hoboken, NJ, USA, 2013.

31. Liu, L.; Zhao, X.; Yin, F.; Lv, Q. Spatio-temporal clustering of hand, foot and mouth disease at the county level in Sichuan province, China, 2008–2013. *Epidemiol. Infect.* **2015**, *143*, 831–838. [CrossRef] [PubMed]
32. BMBS. Beijing Municipal Bureau of Statistics. Available online: http://www.bjstats.gov.cn/esite/ (accessed on 15 September 2015).
33. Data Sharing Infrastructure of Earth System Science. Base Map of Beijing City, China. Available online: http://www.geodata.cn/Portal/ (accessed on 17 July 2015).
34. Ministry of Health of China. Incidence of Notifiable Infectious Disease in China. Available online: www.moh.gov.cn/public files (accessed on 27 September 2015).
35. GIS Geography. Spatial Autocorrelation and Moran's I in GIS. Available online: http://gisgeography.com/spatial-autocorrelation-moran-i-gis/ (accessed on 27 August 2015).
36. Getis, A. Spatial interaction and spatial autocorrelation: Across-product approach. *Environ. Plan. A* **1991**, *23*, 1269–1277. [CrossRef]
37. How Spatial Autocorrelation: Moran's I (Spatial Statistics) Works. Available online: http://webhelp.esri.com/arcgisdesktop/9.3/index.cfm?TopicName=How%20Spatial%20Autocorrelation:%20Moran's%20I %20(Spatial%20Statistics)%20works (accessed on 28 January 2015).
38. Spatial Autocorrelation Statistic (Moran's *I*) (Spatial Statistic). Available online: webhelp.esri.com/arcgisdesktop/9.2/index.cfm?TopicName=Spatial Autocorrelation (accessed on 23 November 2015).
39. Anselin, L. Local indicators of spatial association-LISA. *Geogr. Anal.* **1995**, *27*, 93–115. [CrossRef]
40. ArcGIS Software for Geographic Study Version 10.1. Available online: www.ersi.com (accessed on 27 August 2015).
41. Zulu, L.C.; Kalipeni, E.; Johannes, E. Analyzing spatial clustering and the spatiotemporal nature and trends of HIV/AIDS prevalence using GIS: The case of Malawi, 1994–2010. *BMC Infect. Dis.* **2014**, *14*, 285. [CrossRef] [PubMed]
42. Anselin, L.; Getis, O. Spatial statistical analysis and geographic information systems. *Ann. Reg. Sci.* **1992**, *26*, 19–33. [CrossRef]
43. Getis, A.; Ord, J.K. The analysis of spatial association by use of distance statistics. *Geogr. Anal.* **1992**, *24*, 189–206. [CrossRef]
44. Getis, A.; Ord, J.K. Local spatial statistics: An overview. In *Spatial Analysis: Modeling in GIS Environment*; Longley, P., Batty, M., Eds.; John Wiley & Sons Inc.: New York, NY, USA, 1996; pp. 261–278.
45. Mitchell, A. Spatial measurements & statistics. In *The ESRI Guide to GIS Analysis*; ESRI Press: Redlands, CA, USA, 2005; Volume 2.
46. Peeters, A.; Zude, M.; Käthner, J.; Ünlü, M.; Kanber, R.; Hetzroni, A. Getis-ord's hot- and cold-spot statistics as a basis for multivariate spatial clustering of orchard tree data. Comput. *Electron. Agric.* **2015**, *111*, 140–150. [CrossRef]
47. Saxena, R.; Nagpal, B.N.; Das, M.K.; Srivastava, A.; Gupta, S.K. A spatial statistical approach to analyze malaria situation at micro level for priority control in Ranchi district, Jharkhand. *Indian J. Med. Res.* **2012**, *136*, 776–782. [PubMed]
48. Kulldorff, M.; Feuer, E.J.; Miller, B.A.; Freedman, L.S. Breast cancer clusters in the northeast United States: A geographic analysis. *Am. J. Epidemiol.* **1997**, *146*, 161–170. [CrossRef] [PubMed]
49. SaTScan. Available online: www.satscan.org (accessed on 23 November 2015).
50. Zhang, W.; Wang, L.; Fang, L.; Ma, J.; Xu, Y.; Jiang, J.; Hui, F.; Wang, J.; Liang, S.; Yang, H.; *et al.* Spatial analysis of malaria in Anhui province, China. *Malar. J.* **2008**, *7*. [CrossRef] [PubMed]
51. Kleinman, K.P.; Abrams, A.M.; Kulldorff, M.; Platt, R. A model-adjusted space-time scan statistic with an application to syndromic surveillance. *Epidemiol. Infect.* **2005**, *133*, 409–419. [CrossRef] [PubMed]
52. SaTScan Software for Spatial Temporal Analysis. Software for the Spatial and Space-Time Scan Statistics. Available online: www.satsean.org (accessed on 9 June 2015).
53. Kulldorff, M. A spatial scan statistic. *Commun. Stat. Theory Methods* **1997**, *26*, 1481–1496. [CrossRef]
54. Kulldorff, M. SaTScan User Guide for Version 9.4. Available online: http://www.satscan.org/cgi-bin/satscan/register.pl/SaTScan_Users_Guide.pdf?todo=process_userguide_download (accessed on 23 November 2015).
55. Deng, T.; Huang, Y.; Yu, S.; Gu, J.; Huang, C.; Xiao, G.; Hao, Y. Spatial-temporal clusters and risk factors of hand, foot, and mouth disease at the district level in Guangdong Province, China. *PLoS ONE* **2013**, *8*, e56943. [CrossRef] [PubMed]

56. Public Health England, United Kingdom. Scarlet Fever: Second Year of High Activity. Available online: https://www.gov.uk/government/news/scarlet-fever-second-year-of-high-activity (accessed on 9 July 2015).
57. Lamagni, T.; Dennis, J.; George, R.; Efstratiou, A. Analysis of Epidemiological Patterns during a Century of Scarlet Fever. 2008. Available online: www.hpa.org.uk/webc/hpawebfile/hpaweb (accessed on 27 August 2015).

# 5

# Urban Natural Environments, Obesity and Health-Related Quality of Life among Hispanic Children Living in Inner-City Neighborhoods

**Jun-Hyun Kim *, Chanam Lee and Wonmin Sohn**

Academic Editors: Harry Timmermans, Astrid Kemperman and Pauline van den Berg

Department of Landscape Architecture and Urban Planning; Texas A&M University, College Station, TX 77843, USA; clee@arch.tamu.edu (C.L.); wonmin.sohn@tamu.edu (W.S.)

* Correspondence: jhkim@arch.tamu.edu.

**Abstract:** Although a substantial body of literature has provided evidence supporting the positive effects of natural environments on well-being, little has been known about the specific spatial patterns of urban nature in promoting health-related quality of life (HRQOL) among children. This study assessed the association that the urban natural environment measured by landscape spatial patterns may have with obesity and HRQOL among Hispanic children. Ninety-two 4th and 5th grade students were recruited from Houston, Texas, and the Pediatric Quality of Life Inventory (PedsQL) was used to capture the children's HRQOL. The quality of urban natural environments was assessed by quantifying the landscape spatial patterns, using landscape indices generated by Geographic Information Systems and remote sensing. From the bivariate analyses, children's body mass index showed a significantly negative association with their HRQOL. After controlling for socio-demographic factors, the results revealed that larger and more tree areas were positively correlated with children's HRQOL. In addition, those children living in areas with tree patches further apart from each other showed higher HRQOL. This research adds to the current multi-disciplinary area of research on environment-health relationships by investigating the roles of urban greeneries and linking their spatial structures with children's HRQOL.

**Keywords:** health-related quality of life; urban natural environments; children; landscape spatial patterns; obesity; green space; GIS; remote sensing; landscape ecology

## 1. Introduction

Studies connecting neighborhood environments with public health outcomes have grown rapidly in recent years. Earlier works have focused on how natural and well-designed built environments can support physical activities such as walking and bicycling [1–14]. In addition, many previous studies have examined the relationship between urban natural environments and mental health condition [15,16]. Researchers found that urban greeneries can contribute to public health, not only by promoting outdoor activities, but also by helping people relieve daily stresses through their well-documented restorative effects [15,17,18]. However, although a substantial body of literature has provided evidence supporting the effects of the natural environment on the well-being of adult populations, the influence of the natural environment on the physical and mental health outcomes of children or adolescents has not yet been fully investigated. In addition, so far, little is known about the role of urban natural environments shaping landscape spatial patterns in promoting health-related quality of life (HRQOL), especially among children and adolescents. The effects of the natural and built environments on children's health require more investigation because the results of childhood environmental interactions will likely differ from those of adult populations [19].

A number of previous studies have focused on the relationship between the natural environment and children's mental health. Wells and Evans [20] examined the effects of nearby natural environments in rural residential areas on children's psychological well-being. With the 337 rural 3rd to 5th grade children studied, they found that living near nature could offer moderating effects on the impacts of stressful life events, and also improve cognitive functioning. In addition, children living in greener residential environments showed lower levels of stress from life events, while children living in environments with little nearby nature reported higher levels of stress. Kuo and Faber Taylor [21] examined the influence of natural environments on Attention Deficit Hyperactivity Disorder (ADHD) in children. They found that ADHD symptoms were significantly reduced during green outdoor activities than in activities conducted in either built outdoor or indoor settings. In addition, this result appeared consistent across a wide range of characteristics including individual demographic factors, residential settings, and ADHD case severity. Amoly and colleagues [22] also observed 2111 children from 7 to 10 years of age and examined the associations among green spaces, children's behavioral development, and symptoms of ADHD. By using the Normalized Difference Vegetation Index (NDVI) to examine green spaces, they found that both the accessibility and usage times of green spaces surrounding children's residential areas were associated with decreased behavioral problems, including ADHD. Similar cross-sectional studies showed how urban green spaces could play a pivotal role in soothing children's hyperactivity and behavioral problems [23,24]. A small number of studies emphasized the effects of the built environment on children's HRQOL. After reviewing extensive literature on pediatrics, Sherman and colleagues [25] synthesized certain components of the physical environment that influence children's mental health, including their psychological, emotional, and cognitive functioning. They found access to nature to be beneficial in improving children's HRQOL. Although many pediatric researchers have recognized the potential impacts of nature on children's HRQOL, it has not yet fully been measured and a significant knowledge gap still exists.

Childhood obesity caused by a sedentary lifestyle is strongly associated with a poor quality of life and low level of well-being [26–29]. The majority of studies on childhood and adolescent quality of life have focused on groups with chronic health conditions or comorbidities. Several studies have attempted to measure the HRQOL of obese children or adolescents [26–33]. Schwimmer and colleagues [32] recruited three groups of children and adolescents: an obese group, a healthy group, and a group of children and adolescents diagnosed with cancer. Of these groups, obese children and adolescents showed significantly lower values of HRQOL than did healthy children and adolescents. Obesity in a child or adolescent increased the likelihood of an impaired HRQOL 5.5 times more than a healthy child or adolescent, and 1.3 times more than a cancer patient. In addition, the authors stated that measuring HRQOL of Hispanic children would be important due to their high risk of obesity. Tyler and colleagues [34] supported the conclusion that obese Mexican-American children showed lower HRQOL than other ethnicity groups. Studies involving obese adults have shown that they tend to have a lower HRQOL than non-obese adults [35,36], but more empirical studies are necessary that examine HRQOL and obese children [32]. Since children are one of the most vulnerable population groups as they may not be ready to make informed health-related choices by themselves, more studies are required to understand the influence of obesity on their quality of life. Because evidence suggests that childhood obesity could be related to many psychological and social factors [37], the multidimensional and comprehensive structure of HRQOL makes it a useful tool for assessing and enhancing childhood health conditions [32,38].

Based on these findings, the main purpose of this research is to examine the association between urban natural environments measured by landscape spatial patterns and HRQOL among Hispanic children. Since it is widely known that natural environments in neighborhoods provide positive effects on improving the mental health of adults [18,39–41] and children [20,21], it was hypothesized that children's HRQOL would be positively associated with the quality of their neighborhood environments and landscape spatial patterns. In addition, this research hypothesized that children's HRQOL would be negatively correlated with their levels of obesity.

## 2. Methods

### 2.1. *Study Location and Sample*

To obtain reliable self-reported data [42], this research recruited ninety-two 4th and 5th grade students (9 to 11 years old) from five elementary schools in the East End district of Houston, Texas. The East End district is located on the east side of the Houston downtown area. The rationale in choosing this district was as follows. First, according to 2010 Census data [43], a significant proportion of this area is Hispanic (85%) with children younger than 18 years of age (27%). Second, 32% of East End households reported earning less than $ 20,000 in 2010, and 30% of Hispanics in the district reported living in poverty. In addition, 48% of residents who are 25 years and older had no high school diploma, and only 12% had a college degree or higher. Finally, the district offers diverse physical environmental settings including different types of parks, land uses, and housing types. As the neighborhoods in this district were established in the early 1910s, many mature trees have shaped the landscape spatial patterns.

### 2.2. *Data Collection and Measurement*

The consent forms, protocols, and data collection for this study were approved by the Institutional Review Board (IRB) at both the University of Houston and Texas A&M University (2008-0092). Our IRB approval process reviewed every item of this study including all instructions of the Declaration of Helsinki. A research staff administered surveys to participating children at each school site, capturing all of the study variables for this research. This study used the specific study eligibility criteria to recruit participants: children being of Hispanic origin who can speak and read in either English or Spanish, and children who do not have any medical restrictions limiting their physical activity. All subjects voluntarily signed their informed consent for participating in this research.

#### 2.2.1. Health-Related Quality of Life (Dependent Variable)

To evaluate children's HRQOL, this research used the Pediatric Quality of Life Inventory (PedsQL$^{TM}$) 4.0 generic core scale developed by Varni [38,44–48]. According to Varni [38], the PedsQL is a modular instrument used to assess HRQOL among children and adolescents ages 2 to 18. The PedsQL consists of parallel child self-reports and parent proxy reports in essentially identical forms. Separate reports from parents and children are used because children's self-reports are designed to assess perceptions of their internal states, while parents' reports reflect their children's observable behaviors. This instrument has been tested extensively and used in many HRQOL studies [29,32–34]. Most questions regarding perceptions related to health conditions use a five-point Likert scale (0 = never a problem, 1 = almost never a problem, 2 = sometimes a problem, 3 = often a problem, and 4 = almost always a problem). The 23-item PedsQL 4.0 generic core scale is classified into four sub-factors, including physical functioning (8 items), emotional functioning (5 items), social functioning (5 items), and school functioning (5 items). To facilitate the description of children's HRQOL (higher scores show better HRQOL), after collecting all of the questionnaires, each item was reverse-scored and linearly transformed on a zero to 100 scale (0 = 100, 1 = 75, 2 = 50, 3 = 25, 4 = 0). Then, a total scale score calculated by the mean of all 23 items was determined to indicate a summary of the child's HRQOL. Across the ages, the total scale score of self-reporting and proxy-reporting was shown to approach a high score Cronbach alpha reliability coefficient of 0.90 [38]. Previous research has shown that the PedsQL can detect HRQOL differences between healthy and unhealthy children [32,38,44]. The PedsQL was selected for this research because of its high reliability and ease of use for both participants and researchers.

#### 2.2.2. Landscape Spatial Patterns (Independent Variables)

To quantify the quality of the urban natural environments in children's neighborhoods, this research utilized the patch-corridor-matrix model (P-M model), which has been widely used in the landscape ecology field to measure landscape spatial patterns [49]. To classify land cover, 1m color infrared (CIR) high-resolution Digital Orthophoto Quarter Quadrangle (DOQQ) images from the National Agriculture Imagery Program (NAIP) were obtained from the Texas Natural Resources Information System (TNRIS) [50]. To objectively measure different land cover types, Geographic Information Systems (GIS; ArcGIS Version 9.2; ESRI, Redlands, CA, USA) and remote sensing (ENVI Version 4.3, geospatial image analyzing software; ITT Visual Information Solutions, White Plains, New York, NY, USA) were used. Based on light spectral similarities with the unsupervised classification process, 40 different land cover classes were generated. Those classes were then grouped into three main land cover types: grass areas, trees/forests, and developed/impervious areas. To remove isolated pixels affecting the landscape index values, post-classification processes were conducted; these included sieving, clumping, and filtering [51,52].

For measuring landscape spatial patterns surrounding a participant's home, both half-mile and quarter-mile airline buffers (Figure 1) were generated and analyzed using various landscape indices obtained from FRAGSTATS 3.3, a spatial pattern analysis software program developed by McGarigal and Marks [53]. The buffer distance of a half-mile, approximately 800 m, around homes was selected based on the reported distance that both adults and children would likely be willing to walk from previous research [54–57]. Based on previous studies, a quarter-mile buffer, approximately 400 m, was also added to assess the more proximate home neighborhood environments around each child's home [58,59]. In addition, this research tested the associations between the children's HRQOL and landscape spatial patterns of urban forests and trees within the half-mile and quarter-mile street network buffers. However, the PedsQL scores were not associated with any landscape indices captured within these buffer types.

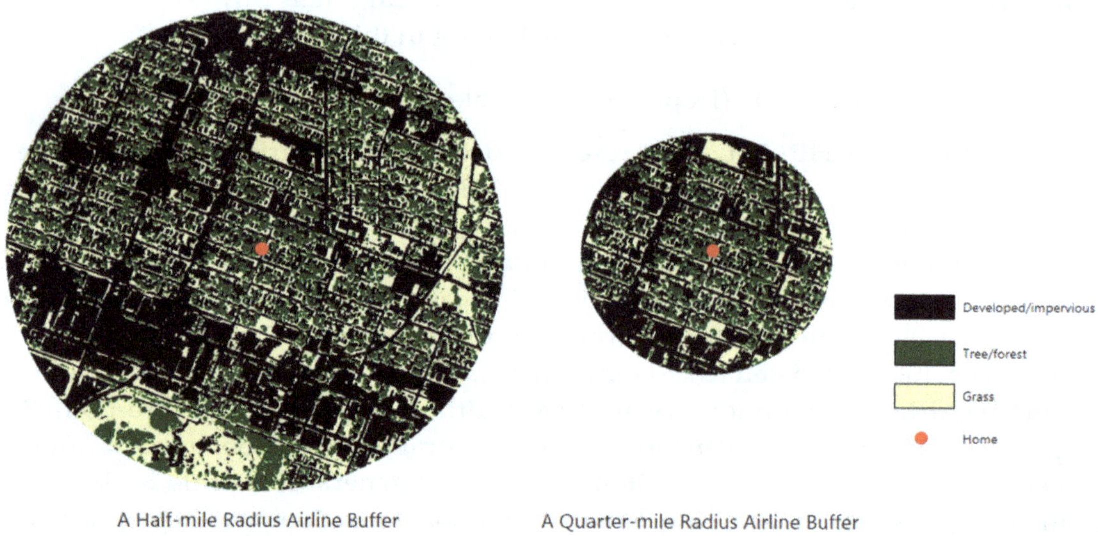


**Figure 1.** Different spatial settings to measure landscape spatial patterns.

From the greater body of landscape indices that measure landscape spatial patterns, this research selected six landscape indices: Percentage of Landscape (PLAND), Number of Patches (NP), Mean Patch Size (MPS), Mean Shape Index (MSI), Mean Nearest Neighbor Distance (MNN), and Patch Cohesion Index (COHESION) (Table 1). To select the appropriate landscape indices, we reviewed the previous literature and established guidelines/criteria addressing both landscape ecological and public health perspectives [12,39,49,60–65]. Each landscape index quantified various aspects of the urban natural environment, including configurations such as size and the existence of tree patches, shapes of patch boundaries, distances between single patches, and connectivity.

### 2.2.3. Control Variables

Control variables including socio-demographic, children's Body Mass Index (BMI), physical activity, and environmental perception variables were captured from the surveys. To collect the children's BMI data, we objectively collected students' heights and weights using a scale (TANITA TBT 310) and portable stationmaster (Seca 213). Each participant's height and weight was measured twice, and the average of the two assessments was used for the final value. Participants' heights and weights were then used to calculate their BMI values. The Quetelet's index (weight in kg/height $m^2$) was used to compute the BMI. The guidelines provided by the Center for Disease Control and Prevention (CDC) [66] were applied to analyze age- and gender-specific BMI z-scores. The Physical Activity Questionnaire for Older Children (PAQ-C) [67] was used to capture the children's physical activity levels in different situations including both weekday and weekend days, and sedentary activity patterns. Participants' neighborhood environmental perceptions were captured regarding accessibility, safety, comfort, attractiveness, and satisfaction, using a survey based on previously-developed instruments [68–71] and the relevant literature [1,57,72].

**Table 1.** Selected landscape indices, acronyms, formulas, and descriptive statistics of landscape spatial patterns.

| Criteria | Landscape Indices (Acronym) | Formula [a] | Description | Unit (Range) | Mean * | Std. Dev. |
|---|---|---|---|---|---|---|
| *Size & Existence & Fragmentation* | Percentage of landscape (PLAND) | $\sum_{j=1}^{a} a_{ij}/A \times 100$ | Higher PLAND values indicate larger patch sizes. | % | 25.99 | 7.29 |
| | Number of patches (NP) | $n_i$ | Higher NP values indicate more number of patches and fragmented condition. | Count | 3809.29 | 1211.09 |
| | Mean Patch size (MPS) | $\sum_{j=1}^{n} a_{ij}/n_i$ | Lower MPS values indicate more fragmented conditions. | Square-meter (MPS ⩾ 0, without limit) | 162.54 | 128.06 |
| *Shape* | Mean shape index (MSI) | $\left[\sum_{j=1}^{n} (0.25p_{ij}/\sqrt{a_{ij}})\right]/n_i$ | Higher MSI values indicate more irregular shapes. | None (MSI ⩾ 1, without limit) | 1.25 | 0.03 |
| *Distance (Isolation)* | Mean nearest neighbor distance (MNN) | $\sum_{j=1}^{a} h_{ij}/n_i$ | Higher MNN values indicate more isolated patterns. | Meter | 2.87 | 0.39 |
| *Connectivity* | Patch cohesion index (COHESION) | $\left(1-\sum_{j=1}^{n} p_{ij}/\sum_{j=1}^{n} (p_{ij}\sqrt{a_{ij}})\right) \times \left(1-1/\sqrt{A}\right)^{-1} \times 100$ | Higher COHESION values indicate more connected patterns. | % | 97.78 | 1.15 |

Notes: This table is adopted and revised form Kim *et al.* [12]. $n_i$ = number of patches in the landscape of patch type I; $a_{ij}$ = area ($m^2$) of patch ij; A = total landscape area ($m^2$); $p_{ij}$ = perimeter of patch ij; $h_{ij}$ = distance (m) from patch ij to nearest neighboring patch of the same type, based on edge-to-edge distance. * Mean values of half-mile airline buffers; [a] See McGarigal and Marks [53] for more details.

## 2.3. *Data Analysis*

The data analysis focused on assessing the correlations between landscape spatial patterns of urban nature and children's HRQOL. First, descriptive statistics were performed to understand the respondents' socio-demographic characteristics, overall physical activity patterns, BMI, HRQOL, and landscape spatial patterns computed by landscape indices. Then the standard diagnostic testing was performed to identify key variables and outliers. In addition, this study tested for the potential multicollinearity problems among the independent variables, especially with regards to the various landscape indices. Once descriptive analyses were completed, bivariate analyses were conducted to detect associations between independent variables (landscape spatial patterns) and the dependent variable (children's HRQOL). Correlations among children's BMI and HRQOL were also evaluated.

A Principal Component Analysis (PCA) was performed to reduce the number of environmental perception variables. The criterion for selecting the appropriate number of factors was above the 0.5 value of the Kaiser-Myer-Olkin (KMO) and Bartlett's test of sphericity with the varimax rotation [73,74]. This research also suppressed absolute values of factor loading less than 0.4 [75]. Finally, statistical models were estimated using a series of multiple regression analyses to predict children's HRQOL, using the independent variables of the landscape indices and controlling for BMI, physical activity, factor scores of environmental perception, and socio-demographic variables. Separate models were estimated using the GIS variables captured at different buffer sizes. We used SPSS 19.0 (IBM Corp., Armonk, NY, USA) to analyze the data.

## 3. Results

### 3.1. Characteristics of the Participants

Table 2 shows the socio-demographic characteristics of the participants. The proportion of the children's genders was approximately 40% boys and 60% girls. Their ages ranged between 9 and 11 years old. In terms of ethnicity, the Hispanic group was dominant with over 80%; the rest of the respondents answered "don't know" to queries regarding their ethnicities. The majority of the participants was born in the U.S. and living with both parents. The average age of the respondents' mothers was about 36 years old. Approximately 65% of the mothers were married. The majority of the mothers were employed, followed by those identifying as homemakers; only about 20% had at least a college degree. The children lived in households with an average of five people. The majority had at least two TVs and two cars per household. About 58% of the children were categorized as overweight or obese, and their mean BMI z-score was 0.90.

**Table 2.** Socio-demographic characteristics of child respondents ($N = 92$).

| Variables | Freq. (%) | Variables | Freq. (%) | Variables | Freq. (%) |
|---|---|---|---|---|---|
| **Gender** | | **Mother's age** | Value | **People in Household** | Value |
| Boy | 35 (38.0%) | Mean | 35.98 | Mean | 5.11 |
| Girl | 57 (62.0%) | SD | 7.17 | SD | 1.60 |
| **Age** | | **Mother's marital status** | | **# of TVs at Home** | |
| 9 | 21 (22.8%) | Single, never married | 13 (14.1%) | 0 | 1 (1.1%) |
| 10 | 50 (54.4%) | Married | 60 (65.2%) | 1 | 9 (9.8%) |
| 11 | 21 (22.8%) | Living with partner | 7 (7.6%) | 2 | 22 (23.9%) |
| **Grade** | | Divorced, widow, separated | 11 (12.0%) | 3 + | 60 (65.2%) |
| 4th | 43 (46.7%) | Other | 1 (1.1%) | | |
| 5th | 49 (53.3%) | **Mother's employment** | | **Number of cars per household** | |
| **Ethnicity** | | Employed for wages | 39 (42.4%) | 0 | 5 (5.4%) |
| Hispanic | 76 (82.6%) | Self-employed | 6 (6.5%) | 1 | 33 (35.9%) |
| Don't know | 16 (17.4%) | Out of work for more than 1 year | 5 (5.4%) | 2 | 42 (45.7%) |
| **Country Born** | | Our of work for less than 1 year | 6 (6.5%) | 3 + | 12 (13.0%) |
| US | 75 (81.5%) | A homemaker | 33 (35.9%) | **Weight Status** [a] | |
| Mexico | 12 (13.0%) | A student | 1 (1.1%) | Obese or overweight | 53 (57.6%) |
| Central America | 3 (3.3%) | Retired | 1 (1.1%) | Normal or underweight | 39 (42.4%) |
| Missing | 2 (2.2%) | Unable to work | 1 (1.1%) | **BMI** | Value |
| **Guardians** | | **Mother's education** | | Mean | 21.69 |
| Mom only | 15 (16.3%) | Elementary to 6th | 15 (16.3%) | SD | 5.53 |
| Dad only | 0 (0.0%) | 7th–8th | 12 (13.1%) | **BMI z-score** | Value |
| Mom and Dad | 70 (76.1%) | 9th–12th | 45 (48.9%) | Mean | 0.90 |
| Parent and Step-parent | 6 (6.5%) | College degree or higher | 15 (16.3%) | SD | 1.21 |
| Missing | 1 (1.1%) | Vocational/technical | 5 (5.4%) | | |

[a] obese or overweight = at or above the 85th percentile of body mass index (BMI) values; Normal or underweight = less than the 85th percentile of BMI values.

According to their physical activity patterns, children showed a relatively low frequency of walking to utilitarian destinations including schools, parks, friends' houses, and stores or shops. About 62% did not walk to school during a typical week. In addition, over 40 percent of children did not walk to a park on a weekly basis. More than 65% participated in physical education or athletics classes only once or twice a week, and only about 30% of children were engaged in physical activities twice or three times on a regular weekend day. As a result of their sedentary patterns, children reported longer amounts of time watching TV during a typical weekend, as compared to weekdays. About 38% of children watched TV more than three hours per weekday, while approximately 52% spent more than three hours watching TV on the weekends. Overall, respondents shared a generally positive environmental perception of and satisfaction with their neighborhoods. In using a five-point Likert scale, approximately 51% of children believed they had adequate accessibility to recreational or playable places, and more than 58% were satisfied with walking or biking conditions in and around their neighborhoods.

### *3.2. Children's HRQOL*

Table 3 represents the children's HRQOL measured by both the children's self-report and their mothers' proxy PedsQL surveys. The children's self-reported PedsQL scores were higher than the scores assessed by their mothers. This result supports the findings of previous research [31]. The total PedsQL score was composed of two main health summary scores: the physical and psychosocial health summary scores; both summary scores were higher in the children's self-report surveys as compared to the mothers' proxy surveys. In the children's self-report surveys, boys reported higher total PedsQL scores (1799.29) than girls (1726.79), while girls had higher total PedsQL scores (1664.91) than boys (1624.24) in the mothers' proxy reports. However, a child's gender did not play a statistically significant role in either survey.

**Table 3.** Children's health-related quality of life (HRQOL) Assessed by the pediatric quality of life (PedsQL) Surveys.

| Surveys | Statistics | Mean PedsQL Score | Total PedsQL Score | Physical Health Summary Score | Psychosocial Health Summary Score |
|---|---|---|---|---|---|
| Child self-report survey | Mean | 76.35 | 1754.67 | 632.61 | 1122.25 |
| | Std. Deviation | 13.32 | 307.86 | 131.96 | 214.72 |
| Mother proxy survey | Mean | 71.44 | 1650.00 | 554.62 | 1093.33 |
| | Std. Deviation | 16.41 | 377.75 | 187.52 | 237.37 |

Total of mean PedsQL score = 100, total of the total PedsQL score = 2300, total of the physical health summary score = 800, total of the psychosocial health summary score = 1500.

The results of the bivariate analyses of the children's BMI and HRQOL indicated that the children's BMI z-scores were negatively correlated to their HRQOL. According to the unadjusted bivariate analysis results from the children's self-report PedsQL surveys, there was a negatively significant relationship between the children's BMI z-scores and their physical health summary scores ($p < 0.05$), while the total PedsQL scores were not significantly associated with their BMI. On the other hand, there were slightly more statistically significant correlations between the children's BMI z-scores and their total and psychosocial health summary scores from the mothers' proxy surveys (Table 4).

**Table 4.** Correlations between children's BMI and HRQOL.

| Variables | Child Self-Report Survey | | Mother Proxy Survey | |
|---|---|---|---|---|
| | Beta | Sig. | Beta | Sig. |
| Total PedsQL score | −0.146 | 0.166 | −0.248 | 0.019 * |
| Physical health summary score | −0.227 | 0.030 * | −0.165 | 0.117 |
| Psychosocial health summary score | −0.068 | 0.520 | −0.269 | 0.010 * |

Notes: Dependent variable: Children's BMI z-score. * $p < 0.05$.

### 3.3. Correlation between Children's HRQOL and Landscape Spatial Patterns

This research estimated two multivariate regression models to measure the association between landscape spatial patterns and children's HRQOL (Table 5). The first model (HA) was based upon the half-mile radius airline buffer measures, and the other model (QA) was based upon the quarter-mile radius airline buffer measures. The HA model explained about 43% of the variance in the children's HRQOL measured by the self-reported PedsQL scores. The children's HRQOL was negatively associated with their age ($p < 0.05$). Their HRQOL was positively correlated with the mother's employment status, but the p value was .061. The children's HRQOL was negatively correlated with their BMI z-scores ($p = 0.01$) and weekend television watching times ($p < 0.05$), while more weekend physical activity time was more likely to lead to a higher HRQOL value ($p < 0.01$). Among the variables related to neighborhood environmental perceptions, neighborhood disorder ($p < 0.01$) and walking barriers ($p < 0.05$) were negatively related to children's HRQOL, whereas accessibility to schools and open spaces to play ($p < 0.05$) and the existence of parks in their neighborhoods ($p < 0.01$) were positively associated with their HRQOL. Regarding the landscape spatial pattern variables, PLAND ($p < 0.05$), NP ($p < 0.05$), and MNN ($p < 0.01$) were positively correlated with the children's HRQOL. Other landscape indices (MPS and MSI) showed a negative relationship with the respondent's HRQOL, but only of marginal significance ($0.05 < p < 0.1$). These results indicated that children having larger (higher PLAND) and more (higher NP) tree/forest areas within a half-mile of their homes were likely to have higher HRQOL. In addition, longer distances between tree patches (higher MNN) were positively associated with higher HRQOL.

The QA model using the quarter-mile radius airline buffer measured the children's HRQOL to be approximately 42%. Among the socio-demographic variables, only age was negatively associated with the children's PedsQL scores. Similar to the results of the HA model, the children's BMI z-scores ($p < 0.01$) and weekend television watching times ($p < 0.01$) were negatively related to their HRQOL, whereas their physical activity times during a regular weekend day ($p < 0.01$) were positively correlated with their HRQOL. From the neighborhood environment perception variables, neighborhood disorder ($p < 0.01$) was negatively associated with the children's HRQOL. However, accessibility to schools and open spaces to play ($p < 0.05$) and the existence of parks in the neighborhood ($p < 0.05$) were both positively correlated with improved HRQOL. Among the selected landscape indices, NP ($p < 0.05$) and MNN ($p < 0.01$) showed significantly positive relationships with the children's HRQOL, which was consistent with the HA model results. The QA model results suggest that children living in areas with more tree patches and longer distances between the patches were likely to report higher HRQOL.

**Table 5.** Landscape Indices Correlated with children's HRQOL: Final Regression Model Results Based on the Half-Mile Airline (HA) and Quarter-mile Airline (QA) Buffer Measures.

| **Half-Mile Airline Buffer: Model HA** | | | **Quarter-Mile Airline Buffer: Model QA** | | |
|---|---|---|---|---|---|
| **Variables** | **Beta** | **Sig.** | **Variables** | **Beta** | **Sig.** |
| **Socio-Demographic Factors** | | | **Socio-Demographic Factors** | | |
| Children's age | −0.189 | 0.041 * | Children's age | −0.243 | 0.010 * |
| Children's gender [a] | 0.202 | 0.038 * | Children's gender [a] | 0.182 | 0.061 |
| Mother's employment status [b] | 0.170 | 0.061 | Mother's employment status [b] | 0.139 | 0.142 |
| **BMI and Physical Activity (PA)** | | | **BMI and Physical Activity (PA)** | | |
| Children's BMI z-score | −0.239 | 0.010 * | Children's BMI z-score | −0.247 | 0.008 ** |
| Total weekend PA times | 0.392 | 0.000 ** | Total weekend PA times | 0.346 | 0.000 ** |
| Total TV watching hours during the weekend | −0.225 | 0.015 * | Total TV watching hours during the weekend | −0.252 | 0.005 ** |

**Table 5.** *Cont.*

| Half-Mile Airline Buffer: Model HA | | | Quarter-Mile Airline Buffer: Model QA | | |
|---|---|---|---|---|---|
| **Variables** | **Beta** | **Sig.** | **Variables** | **Beta** | **Sig.** |
| **Neighborhood Environmental Perceptions** | | | **Neighborhood Environmental Perceptions** | | |
| Accessibility for utilitarian walking | 0.163 | 0.065 | Accessibility to playgrounds and streets to play | 0.171 | 0.059 |
| Unattractiveness in walking conditions | −0.142 | 0.099 | Neighborhood disorder | −0.283 | 0.002 ** |
| Neighborhood disorder | −0.279 | 0.002 ** | Accessibility to schools and open spaces to play | 0.194 | 0.041 * |
| Walking barriers | −0.201 | 0.035 * | Park existence in neighborhoods | 0.251 | 0.015* |
| Accessibility to schools and open spaces to play | 0.198 | 0.042 * | | | |
| Park existence in neighborhoods | 0.298 | 0.005 ** | | | |
| **Landscape Spatial Patterns** | | | **Landscape Spatial Patterns** | | |
| PLAND | 0.357 | 0.023 * | PLAND | 0.255 | 0.069 |
| NP | 0.382 | 0.016 * | NP | 0.385 | 0.020 * |
| MPS | −0.299 | 0.072 | MPS | N/S | N/S |
| MSI | −0.191 | 0.089 | MSI | N/S | N/S |
| MNN | 0.608 | 0.001 ** | MNN | 0.536 | 0.004 ** |
| COHESION | N/S | N/S | COHESION | N/S | N/S |
| (Constant: Coeff. = 3060.916 **) | | | (Constant: Coeff. = 2848.380 **) | | |
| N = 92/Sig. < 0.000 / Adj. $R^2$ = 0.431 | | | N = 92/Sig. < 0.000 / Adj. $R^2$ = 0.423 | | |

Dependent Variable: Child Self-report Total PedsQL Score. Dummy variables: [a] 0 = girl, 1 = boy, [b] 0 = unemployment, 1 = employment. Abbreviations: PLAND, percentage of landscape; NP, number of patches; MPS, mean patch size; MSI, mean shape index; MNN, mean nearest neighbor distance; COHESION, patch cohesion index; N/S, not significant. More detailed information about each landscape index in Table 1. * $p < 0.05$; ** $p < 0.01$.

## 4. Discussion

This research assessed the correlations between landscape spatial patterns and HRQOL among Hispanic children. Although a number of researchers reported that there were significant correlations between the natural environment and mental health conditions [15,17,18,39–41,65,76], only a few studies have attempted to assess relationships related to objectively-measured landscape spatial patterns of urban natural environments. This research expands the existing knowledge regarding those relationships. The results revealed that larger sizes (PLAND) and a greater number (NP) of forests or tree areas in urban neighborhoods were positively associated with Hispanic children's HRQOL. In addition, longer distances between tree/forest patches (MNN) were positively associated with children's HRQOL. These results may relate to safety concerns about urban natural environments. A sense of safety in open spaces has been found to be significantly related to HRQOL [77]. In previous studies, a sense of safety was associated with landscape structure [63,78,79], and they found more open spaces and clearer edge conditions without dense understories would be considered to be safer than closed configurations [63,80]. Thus, the MNN's positive association with HRQOL can be explained by the fact that higher MNN values typically represent landscape spatial patterns with sufficient openings between patches. In addition, this research found that more regularly shaped landscape patterns (lower MSI) were positively associated with children's HRQOL. Although it was a marginal significance between MSI and children's HRQOL, this conclusion also supports the notion that clearly delineated edges contribute to children's higher HRQOL.

The final models for this study showed consistent results with those of previous studies indicating negative relationships among BMI, physical activity, and HRQOL [35–37]. The level of physical activity and overall physical health strongly affect mental health conditions and human well-being. Many researchers have documented that being more active could provide a better sense of mental well-being [37,81]. In addition, previous research has reported that obese children and adults present significantly lower values of HRQOL than those of normal weight [32,35,36]. However, few studies have considered the influence of urban natural environments on children's HRQOL and BMI. The results of this research have confirmed that there are significant correlations between children's BMI, level of physical activity, and HRQOL as they relate to neighborhood landscape spatial patterns [29,31,82–86].

This study has several limitations. First, because this study captured a large number of variables through labor-intensive recruitment and data collection methods to ensure completeness and accurate matches of child-mother pairs, only a relatively small number of samples ($n$ = 92 pairs) could be recruited; therefore, not all potential confounding factors could be considered in the multivariate modeling. Second, the study focused on an inner-city district with a high proportion of minorities and economically challenged populations; therefore, the results are not generalizable to other settings or populations. Third, this research did not objectively measure children's physical activity; conversely, children's BMI and landscape spatial patterns were evaluated using objective measurements. Finally, to examine the landscape spatial patterns of urban forests and trees, this research used DOQQ imagery. This imagery only allows for the analysis of two-dimensional landscape patterns determined by different land cover types. To understand the full range of layers in landscape structure, more advanced media such as LIDAR (Light Detection and Ranging) satellite imagery should be considered. Future studies should consider a larger sample size, and additional and more diverse population groups and community settings. Despite these limitations, to our knowledge, this is one of the first empirical studies assessing associations between children's HRQOL, BMI, and landscape spatial patterns shaped by urban trees and forests, captured objectively through landscape indices.

## 5. Conclusions

This research adds to the current multi-disciplinary research on environment-health relationships by investigating the roles of urban greeneries and linking their spatial structures to Hispanic children's HRQOL. Hispanic populations possess more serious conditions related to physical inactivity and obesity, and they are more likely to be economically challenged, and suffer from poverty and limited access to healthcare. Many previous studies documented that Hispanic children and adolescents were more vulnerable to health conditions such as a high rate of sedentary patterns, obese conditions, and an impaired quality of life than other ethnicity groups [32,34]. The better quality of urban neighborhood environments with more green spaces would be a feasible means of promoting physical activity and improving quality of life by incorporating the practice into children's daily routines. One of the critical goals in urban and landscape planning is improving the quality of life of community members. Because landscape spatial patterns provide one of the most important factors shaping neighborhood environments and have been shown to improve psychological and physical health, it should be considered as a significant neighborhood component requiring proper planning and management. To improve the quality of life among children, it is essential to understand the specific roles of various natural and built environment elements within their neighborhoods. Further, to promote quality of life among children living in urban communities, continued multi-disciplinary efforts are needed to incorporate urban nature and ecological planning considerations into the decision-making processes in the urban planning and public health fields.

**Acknowledgments:** This research was supported by the Robert Wood Johnson Foundation's doctoral dissertation award through the Active Living Research program (#65536), and the Nurture by Steelcase Dissertation Grant from the Center for Health Systems and Design at Texas A&M University. Data collection of this paper was supported by the Urban Hispanic Perceptions of Environment and Activity among Kids (UH-PEAK) research funded by Robert Wood Johnson Foundation's Active Living Research program (#63755). Co-Principal Investigators are Norma Olvera, and Chanam Lee. Authors specially thank Norma Olvera, Stephanie Kellam and undergraduate students from the University of Houston for their assistance with data collection. The open access publishing fees for this article have been covered by the Texas A&M University Online Access to Knowledge (OAK) Fund, supported by the University Libraries and the Office of the Vice President for Research.

**Author Contributions:** Jun-Hyun Kim conceived of the study and led data analysis and manuscript writing. Chanam Lee provided conceptual advice, participated in the development of this study, and edited the manuscript.

Wonmin Sohn contributed to reviewing and summarizing the existing literature, and revised the manuscript. All authors made significant contributions to this research and have approved the final manuscript.

## Abbreviations

| | |
|---|---|
| HRQOL | Health-related quality of life |
| ADHD | Attention Deficit Hyperactivity Disorder |
| PedsQL$^{TM}$ | Pediatric Quality of Life Inventory |
| DOQQ | Digital Orthophoto Quarter Quadrangle |
| GIS | Geographic Information Systems |
| PLAND | Percentage of Landscape |
| NP | Number of Patches |
| MPS | Mean Patch Size |
| MSI | Mean Shape Index |
| MNN | Mean Nearest Neighbor Distance |
| COHESION | Patch Cohesion Index |
| BMI | Body Mass Index |
| HA | Half-mile airline |
| QA | Quarter-mile airline |
| NDVI | Normalized Difference Vegetation Index |

## References

1. Jago, R.; Baranowski, T.; Baranowski, J.C. Observed, GIS, and self-reported environmental features and adolescent physical activity. *Am. J. Health Promotion* **2006**, *20*, 422–428. [CrossRef] [PubMed]
2. Brownson, R.C.; Hoehner, C.M.; Day, K.; Forsyth, A.; Sallis, J.F. Measuring the built environment for physical activity: State of the science. *Amer. J. Prev. Med.* **2009**, *36*, 99–123. [CrossRef] [PubMed]
3. Giles-Corti, B.; Broomhall, M.H.; Knuiman, M.; Collins, C.; Douglas, K.; Ng, K.; Lange, A.; Donovan, R.J. Increasing walking: How important is distance to, attractiveness, and size of public open space? *Am. J. Prev. Med.* **2005**, *28*, 169–176. [CrossRef] [PubMed]
4. Giles-Corti, B.; Donovan, R.J. Relative influences of individual, social environmental, and physical environmental correlates of walking. *Am. J. Public Health* **2003**, *93*, 1583–1589. [CrossRef] [PubMed]
5. King, A.C.; Stokols, D.; Talen, E.; Brassington, G.S.; Killingsworth, R. Theoretical approaches to the promotion of physical activity: Forging a transdisciplinary paradigm. *Am. J. Prev. Med.* **2002**, *23*, 15–25. [CrossRef]
6. Lee, C.; Moudon, A.V. Physical activity and environment research in the health field: Implications for urban and transportation planning practice and research. *J. Plan. Lit.* **2004**, *19*, 147–181. [CrossRef]
7. Moudon, A.; Lee, C.; Cheadle, A.D.; Garvin, C.; Johnson, D.B.; Schmid, T.L.; Weathers, R.D. Attributes of environments supporting walking. *Amer. J. Health Promotion* **2007**, *21*, 448–459. [CrossRef]
8. Moudon, A.V.; Lee, C.; Cheadle, A.D.; Garvin, C.; Johnson, D.; Schmid, T.L.; Weathers, R.D.; Lin, L. Operational definitions of walkable neighborhood: Theoretical and empirical insights. *J. Phys. Act. Health* **2006**, *3*, 99–117.
9. Saelens, B.E.; Handy, L.S. Built environment correlates of walking: A review. *Med.Sci. Sport. Exerc.* **2008**, *40*, 550–566. [CrossRef] [PubMed]
10. Saelens, B.E.; Sallis, J.F.; Black, J.B.; Chen, D. Neighborhood-based differences in physical activity: An environment scale evaluation. *Am. J. Public Health* **2003**, *93*, 1552–1558. [CrossRef] [PubMed]
11. Tilt, J.H.; Unfried, T.M.; Roca, B. Using objective and subjective measures of neighborhood greenness and accessible destinations for understanding walking trip and bmi in Seattle, Washington. *Am. J. Health Promot.* **2007**, *21*, 371–379. [CrossRef] [PubMed]
12. Kim, J.-H.; Lee, C.; Olvara, N.E.; Ellis, C.D. The role of landscape spatial patterns on obesity in hispanic children residing in inner-city neighborhoods. *J. Phys. Act. Health* **2014**, *11*, 1449–1457. [CrossRef] [PubMed]

13. Booth, K.M.; Pinkston, M.M.; Poston, W.S.C. Obesity and the built envrionment. *J. Am. Diet. Assn.* **2005**, *105*, 110–117. [CrossRef] [PubMed]
14. Norman, G.J.; Nutter, S.K.; Ryan, S.; Sallis, F.S.; Calfas, K.J.; Patrick, K. Community design and access to recreational facilities as correlates of adolescent physical activity and body-mass index. *J. Phys. Act. Health* **2006**, *3*, 118–128.
15. Kaplan, R. Impact of urban nature: A theoretical analysis. *Urban Ecol.* **1984**, *8*, 189–197. [CrossRef]
16. Kaplan, S.; Kaplan, R. Health, supportive environments, and the reasonable person model. *Am. J. Public Health* **2003**, *93*, 1484–1489. [CrossRef] [PubMed]
17. Hartig, T.; Mang, M.; Evans, G.W. Restorative effects of natural environment experiences. *Environ. Behav.* **1991**, *23*, 3–26. [CrossRef]
18. Ulrich, R.S.; Simons, R.F.; Losito, B.D.; Fiorito, E.; Miles, M.A.; Zelson, M. Stress recovery during exposure to natural and urban environments. *J. Environ. Psychol.* **1991**, *11*, 201–230. [CrossRef]
19. Jackson, R.J.; Tester, J. Environment shapes health, including children's mental health. *J. Am. Acad. Child Adolesc. Psy.* **2008**, *47*, 129–131. [CrossRef] [PubMed]
20. Wells, N.M.; Evans, G.W. Nearby nature: A buffer of life stress among rural children. *Environ. Behav.* **2003**, *35*, 311–330. [CrossRef]
21. Kuo, F.E.; Faber Taylor, A. A potential natural treatment for attention-deficit/hyperactivity disorder: Evidence from a national study. *Am. J Public Health* **2004**, *94*, 1580–1586. [CrossRef] [PubMed]
22. Amoly, E.; Dadvand, P.; Forns, J.; López-Vicente, M.; Basagaña, X.; Julvez, J.; Alvarez-Pedrerol, M.; Nieuwenhuijsen, M.J.; Sunyer, J. Green and blue spaces and behavioral development in barcelona schoolchildren: The breathe project. *Environ. Health Perspect.* **2014**, *122*, 1351–1358. [CrossRef] [PubMed]
23. Balseviciene, B.; Sinkariova, L.; Grazuleviciene, R.; Andrusaityte, S.; Uzdanaviciute, I.; Dedele, A.; Nieuwenhuijsen, M.J. Impact of residential greenness on preschool children's emotional and behavioral problems. *Int. J. Environ. Res. Public Health* **2014**, *11*, 6757–6770. [CrossRef] [PubMed]
24. Markevych, I.; Tiesler, C.M.T.; Fuertes, E.; Romanos, M.; Dadvand, P.; Nieuwenhuijsen, M.J.; Berdel, D.; Koletzko, S.; Heinrich, J. Access to urban green spaces and behavioural problems in children: Results from the giniplus and lisaplus studies. *Environ. Int.* **2014**, *71*, 29–35. [CrossRef] [PubMed]
25. Sherman, S.A.; Shepley, M.M.; Varni, J.W. Children's environments and health-related quality of life: Evidence informing pediatric healthcare environmental design. *Child. Youth Environ.* **2005**, *15*, 186–223.
26. De Beer, M.; Hofsteenge, G.H.; Koot, H.M.; Hirasing, R.A.; Delemarre-van de Waal, H.A.; Gemke, R. Health-related-quality-of-life in obese adolescents is decreased and inversely related to BMI. *Acta Paediat.* **2007**, *96*, 710–714. [CrossRef] [PubMed]
27. Ottova, V.; Erhart, M.; Rajmil, L.; Dettenborn-Betz, L.; Ravens-Sieberer, U. Overweight and its impact on the health-related quality of life in children and adolescents: Results from the european kidscreen survey. *Qual. Life Res.* **2012**, *21*, 59–69. [CrossRef] [PubMed]
28. Perry, T.T.; Moore, P.C.; Redwine, K.M.; Robbins, J.M.; Weber, J.L. Physical activity, screen time and pediatric health-related quality of life in Mississippi delta. *Open J. Prev. Med.* **2012**, *2*, 105–111. [CrossRef]
29. Tsiros, M.D.; Olds, T.; Buckley, J.D.; Grimshaw, P.; Brennan, L.; Walkley, J.; Hills, A.P.; Howe, P.R.C.; Coates, A.M. Health-related quality of life in obese children and adolescents. *Int. J. Obesity* **2009**, *33*, 387–400. [CrossRef] [PubMed]
30. Pinhas-Hamiel, O.; Singer, S.; Pilpel, N.; Fradkin, A.; Modan, D.; Reichman, B. Health-related quality of life among children and adolescents: Associations with obesity. *Int. J. Obesity* **2005**, *30*, 267–272. [CrossRef] [PubMed]
31. Zeller, M.H.; Modi, A.C. Predictors of health-related quality of life in obese youth. *Obesity* **2006**, *14*, 122–130. [CrossRef] [PubMed]
32. Schwimmer, J.B.; Burwinkle, T.M.; Varni, J.W. Health-related quality of life of severely obese children and adolescents. *J. Am. Med. Assn.* **2003**, *289*, 1813–1819. [CrossRef] [PubMed]
33. Williams, J.; Wake, M.; Hesketh, K.; Maher, E.; Waters, E. Health-related quality of life of overweight and obese children. *J. Am. Med. Assn.* **2005**, *293*, 70–76. [CrossRef] [PubMed]
34. Tyler, C.; Johnston, C.A.; Fullerton, G.; Foreyt, J.P. Reduced quality of life in very overweight mexican american adolescents. *J. Adolesc. Health* **2007**, *40*, 366–368. [CrossRef] [PubMed]
35. Fontaine, K.R.; Bartlett, S.J. Estimating health-related quality of life in obese individuals. *Dis. Manag. Health Outcomes* **1998**, *3*, 61–70. [CrossRef]

36. Kolotkin, R.L.; Head, S.; Hamilton, M.; Tse, C.-K.J. Assessing impact of weight on quality of life. *Obesity Res.* **1995**, *3*, 49–56. [CrossRef]
37. Banis, H.T.; Varni, J.W.; Wallander, J.L.; Korsch, B.M.; Jay, S.M.; Adler, R.; Garciatemple, E.; Negrete, V. Psychological and social adjustment of obese children and their families. *Child Care Health Develop.* **1988**, *14*, 157–173. [CrossRef]
38. Varni, J.W.; Seid, M.; Kurtin, P.S. PedsQL$^{TM}$ 4.0: Reliability and validity of the Pediatric Quality of Life Inventory$^{TM}$ Version 4.0 Generic Core Scales in healthy and patient populations. *Med. Care* **2001**, *39*, 800–812. [CrossRef] [PubMed]
39. Hartig, T.; Johansson, G.; Kylin, C. Residence in the social ecology of stress and restoration. *J. Soc. Issues* **2003**, *59*, 611–636. [CrossRef]
40. Kaplan, S. The restorative benefits of nature: Toward an integrative framework. *J. Environ. Psychol.* **1995**, *15*, 169–182. [CrossRef]
41. Sugiyama, T.; Leslie, E.; Giles-Corti, B.; Owen, N. Associations of neighbourhood greenness with physical and mental health: Do walking, social coherence and local social interaction explain the relationships? *J. Epidemiol. Community Health* **2008**. [CrossRef]
42. Welk, G.J.; Corbin, C.B.; Dale, D. Measurement issues in the assessment of physical activity in children. *Res. Quart. Exerc. Sport* **2000**, *71*, 59–73. [CrossRef] [PubMed]
43. U.S. Census Bureau. American Community Survey 5-Year Estimates. Available online: http://www.census.gov/programs-surveys/acs/ (accessed on 12 July 2015).
44. Varni, J.W.; Seid, M.; Rode, C.A. The PedsQL$^{TM}$: Measurement model for the pediatric quality of life inventory. *Med. Care* **1999**, *37*, 126–139. [CrossRef] [PubMed]
45. Chan, K.S.; Mangione-Smith, R.; Burwinkle, T.M.; Rosen, M.; Varni, J.W. The PedsQL$^{TM}$: Reliability and validity of the short-form generic core scales and asthma module. *Med. Care* **2005**, *43*, 256–265. [CrossRef] [PubMed]
46. Varni, J.W.; Burwinkle, T.M.; Seid, M.; Skarr, D. The PedsQL$^{TM}$ 4.0 as a pediatric population health measure: Feasibility, reliability, and validity. *Ambul. Pediatr.* **2003**, *3*, 329–341. [CrossRef]
47. Varni, J.W.; Seid, M.; Knight, T.S.; Uzark, K.; Szer, I.S. The PedsQL$^{TM}$ 4.0 generic core scales: Sensitivity, responsiveness, and impact on clinical decision-making. *J. Behav. Med.* **2002**, *25*, 175–193. [CrossRef] [PubMed]
48. Swallen, K.C.; Reither, E.N.; Haas, S.A.; Meier, A.M. Overweight, obesity, and health-related quality of life among adolescents: The national longitudinal study of adolescent health. *Pediatrics* **2005**, *115*, 340–347. [CrossRef] [PubMed]
49. Forman, R.T.T. *Land Mosaics: The Ecology of Landscapes and Regions*; Cambridge University Press: Cambridge, UK, 1995.
50. TNRIS. Texas Natural Resources Information System. Available online: https://tnris.org/data-download/#!/county/Harris (accessed on 2 September 2014).
51. Gong, P.; Mahler, S.A.; Biging, G.S.; Newburn, D.A. Vineyard identification in an oak woodland landscape with airborne digital camera imagery. *Int. J. Remote Sens.* **2003**, *24*, 1303–1315. [CrossRef]
52. Mas, J.-F.; Gao, Y.; Pacheco, J.A.N. Sensitivity of landscape pattern metrics to classification approaches. *For. Ecol. Manag.* **2010**, *259*, 1215–1224. [CrossRef]
53. McGarigal, K.; Marks, B.J. *Spatial Pattern Analysis Program for Quantifying Landscape Structure*; Gen. Tech. Rep. PNW-GTR-351; U.S. Department of Agriculture, Forest Service, Pacific Northwest Research Station: Portland, OR, USA, 1995.
54. Ewing, R. Beyond density, mode choice, and single purpose trips. *Transp. Quart.* **1995**, *49*, 15–24.
55. Lee, C.; Moudon, A.V. The 3Ds + R: Quantifying land use and urban form correlates of walking. *Transp. Res. Pt. D: Trans. Environ.* **2006**, *11*, 204–215. [CrossRef]
56. Lee, C.; Moudon, A.V.; Courbois, J.-Y.P. Built environment and behavior: Spatial sampling using parcel data. *Ann. Epidemiol.* **2006**, *16*, 387–394. [CrossRef] [PubMed]
57. Timperio, A.; Crawfore, D.; Telford, A.; Salmon, J. Perceptions about the local neighborhood and walking and cycling among children. *Prev. Med.* **2004**, *38*, 39–47. [CrossRef] [PubMed]
58. McMillan, T.E. The relative influence of urban form on a child's travel mode to school. *Transp. Res. Part A: Policy Pract.* **2007**, *41*, 69–79. [CrossRef]

59. Kaczynski, A.T.; Henderson, K.A. Environmental correlates of physical activity: A review of evidence about parks and recreation. *Leisure Sci.* **2007**, *29*, 315–354. [CrossRef]
60. Dramstad, W.E.; Olson, J.D.; Forman, R.T.T. *Landscape Ecology Principles In Landscape Architecture And Land-Use Planning*; Island Press: Washington, DC, USA, 1996.
61. Forman, R.T.T. Some general principles of landscape and regional ecology. *Landscape Ecol.* **1995**, *10*, 133–142. [CrossRef]
62. Shafer, C. Beyond park boundaries. In *Landscape Planning and Ecological Networks*; Cook, E.A., van Lier, H.N., Eds.; Elsevier: New York, NY, USA, 1994; pp. 201–224.
63. Jorgensen, A.; Hitchmough, J.; Calvert, T. Woodland spaces and edges: Their impact on perception of safety and preference. *Landscape Urban Plan.* **2002**, *60*, 135–150. [CrossRef]
64. Mitchell, R.; Popham, F. Greenspace, urbanity and health: Relationships in england. *J. Epidemiol. Community Health* **2007**, *61*, 681–683. [CrossRef] [PubMed]
65. Ulrich, R.S. View through a window may influence recovery from surgery. *Science* **1984**, *224*, 420–421. [CrossRef] [PubMed]
66. CDC. Percentile Data Files with Lms Values. Available online: http://www.cdc.gov/growthcharts/percentile_data_files.htm (accessed on 22 September 2013).
67. Crocker, P.R.; Bailey, D.A.; Faulkner, R.A.; Kowalski, K.C.; McGrath, R. Measuring general levels of physical activity: Preliminary evidence for the physical activity questionnaire for older children. *Med. Sci. Sport. Exerc.* **1997**, *29*, 1344–1349. [CrossRef]
68. Brownson, R.C.; Chang, J.J.; Eyler, A.A.; Ainsworth, B.E.; Kirtland, K.A.; Saelens, B.E.; Sallis, J.F. Measuring the environmental friendliness toward physical activity: A comparison of the reliability of 3 questionnaires. *Amer. J. Public Health* **2004**, *94*, 473–483. [CrossRef]
69. Forsyth, A.; Schumitz, K.H.; Oakes, M. Twin Cities Walking Survey. Available online: http://www.Activelivingresearch.Org/node/10619 (accessed on 6 December 2008 ).
70. SIP 4–99 Research Group. Environmental Supports for Physical Activity Questionnaire. Available online: http://prevention.sph.sc.edu/tools/docs/Env_Supports_for_PA.pdf (accessed on 6 December 2008).
71. Telford, A.; Salmon, J.; Jolley, D.; Crawford, D. Reliability and validity of physical activity questionnaires for children: The children's leisure activities study survey (class). *Pediatr. Exerc. Sci.* **2004**, *16*, 64–78.
72. Hume, C.; Ball, K.; Salmon, J. Development and reliability of a self-report questionnaire to examine children's perceptions of the physical activity environment at home and in the neighborhood. *Int. J. Behav. Nutr. Phys. Activ.* **2006**. [CrossRef] [PubMed]
73. Hutcheson, G.; Sofroniou, N. *The Multivariate Social Scientist: Introductory Statistics Using Generalized Linear Models*; Sage: London, UK, 1999.
74. Kaiser, H.F. An index of factorial simplicity. *Psychometrika* **1974**, *39*, 31–36. [CrossRef]
75. Stevens, J.P. *Applied Multivariate Statistics for the Social Sciences*, 2nd ed.; Erlbaum: Hillsdale, NJ, USA, 1992.
76. Kaplan, R.; Kaplan, S. *The Experience of Nature: A Psychological Perspective*; Cambridge University Press: Cambridge, UK, 1989.
77. Parra, D.C.; Gomez, L.F.; Sarmiento, O.L.; Buchner, D.; Brownson, R.; Schimd, T.; Gomez, V.; Lobelo, F. Perceived and objective neighborhood environment attributes and health related quality of life among the elderly in Bogota, Colombia. *Soc. Sci. Med.* **2010**, *70*, 1070–1076. [CrossRef] [PubMed]
78. Schroeder, H.; Anderson, L. Perception of personal safety in urban recreation sites. *J. Leisure Res.* **1985**, *16*, 178–194.
79. Sugiyama, T.; Ward Thompson, C. Associations between characteristics of neighbourhood open space and older people's walking. *Urban For. Urban Green.* **2008**, *7*, 41–51. [CrossRef]
80. Ulrich, R.S. Human responses to vegetation and landscapes. *Landscape Urban Plan.* **1986**, *13*, 29–44. [CrossRef]
81. Fox, K.R. The influence of physical activity on mental well-being. *Public Health Nutr.* **1999**, *2*, 411–418. [CrossRef] [PubMed]
82. Fullerton, G.; Tyler, C.; Johnston, C.A.; Vincent, J.P.; Harris, G.E.; Foreyt, J.P. Quality of life in mexican-american children following a weight management program. *Obesity* **2007**, *15*, 2553–2556. [CrossRef] [PubMed]
83. Kolotkin, R.L.; Zeller, M.; Modi, A.C.; Samsa, G.P.; Polanichka Quinlan, N.; Yanovski, J.A.; Bell, S.K.; Maahs, D.M.; Gonzales de Serna, D.; Roehrig, H.R. Assessing weight-related quality of life in adolescents. *Obesity* **2006**, *14*, 448–457. [CrossRef] [PubMed]

84. Sawyer, M.G.; Harchak, T.; Wake, M.; Lynch, J. Four-year prospective study of bmi and mental health problems in young children. *Pediatrics* **2011**, *128*, 677–684. [CrossRef] [PubMed]
85. Wallander, J.L.; Taylor, W.C.; Grunbaum, J.A.; Franklin, F.A.; Harrison, G.G.; Kelder, S.H.; Schuster, M.A. Weight status, quality of life, and self-concept in african american, hispanic, and white fifth-grade children. *Obesity* **2009**, *17*, 1363–1368. [CrossRef] [PubMed]
86. Zeller, M.H.; Modi, A.C. Development and initial validation of an obesity-specific quality-of-life measure for children: Sizing me up. *Obesity* **2009**, *17*, 1171–1177. [CrossRef] [PubMed]

6

# Assessment on Distributional Fairness of Physical Rehabilitation Resource Allocation: Geographic Accessibility Analysis Integrating Google Rating Mechanism

**Hui-Ching Wu [1,2,†], Ming-Hseng Tseng [3,†] and Chuan-Chao Lin [4,5,*]**

[1] Department of Medical Sociology and Social Work, Chung Shan Medical University, Taichung 402367, Taiwan; graciewu@csmu.edu.tw
[2] Social Service Section, Chung Shan Medical University Hospital, Taichung 402367, Taiwan
[3] Department of Medical Informatics, Chung Shan Medical University, Taichung 402367, Taiwan; mht@csmu.edu.tw
[4] School of Medicine, Chung Shan Medical University, Taichung 402367, Taiwan
[5] Department of Physical Medicine and Rehabilitation, Chung Shan Medical University Hospital, Taichung 402367, Taiwan
* Correspondence: t01062@csmu.edu.tw.
† These authors contributed equally to this work.

**Abstract:** Identifying and treating co-existing diseases are essential in healthcare for the elderly, while physical rehabilitation care teams can provide interdisciplinary geriatric care for the elderly. To evaluate the appropriateness of demand and supply between the population at demand and physical rehabilitation resources, a comparative analysis was carried out in this study. Our study applied seven statistical indices to assess five proposed methods those considered different factors for geographic accessibility analysis. Google ratings were included in the study as a crucial factor of choice probability in the equation for calculating the geographic accessibility scores, because people's behavioral decisions are increasingly dependent on online rating information. The results showed that methods considering distances, the capacity of hospitals, and Google ratings' integrally generated scores, are in better accordance with people's decision-making behavior when they determine which resources of physical rehabilitation to use. It implies that concurrent considerations of non-spatial factors (online ratings and sizes of resource) are important. Our study proposed an integrated assessment method of geographical accessibility scores, which includes the spatial distribution, capacity of resources and online ratings in the mechanism. This research caters to countries that provide citizens with a higher degree of freedom in their medical choices and allows these countries to improve the fairness of resource allocation, raise the geographic accessibilities of physical rehabilitation resources, and promote aging in place.

**Keywords:** physical rehabilitation; elderly; geographic accessibility; resources allocation; spatial inequality; medical geology

---

## 1. Introduction

### 1.1. *Physical Rehabilitation Resources and Active Aging*

The World Health Organization (WHO) proposed a policy framework for active aging in 2002, emphasizing that active aging is a process wherein aging is guided by policies. By providing the elderly with the best opportunities in pursuit of health, social participation, and a safe environment, their quality of life can be effectively promoted [1]. Therefore, the crucial implication active aging is to help the elderly achieve the stage of successful aging. Phelan et al. [2] pointed out that the elderly believe successful aging involves the integration of multi-faceted health conditions, including physical, functional, psychological, and social abilities. In addition to medical services, social activities that increase mental flexibility and connection to support networks that strengthen health also promote the quality of life of the elderly. Due to the physical limitations of the elderly, the geographical accessibility of physical rehabilitation resources affects their ability to use community care resources and reflects fairness in the design of the resource allocation policy.

Identifying and treating co-existing diseases are essential in the healthcare for the elderly, while physical rehabilitation care teams can provide high-quality and interdisciplinary geriatric care for the elderly [3–5]. Board-certificated physiatrists are practitioners who complete their training in physical medicine and rehabilitation residency and pass the national examinations. They possess the professional knowledge to diagnose and treat many diseases of the elderly. Research has shown that with the intervention of physiatrists, the elderly enjoy better functional recovery from injuries and illnesses [6]. With the rapid growth of the elderly population, Taiwan is about to become a super-aged society. Every older person in Taiwan has the same health insurance. The current healthcare system in Taiwan, known as National Health Insurance (NHI), was instituted in 1995. NHI is a single-payer compulsory social insurance plan that centralizes the disbursement of healthcare funds. The system promises equal access to healthcare for all citizens, and the population coverage has reached 99% [7]. The National Health Insurance of Taiwan covers medical insurance for 99% of the population. People are free to choose from medical centers, community hospitals, and specialist clinics when they look for treatments. The integrated medical specialist teams led by physiatrists and supported by physiotherapists, occupational therapists, speech therapists, nurses, nutritionists, and orthotists can provide interdisciplinary physical rehabilitation care in appropriate environments with proper equipment and provide comprehensive care for the elderly [8,9].

### *1.2. Accessibility Assessment of Elderly Physical Rehabilitation Resources*

Some studies addressed the perception of accessibility of elderly physical resources, such as reports by clinic managers versus actual accessibility in healthcare clinics for persons using wheelchairs [10], problems of access to primary care [11], people with physical disabilities feel they are experiencing difficulty accessing adequate and appropriate primary healthcare services [12]. According to these studies, the transportation factor is important for the elderly to access healthcare resources. Therefore, evaluating the appropriateness of the demand and supply between the population at demand and physical rehabilitation resource is important for policy-making. In a comprehensive review of the literature, studies that address a geographic accessibility assessment of elderly physical rehabilitation resources are rare.

A geographic accessibility assessment could provide a fair distribution in allocating healthcare resources [13–24]. Identifying and treating co-existing diseases are essential in healthcare for the elderly, while the accessibility of physical rehabilitation resources should be taken seriously.

Frail older adults can go to hospitals by their family's vehicles or apply for the governmental rehabilitation bus service. Taiwan's NHI provides a free rehabilitation bus service for those who have a handbook of physical and mental disabilities with moderate or above multiple disabilities including limbs, moderate or above visually impaired, vegetative (wheelchair accessible), and extremely severely disabled vital organs [25]. For those who could not pay the premium, the premium is fully subsidized for the households below the poverty line. Or, the NHI can refer those very poor persons to charitable organizations for help. The transportation for older adults is organized by families and the NHI. Therefore, income would not become the main obstacle of transportation, but the accessibility of resources would be an important issue for aging in place.

The discussion of fairness in the distribution of physical rehabilitation resources involves the degree of coordination between the population at demand and service supply as well as distance factors. For frail older adults who need regular and periodic physical rehabilitation, high-geographic accessibility is important to promote aging in place. The use of assessment methods for resource accessibility help examine whether the allocation of physical rehabilitation resources shows inequity due to regional differences. The author of this study attempted to employ geographic accessibility as the assessment method. The investigation analyzed and compared the results drawn from five types of geographic accessibility calculation methods examining the adequacy of physical rehabilitation resource allocation.

In this study, open data of 2020 were retrieved from Taiwan Academy of Physical Medicine and Rehabilitation, and its member list of board-certificated physiatrists and registered clinics was the supply points for the resources. People aged 65 and above in towns were held to be the population at demand. With the aforementioned data, the geographical accessibility of rehabilitation resources for the population at demand in towns were examined. For the presentation of analytical data, assessments focused on the data of county/city levels, which were aggregated from the data of town levels. Therefore, the counties/cities over the island that need to be prioritized for the improvement of resource accessibilities at physical rehabilitation points are pointed out in this study to present the problems in the appropriateness of demand and supply between geographical locations and the density distribution of population at demand. The research results are expected to turn into references for relevant administrative and management departments when they formulate resource allocation policies of rehabilitation resources.

The current distribution of the population at demand and physical rehabilitation resources, population at demand to physical rehabilitation resources ratio, and the service load of rehabilitation hospitals were examined in the study. The investigation helped consider how to increase the accessibilities of physical rehabilitation for the elderly by assisting them to look for treatment at clinics nearest to their homes and reduce traffic obstacles they may encounter to promote their health. The author explored the following issues:

1. The spatial distribution of the population at demand and number of physical rehabilitation resources in towns.
2. To carry out a comparative analysis on geographical accessibility scores of physical rehabilitation resources with five calculation methods based on different decision-making considerations and choice probabilities.
3. To suggest follow-up improvements of policies based on the differences in densities of physical rehabilitation resources in counties/cities.

## 2. Materials and Methods

### *2.1. Data Collection: Study Area and Datasets*

The geographical area covered by the analysis in this study includes 19 counties/cities and 349 towns on the main island of Taiwan. Information about board-certificated physiatrists was retrieved from the open data of Taiwan Academy of Physical Medicine and Rehabilitation in 2020 [26]. Information about the population aged 65 and above in towns was retrieved from the database of Department of Household Registration, Ministry of the Interior, which was released in March 2020 [27].

The convenience of transportation is an important factor that determines senior citizens' access to community care resources. However, to examine the differences in convenience of transportation in counties/cities, we will have to consider the types of vehicles, frequencies of running and travel time, fare policies, as well as fare subsidy policies of counties/cities. Due to the scarcity or low credibility of relevant data, it is infeasible to include such information in the analysis of road network data. In the evaluation of factors that affect geographic accessibility, the study took reference from the research method of Page et al. [28]. While retrieving data for the analysis of transportation influencing factors, the road network data in government open data representing actual route distances were adopted instead of the traditional map distances (the linear distance between two points) to reduce the error. As for map data, numerical maps were taken from the Ministry of Transportation and Communications [29]. The ArcGIS application, which adopts geographic information systems, was used to calculate geographic accessibility by a geography information system (GIS)-based network analysis. As the geographic accessibility analysis focused on the convenience of users' mobility, if the data of supply points in the main island and outlying islands of Taiwan are mixed and assessed collectively, the issues in traffic and geographic distance will produce deviations in resource accessibility assessment. Therefore, the study area was limited to the main island of Taiwan.

To define the searching area of physiatrist resources, registered specialist clinics were listed and filtered in this study, according to the Taiwan Academy of Physical Medicine and Rehabilitation. As these resource data only list service units, we had to search for the addresses of every service unit before converting the addresses to coordinates by geocoding applications. Next, with the use of the geography information system (GIS), the latitudes and longitudes of the locations of every resource were positioned in the TWD97 2-degree transverse Mercator coordinate system. Cartographic visualization was employed to test the accuracy of every coordinated point and reduce location error. Finally, the cartographic data of physiatrist resources were produced. As of March 2020, there were 688 physiatrist service points in the main island of Taiwan, while there were 1140 physiatrists in total.

Preliminary investigations in this study indicated that there were 3,618,878 people aged 65 and above on the main island of Taiwan as of March 2020. As there were 1140 physiatrists in total, it means that for every 10,000 elderly people there were 3.15 physiatrists on average. As the number is close to the population of towns, the weighted center point of towns (generated by the weighed calculation of population in villages) would represent the center point of people in demand for resources.

### 2.2. *Measuring Geographic Accessibility to Elderly Physical Rehabilitation Resources*

The geographic accessibility of resources is a critical basis for considering resource allocation. A main method to analyze resource accessibility is to calculate the ratio of resources allocated (amount and spatial distribution) to the population at demand.

At present, Taiwan's policy formulation relies on the regional average method in weighing medical resource accessibility. Taking each administrative region as a unit, the number of hospitals, medical personnel, and hospital beds per 10,000 (or per 100,000) people in the region is calculated and becomes a potential accessibility indicator for the framework of accessibility to medical resources [30]. In terms of the assessment of medical resources, the method assumes that the administrative region equal to the activity space where people utilize medical resources and distances does not bring about differences in the usage of medical resources within the region. However, patients can seek treatments by crossing into different administrative regions in reality. This characteristic of spatial mobility is not taken into consideration in the regional average method, and this is where problems arise [15]. The method was identified as method A0 in this study, with Equation (1) as follows:

$$A_{i,0} = \frac{\sum_{j \in D_i} S_j}{P_i} \tag{1}$$

where $A_i$ is the geographic accessibility score of a town $i$ and implies the average amount of supply point resources enjoyed by each person in demand in the region of the town $i$; $\sum_{j \in D_i} S_j$ represents the amount of supply point resources in region $i$ of the town; $P_i$ represents the population at demand aged 65 or above in the region $i$ of towns.

Luo and Wang [14] proposed the two-step floating catchment area method, which breaks the aforementioned limitations caused by setting administrative regions as activity areas. Not only does the research method consider the possibilities of cross-region healthcare utilization by people, but it also sets a reasonable range of seeking treatment and, in turn, assesses the spatial accessibility of medical resources. The two-step floating catchment area method is primarily divided into two stages [19,21,31]. In stage one, the service loads of each service provider of resources are calculated. In stage two, the ratios of resources that can be reached by each location of the population at demand are calculated to assess the geographic accessibility scores of resources [20].

The three-step floating catchment area method [32] is an advanced and improved search method derived from the two-step floating catchment area method. The new method evaluates different choice probabilities of the population at demand when people approach nearby locations of medical resources. The effects of hospitals' capacities and travel distances on the utilization behavior of medical resources are specifically taken into consideration. The concept of this method is to calculate the probability of seeking treatment, which represents the probability of each patient to visit different hospitals through distance weighting and hospital capacities. The probability of seeking treatment is then used to estimate the average ability of the medical resource allocation of each hospital. In the same manner, with the hospitals' capacities and distances from the served regions, the probability of each region in demand to visit different hospitals is calculated. According to the choice probabilities, the average ability of the medical resource allocation of each hospital will be allocated to the region in demand appropriately, wherein we obtain the distribution situation of geographical accessibility to medical resources in the research area.

The calculation of geographical accessibility proposed in this study originates from the calculus concepts of the three-step floating catchment area method. Equations (2)–(5) are as follows:

$$A_{i,1} = \sum\nolimits_{r=1\sim h} \sum_{j\in D_r} \frac{S_j * f(d_{ij})}{\sum_{r=1\sim h} \sum_{k\in D_r} P_k * f(d_{jk})} \tag{2}$$

$$A_{i,2} = \sum\nolimits_{r=1\sim h} \sum_{j\in D_r} \frac{S_j * K_{ij} * f(d_{ij})}{\sum_{r=1\sim h} \sum_{k\in D_r} P_k * K_{jk} * f(d_{jk})} \tag{3}$$

$$A_{i,3} = \sum\nolimits_{r=1\sim h} \sum_{j\in D_r} \frac{S_j * V_{ij} * f(d_{ij})}{\sum_{r=1\sim h} \sum_{k\in D_r} P_k * V_{jk} * f(d_{jk})} \tag{4}$$

$$A_{i,4} = \sum\nolimits_{r=1\sim h} \sum_{j\in D_r} \frac{S_j * K_{ij}^{V} * f(d_{ij})}{\sum_{r=1\sim h} \sum_{k\in D_r} P_k * K_{jk}^{V} * f(d_{jk})} \tag{5}$$

where $A_{i,1}$ is the simplest calculation of the geographical accessibility score of a location at demand $i$ and implies the average amount of supply point resources enjoyed by people at demand in the location at demand $i$; $S_j$ represents the scale of supply at each service point (physiatrist) $j$; $P_k$ represents the size of the elderly population in the location at demand $k$; $d_{ij}$ is the route distance between the location at demand $i$ and the service point $j$; $d_{jk}$ is the route distance between the service point $j$ and the location at demand $k$. In the equations, $f(d_{ij})$ is the distance-decay function, while the search radii of resources in this study are divided into three districts ($r$ = 1~3) according to the respective distance. The first district ($d_{ij} \leq 3$ km) is the area that the elderly can reach on foot in about an hour [24]. The second district (3 km < $d_{ij} \leq 15$ km) is the area that the elderly can reach by driving for about half an hour. The third district (15 km < $d_{ij} \leq 30$ km) is the area that the elderly can reach by driving for about an hour. $f(d_{ij})$ is shown in Equation (6):

$$(d_{ij}) = \left\{ \begin{array}{cc} 1, & d_{ij} \leq 3\text{ km} \\ \frac{3}{d_{ij}}, & 3\text{ km} < d_{ij} \leq 15\text{ km} \\ \frac{15}{(d_{ij})^2}, & 15\text{ km} < d_{ij} \leq 30\text{ km} \\ 0, & d_{ij} > 30\text{ km} \end{array} \right\} \tag{6}$$

$A_{i,2}$ calculates the geographical accessibility score of a location at demand $i$ when $K_{ij}$, which is the different choice probabilities of the population at demand to approach various nearby service points, is taken into consideration. With considerations of the scale of supply at service points $S_j$ and distance decay $d_{ij}$, $K_{ij}$ represents the choice probabilities of the location at demand $i$ to service point $j$ and is expressed in Equation (7):

$$K_{ij} = \sum_{r=1\sim h} \frac{S_j * f(d_{ij})}{\sum_{k \in D_r} S_k * f(d_{ik})} \tag{7}$$

$A_{i,3}$ is a new method of calculation introduced in this study, which calculates the geographical accessibility score of a location at demand $i$ when $V_j$, the overall rating of the location point $j$ summited to Google by ordinary users, is taken into consideration. It represents the crucial decision basis of people when they choose to visit a particular service point in reality. With considerations on rating $V_j$ and distance decay $d_{ij}$, $V_{ij}$ represents the choice probabilities of the location at demand $i$ to service point $j$ and is expressed in Equation (8):

$$V_{ij} = \sum_{r=1\sim h} \frac{V_j * f(d_{ij})}{\sum_{k \in D_r} V_k * f(d_{ik})} \tag{8}$$

$A_{i,4}$ is another new method of calculation introduced in this study. It calculates the geographical accessibility score of a location at demand $i$ while integrating the factors of rating $V_j$, the scale of supply at service points $S_j$, distance decay $d_{ij}$, and the different choice probabilities of the population at demand to approach various nearby service points $K_{ij}^V$. $K_{ij}^V$ is expressed in Equation (9):

$$K_{ij}^V = \sum_{r=1\sim h} \frac{V_j * S_j * f(d_{ij})}{\sum_{k \in D_r} V_k * S_k * f(d_{ik})} \tag{9}$$

The flow of calculation follows Equations (1)–(8). First, we calculated the service load to be provided by each service point of physiatrists to the three districts divided by the distances and within a 30-km search radius of resources (the service load = the total population at demand in towns within a 30-km search radius/scale of service at the particular service point). Finally, we calculated the accumulated service load provided by the service points of physiatrists to each weighted center point of people at demand in towns, while the service points are within a 30-km search radius of the weighted center point. In this way, the accessible ratio of resources at the service points of physiatrists to the population at demand in towns was obtained, which is held to be the geographical accessibility score. Table 1 shows the calculation equations used in this study to evaluate the geographic accessibility scores of physical rehabilitation resources.

**Table 1.** Definition of elderly physical rehabilitation resources geographic accessibility scores.

| Method | Description | Equation | Distance-Decay Function |
|---|---|---|---|
| A0 | Regional average method | $A_{i,0} = \frac{\sum_{j \in D_i} S_j}{P_i}$ | 1 |
| A1 | Two-step floating catchment area method without choice probability | $A_{i,1} = \sum_{r=1\sim h} \sum_{j \in D_r} \frac{S_j * f(d_{ij})}{\sum_{r=1\sim h} \sum_{k \in D_r} P_k * f(d_{jk})}$ | $f(d_{ij}) = \left\{ \begin{array}{ll} 1, & d_{ij} \le 3 \text{ km} \\ \frac{3}{d_{ij}}, & 3 \text{ km} < d_{ij} \le 15 \text{ km} \\ \frac{15}{(d_{ij})^2}, & 15 \text{ km} < d_{ij} \le 30 \text{ km} \\ 0, & d_{ij} > 30 \text{ km} \end{array} \right\}$ |
| A2 | Three-step floating catchment area method with considerations of choice probability $K_{ij}$ | $A_{i,2} = \sum_{r=1\sim h} \sum_{j \in D_r} \frac{S_j * K_{ij} * f(d_{ij})}{\sum_{r=1\sim h} \sum_{k \in D_r} P_k * K_{jk} * f(d_{jk})}$ | |
| A3 | Three-step floating catchment area method with considerations of choice probability $V_{ij}$ | $A_{i,3} = \sum_{r=1\sim h} \sum_{j \in D_r} \frac{S_j * V_{ij} * f(d_{ij})}{\sum_{r=1\sim h} \sum_{k \in D_r} P_k * V_{jk} * f(d_{jk})}$ | |
| A4 | Three-step floating catchment area method with considerations of choice probability $K_{ij}^V$ | $A_{i,4} = \sum_{r=1\sim h} \sum_{j \in D_r} \frac{S_j * K_{ij}^V * f(d_{ij})}{\sum_{r=1\sim h} \sum_{k \in D_r} P_k * K_{jk}^V * f(d_{jk})}$ | |

### 2.3. *Google Rating*

In the era of Web 2.0, consumers increasingly rely on the rating mechanism of online service platforms as crucial factors for decision-making. The online rating mechanism has become an important asset in the digital "reputation" economy [33]. For example, people who prepare to choose a hotel put a high value on the review scores left by tourists on travel information websites Agoda and Tripadvisor.

In 1995, Taiwan implemented the National Health Insurance policy, which provides convenient medical services to citizens. People are free to choose from various hospitals when they look for treatments. Faced with the competition in the free market, hospitals have adopted business models of marketing and branding to attract patients. Hospital rating mechanisms on online platforms, where people provide reviews voluntarily and freely, have emerged as crucial sources of references for patients' healthcare-seeking decisions. Broadly speaking, rating mechanisms include blogging, Facebook, YouTube, and Google's rating mechanism for businesses. Among them, Google Rating is the rating mechanism that performs best in structuring consumers' feelings and is the most recognized by the public [33–35]. Google Rating scores are divided into 1~5 points, representing evaluations ranging from least satisfied to most satisfied.

Based on the open competition in Taiwan's medical market, the high degree of freedom enjoyed by people in seeking treatment, and the multiple choice factors in healthcare-seeking decisions, this study innovates and introduces new methods of calculating the geographic accessibility scores of physical rehabilitation resources, in which Google ratings for businesses is included as a choice factor in the calculation equations. The methods are detailed in the descriptions of methods A3 and A4 or Equations (7) and (8).

### 2.4. *Gini Coefficient*

The Gini coefficient was defined by Italian statistician Corrado Gini based on the Lorenz curve as a measure of income distribution equality within a society [36]. The Gini coefficient can range from 1 to 0, wherein 1 represents complete inequality in people's annual income distribution and 0 represents complete equality in income distribution. Generally speaking, a Gini coefficient below 0.2 indicates highly equitable income distribution, 0.2–0.3 represents equitable income distribution, 0.3–0.4 indicates bearable inequitable income distribution, 0.4–0.6 tends toward serious inequality in income distribution, and above 0.6 indicates high inequality in income distribution [37]. Therefore, when the Gini coefficient is above 0.6, the ruling authority would usually be advised to be on the alert for excessive income inequality within the society, as the situation may lead to social conflicts. Due to its nature, the Gini coefficient is also called the inequality coefficient. With reference to the above-mentioned scaling of the coefficient, this study explains the disparity in the accessible ratio of resources at service points of physiatrists to the population at demand in counties/cities.

The Gini coefficient was used in this study to evaluate the equality of the accessible ratio of service point resources to the population at demand. Therefore, a higher Gini coefficient in a county/city represents a more inequitable distribution of resources at service points to the population at demand. Based on the definition of $y_1 = f(x)$ of the Lorenz curve, the $y$-axis measures the accumulated percentage of the accessible ratio of service point resources in each town, while the $x$-axis measures the accumulated percentage of the population at demand in each town. The Gini coefficient is equal to the area between curve $y_1$ and line $y_2$, divided by the area below line $y_2$. The Equation (10) is as follows [38]:

$$G = \frac{\int_0^1 (y_2 - y_1)dx}{\int_0^1 y_2 dx} = \frac{\int_0^1 (x - f(x))dx}{\int_0^1 x dx} = 2\int_0^1 (x - f(x))dx \quad (10)$$

## 3. Results

### 3.1. Distribution of People at Demand in Towns and Physical Rehabilitation Resources

Table 2 summarizes the results of resource assessments using the regional average method (method A0). The National Health Insurance of Taiwan adopts an ideology of open and free competition concerning the setting up of hospitals. As a result, operating in metropolitan areas to attract clients is the first choice of most physical rehabilitation clinics and physiatrists. Furthermore, teaching hospitals focusing on physical rehabilitation training tend to cluster in metropolitan areas. Consequently, many physiatrists choose to register and practice in the same metropolitan area where they complete their specialist trainings. Therefore, in the six most urbanized municipalities (Taipei City, Kaohsiung City, New Taipei City, Taichung City, Tainan City, and Taoyuan City) where 68.12% of the elderly population live, the density of physiatrists per 10,000 elderly people ranged between 1.55% and 5.29%, while the average density of physiatrists on the main island of Taiwan was 1.80%.

**Table 2.** Summary statistics of 65+ population and physical rehabilitation physicians' scores by administrative districts (method A0).

| Administrative District | $65^+$ Population | $65^+$ Population % | Number of Towns | Number of Physicians | Physicians-to 10,000 Population % |
|---|---|---|---|---|---|
| Yilan County | 76,134 | 2.10% | 12 | 25 | 1.82 |
| Hsinchu County | 71,911 | 1.99% | 13 | 16 | 0.91 |
| Miaoli County | 91,283 | 2.52% | 18 | 21 | 1.13 |
| Changhua County | 205,532 | 5.68% | 26 | 42 | 0.99 |
| Nantou County | 89,157 | 2.46% | 13 | 12 | 0.79 |
| Yunlin County | 127,220 | 3.52% | 20 | 19 | 0.81 |
| Chiayi County | 99,858 | 2.76% | 18 | 13 | 0.92 |
| Pingtung County | 140,607 | 3.89% | 32 | 18 | 0.65 |
| Taitung County | 35,707 | 0.99% | 14 | 8 | 0.54 |
| Hualien County | 55,009 | 1.52% | 13 | 20 | 1.49 |
| Keelung City | 62,020 | 1.71% | 7 | 23 | 3.35 |
| Hsinchu City | 57,138 | 1.58% | 3 | 22 | 3.29 |
| Chiayi City | 42,062 | 1.16% | 2 | 24 | 5.72 |
| Taipei City | 483,523 | 13.36% | 12 | 255 | 5.29 |
| Kaohsiung City | 444,875 | 12.29% | 38 | 143 | 2.52 |
| New Taipei City | 590,644 | 16.32% | 29 | 172 | 2.09 |
| Taichung City | 368,586 | 10.19% | 29 | 141 | 3.66 |
| Tainan City | 299,640 | 8.28% | 37 | 83 | 1.55 |
| Taoyuan City | 277,972 | 7.68% | 13 | 83 | 2.19 |
| Total | 3,618,878 | 100% | 349 | 1140 | 1.80 |

### 3.2. Overview of the Google Ratings of Physical Rehabilitation Hospitals in Towns

Table 3 shows the Google ratings of physical rehabilitation hospitals in towns. Concerning mean values, eight counties/cities had a mean value lower than Taiwan's average value. The eight counties/cities are Hsinchu County, Miaoli County, Yunlin County, Chiayi County, Pingtung County, Chiayi City, New Taipei City, and Taoyuan City. Among them, New Taipei City and Taoyuan City are densely populated and highly urbanized. The two municipalities also have a high number of hospitals and a physiatrist to elderly population ratio higher than Taiwan's average value. However, the mean values of physical rehabilitation institutes' Google ratings in the two municipalities are lower than Taiwan's average value, while the standard deviations are higher than Taiwan's average value. In the digital era, people rely heavily on the rating mechanism of online service platforms as crucial factors for decision-making, and the ratings can alter patients' preference in seeking treatment. They may be more inclined to choose hospitals that have high ratings but greater travel distance comparatively. Therefore, as an innovation, this study introduced methods A3 and A4, which integrated Google ratings into the calculation of choice probabilities affecting geographic accessibility.

**Table 3.** Summary statistics of physical rehabilitation hospitals' Google rating.

| Administrative District | Number of Hospitals | Mean | SD | Min | Max |
|---|---|---|---|---|---|
| Yilan County | 15 | 3.71 | 0.74 ◎ | 2.40 * | 5.00 |
| Hsinchu County | 13 | 3.65 * | 0.47 | 2.60 | 4.60 * |
| Miaoli County | 12 | 3.49 * | 0.63 | 2.50 * | 4.80 * |
| Changhua County | 23 | 3.75 | 0.70 ◎ | 2.50 * | 5.00 |
| Nantou County | 11 | 3.87 | 0.86 ◎ | 2.70 | 5.00 |
| Yunlin County | 11 | 3.69 * | 0.57 | 3.00 | 4.90 |
| Chiayi County | 5 | 3.64 * | 0.21 | 3.40 | 3.90 * |
| Pingtung County | 15 | 3.41 * | 0.58 | 2.50 * | 4.30 * |
| Taitung County | 6 | 3.78 | 0.77 ◎ | 2.50 * | 4.90 |
| Hualien County | 11 | 3.76 | 0.80 ◎ | 2.10 * | 5.00 |
| Keelung City | 11 | 3.85 | 0.55 | 2.80 | 4.50 * |
| Hsinchu City | 12 | 3.79 | 0.67 ◎ | 3.10 | 4.90 |
| Chiayi City | 11 | 3.66 * | 0.49 | 3.20 | 4.50 * |
| Taipei City | 114 | 3.79 | 0.56 | 2.60 | 5.00 |
| Kaohsiung City | 96 | 3.78 | 0.68 ◎ | 2.20 * | 5.00 |
| New Taipei City | 102 | 3.62 * | 0.72 ◎ | 1.90 * | 5.00 |
| Taichung City | 83 | 3.75 | 0.68 ◎ | 2.50 * | 5.00 |
| Tainan City | 57 | 3.74 | 0.69 ◎ | 2.20 * | 5.00 |
| Taoyuan City | 50 | 3.63 * | 0.70 ◎ | 2.00 * | 4.90 |
| Total | 658 | | | | |
| Average | | 3.72 | 0.66 | 2.56 | 4.80 |

Note: 1. *: lower than average. 2. ◎: higher than average.

Some conventional inequality measures are the mean, median, Gini coefficient, maximum and minimum values [39]. The median is the middle number in a sorted list of numbers, with the same amount of numbers below and above. The median is sometimes used as opposed to the mean when there are outliers in a sequence that might skew the average of the values. The median of a sequence can be less affected by outliers than the mean. As physical rehabilitation resources are unequal in Taiwan, especially between urban and rural districts, this study applied these inequality measures to compare accessibility values between methods. Table 4 shows the comparison of all resulting scores of geographic accessibilities. All equations generated a minimum value (Min) of 0, which means that regardless of the calculation method chosen, there exist situations in which no physical rehabilitation resources are reachable within 30 km. The mean value of the results of the regional average method (method A0) is the lowest, but its value of standard deviation and maximum value are the highest among all methods. As the regional average method completely disregards the effectiveness of distance and sets limits on the cross-district usage of resources, the results of the regional average method create an illusion wherein the dispersion of geographic accessibility scores is the highest, with a median at 0 and the highest Gini coefficient.

**Table 4.** Summary statistics of physical rehabilitation resources accessibility scores by methods A0–A4.

| Method | Mean | Median | SD | Min | Max | Median-Mean | Gini Coefficient |
|---|---|---|---|---|---|---|---|
| A0 | 1.80 | 0.00 | 3.32 | 0.00 | 31.74 | −1.80 | 0.53 |
| A1 | 1.87 | 1.56 | 1.58 | 0.00 | 6.27 | −0.31 | 0.05 |
| A2 | 1.89 | 1.20 | 1.72 | 0.00 | 8.96 | −0.69 | 0.10 |
| A3 | 1.91 | 1.16 | 1.96 | 0.00 | 15.06 | −0.75 | 0.15 |
| A4 | 1.90 | 1.21 | 1.74 | 0.00 | 9.13 | −0.69 | 0.11 |

In the results of the two-step floating catchment area method (method A1), the dispersion of geographic accessibility scores is the lowest, with the highest median and the lowest Gini coefficient.

Method A2 represents the results of the traditional three-step floating catchment area method with consideration of choice probabilities related to distances and resource sizes; method A3 represents the results of the new three-step floating catchment area method with considerations of choice probabilities related to distances and Google ratings, and method A4 represents the results of the new three-step floating catchment area method with comprehensive considerations of distances, resource sizes, and Google ratings. The results in Table 4 show that, when comparing the results of methods A2–A4, which considered the choice probabilities of people's healthcare-seeking behavior, and the results of method A1, which disregarded choice probabilities, the results of the former group show higher mean values, standard deviations, and maximum values and had lower median values. The results of methods A2–A4 also show higher dispersions of geographic accessibility scores and higher Gini coefficients.

When comparing the scores of geographic accessibilities in methods A2–A4, which considered the choice probabilities of people's healthcare-seeking behavior, the values generated from method A4 show a tendency to land between the values of methods A3 and A2. With consideration of distance decays, sizes of hospitals, and Google ratings, the standard deviation and maximum value of the results of method A4 are lower than method A3, while the median value is higher than method A3. The dispersion of geographic accessibility scores and the Gini coefficient of method A4 are lower than method A3. It implies that concurrent consideration of non-spatial factors (online ratings and sizes of resource) are in better accordance with people's decision-making behavior when they determine which resources of physical rehabilitation to use compared with the sole consideration of online rating factors.

### *3.3. Assessment of Distribution Inequality of People at Demand in Towns and Physical Rehabilitation Resources*

Table 5 shows the mean values, standard deviations, and median values of geographic accessibility scores of counties/cities, which were evaluated by different calculation methods, while the differences in the resulting values are presented. Method 3 considers distances and Google ratings, whereas method A4 adds resource size factors to the basis of method 3. When the median values and mean values are compared, if the median value is lower than the mean value in a county/city, it implies that more than 50% of the resources have low accessibilities. In methods A3 and A4, only three counties/cities (Hsinchu City, New Taipei City, and Taoyuan City) have median values higher than mean values, which implies that the majority of the towns in the counties/cities enjoy plentiful resources. When we take the next step and compare the median values, 12 counties/cities have higher geographic accessibility scores in method A4 than in method A3, which are marked with "*" next to the median values of method 4. Among the counties/cities with more than 30 towns, Kaohsiung City and Pingtung County have higher accessibility scores in method A4, which means that in counties/cities with vast administrative regions, the resources of medical services are more inclined to concentrate in densely populated areas. Therefore, the people may have access to better medical services, as they would consider the credibility and service sizes of the hospitals and choose to visit hospitals that are farther but larger in size. The median values of method A4 rise due to the above reasons.

**Table 5.** Summary statistics of physical rehabilitation resources accessibility scores by methods A0–A4.

| Administrative District | Number of Towns | *Estimated by 10,000 * Capacity/People* | | | | | | | | | | | | | | |
|---|---|---|---|---|---|---|---|---|---|---|---|---|---|---|---|---|
| | | Method A0 | | | Method A1 | | | Method A2 | | | Method A3 | | | Method A4 | | |
| | | Mean | Median | SD | Mean | Median | SD | Mean | Median | SD | Mean | Median | SD | Mean | Median | SD |
| Yilan County | 12 | 1.82 | 0.00 | 3.33 | 2.43 | 2.52 | 1.73 | 2.27 | 1.72 | 1.90 | 2.27 | 1.72 | 2.00 | 2.28 | 1.75 * | 1.90 |
| Hsinchu County | 13 | 0.91 | 0.00 | 1.96 | 1.67 | 1.64 | 1.13 | 1.45 | 0.91 | 1.42 | 1.41 | 0.89 | 1.39 | 1.44 | 0.91 * | 1.42 |
| Miaoli County | 18 | 1.13 | 0.00 | 1.79 | 1.57 | 1.66 | 1.34 | 1.45 | 1.10 | 1.51 | 1.44 | 1.08 | 1.51 | 1.44 | 1.08 | 1.50 |
| Changhua County | 26 | 0.99 | 0.00 | 1.78 | 1.61 | 1.06 | 0.96 | 1.49 | 0.75 | 1.25 | 1.46 | 0.72 | 1.27 | 1.49 | 0.75 * | 1.25 |
| Nantou County | 13 | 0.79 | 0.00 | 1.06 | 0.83 | 0.81 | 0.74 | 0.92 | 0.50 | 0.94 | 0.89 | 0.51 | 0.90 | 0.92 | 0.52 | 0.93 |
| Yunlin County | 20 | 0.81 | 0.00 | 1.78 | 1.17 | 1.02 | 0.97 | 1.16 | 0.71 | 1.2 | 1.18 | 0.64 | 1.31 | 1.16 | 0.70 * | 1.21 |
| Chiayi County | 18 | 0.92 | 0.00 | 3.02 | 1.59 | 1.64 | 1.05 | 1.33 | 0.95 | 1.11 | 1.30 | 0.85 | 1.48 | 1.33 | 0.95 * | 1.11 |
| Pingtung County | 32 | 0.65 | 0.00 | 1.47 | 0.88 | 0.97 | 0.70 | 0.89 | 0.74 | 0.94 | 0.88 | 0.70 | 0.98 | 0.89 | 0.74 * | 0.93 |
| Taitung County | 14 | 0.54 | 0.00 | 1.38 | 0.55 | 0.02 | 1.25 | 0.87 | 0.03 | 1.76 | 0.88 | 0.03 | 1.81 | 1.00 | 0.03 | 2.14 |
| Hualien County | 13 | 1.49 | 0.00 | 2.50 | 1.63 | 0.78 | 2.03 | 1.66 | 0.82 | 2.04 | 1.65 | 0.82 | 2.06 | 1.66 | 0.82 | 2.04 |
| Keelung City | 7 | 3.35 | 1.83 | 4.07 | 2.67 | 2.52 | 0.58 | 3.45 | 3.45 | 1.32 | 3.38 | 3.23 | 1.72 | 3.45 | 3.40 * | 1.31 |
| Hsinchu City | 3 | 3.29 | 1.92 | 4.15 | 3.41 | 3.88 | 1.16 | 3.26 | 4.03 | 1.80 | 3.29 | 3.91 | 1.9 | 3.26 | 4.03 * | 1.80 |
| Chiayi City | 2 | 5.72 | 5.72 | 0.34 | 4.21 | 4.21 | 0.12 | 4.64 | 4.64 | 0.07 | 4.59 | 4.59 | 0.22 | 4.64 | 4.64 * | 0.06 |
| Taipei City | 12 | 5.29 | 5.43 | 2.45 | 4.74 | 5.00 | 0.86 | 4.73 | 4.66 | 0.44 | 4.84 | 4.67 | 0.63 | 4.70 | 4.64 | 0.43 |
| Kaohsiung City | 38 | 2.52 | 0.47 | 4.73 | 2.06 | 1.78 | 1.55 | 2.11 | 1.52 | 1.72 | 2.29 | 1.45 | 2.63 | 2.12 | 1.54 * | 1.73 |
| New Taipei City | 29 | 2.09 | 1.43 | 3.09 | 2.08 | 2.08 | 1.70 | 2.30 | 2.48 | 2.00 | 2.37 | 2.48 | 2.69 | 2.31 | 2.44 | 2.03 |
| Taichung City | 29 | 3.66 | 2.52 | 6.09 | 3.13 | 2.83 | 1.99 | 3.10 | 3.03 | 1.87 | 3.15 | 2.88 | 1.96 | 3.10 | 3.03 * | 1.89 |
| Tainan City | 37 | 1.55 | 0.00 | 2.93 | 1.68 | 1.28 | 1.41 | 1.67 | 1.00 | 1.51 | 1.65 | 1.01 | 1.54 | 1.67 | 0.99 | 1.51 |
| Taoyuan City | 13 | 2.19 | 2.34 | 1.72 | 2.13 | 2.00 | 1.18 | 2.30 | 2.58 | 1.24 | 2.24 | 2.38 | 1.22 | 2.29 | 2.55 * | 1.24 |
| Total | 349 | | | | | | | | | | | | | | | |
| Average | | 1.80 | 0.00 | 3.32 | 1.87 | 1.56 | 1.58 | 1.89 | 1.20 | 1.72 | 1.91 | 1.16 | 1.96 | 1.90 | 1.21 * | 1.74 |

Note: *: (median by A3) – (median by A4) < 0.

## 4. Discussion

Table 6 compares the regional average method (A0), the two-step floating catchment area method (A1), and the innovative three-step floating catchment area method A4 introduced in this study by the values of "median value minus mean value" and Gini coefficients. When the value of "median value minus mean value" of a county/city is negative, it implies that 50% of the medical resources in its towns have low accessibilities distribution. In addition, the mean values of the counties/cities that are lower than Taiwan's average are marked with "*". In the regional average method (A0), the counties/cities with low accessibilities in 50% of the medical resources in its towns are entirely different from those in methods A1 and A4. When we take a further step and compare the degree of inequality in resource accessibilities of counties/cities using the Gini coefficient, we can see that when we carry out an assessment with the regional average method (A0), as the analysis only included the amount of physical rehabilitation resources within the respective administrative regions, it led to calculation results in which 11 counties/cities fell into the category of resource distribution inequality. When the government allocates resources with reference to the regional average method (A0), it is easy to neglect the effects of distance and cross-district usage of services, and the phenomenon of resource distribution inequality worsens as a result. In method A4, Taitung County is the only place with a negative value of "median value minus mean value" and has a Gini coefficient that represents median inequality. The county belongs to Eastern Taiwan and comprises 14 towns. Despite the vast administrative region, there are only six hospitals and 14 board-certificated physiatrists operating in the county, which makes it the county with the highest inequality in resource accessibilities.

In this study, the geographical accessibility scores are grouped into quintiles and the spatial distributions of the accessibility scores of rehabilitation physicians are clearly presented on maps. The colors from lowest to highest accessibility score are red (0%~20%), orange (21%~40%), green (41%~60%), light blue (61%~80%), and dark blue (81%~100%). Figure 1 shows that red areas (low accessibility) measured by method A0 cover almost the entire island, meaning that many towns' accessibility scores are evaluated as low because their medians are 0.0. Figures 2 and 3 are drawn using method A1 and method A4, respectively. The difference between these two methods is that the latter considers the selection probability of each hospital. Figure 3 shows that the number of high-accessibility towns (light-blue and dark-blue area) is increased compared to Figure 2. This result shows that the distribution of medical service resources tends to concentrate in densely populated areas and downtowns. People may travel farther based on the reputation and service capacity of hospitals to get better medical services.

**Table 6.** Measures of geographic inequality of physical rehabilitation resources accessibility scores by methods A0, A1, A4.

| | | | | | | |
|---|---|---|---|---|---|---|
| | | | | | ***Estimated by 10,000 * Capacity/People*** | |
| **Administrative District** | **Method A0** | | **Method A1** | | **Method A4** | |
| | **Median-Mean** | **Gini Coefficient** | **Median-Mean** | **Gini Coefficient** | **Median-Mean** | **Gini Coefficient** |
| Yilan County | −1.82 * | 0.49 ◎ | 0.09 | 0.08 | −0.53 | 0.11 |
| Hsinchu County | −0.91 | 0.45 ◎ | −0.03 | 0.10 | −0.54 | 0.16 |
| Miaoli County | −1.13 | 0.39 | 0.09 | 0.23 | −0.36 | 0.24 |
| Changhua County | −0.99 | 0.27 | −0.55 * | 0.38 | −0.74 * | 0.28 |
| Nantou County | −0.79 | 0.43 ◎ | −0.02 | 0.16 | −0.40 | 0.25 |
| Yunlin County | −0.81 | 0.61 ◎◎ | −0.16 | 0.24 | −0.46 | 0.29 |
| Chiayi County | −0.92 | 0.81 ◎◎ | 0.06 | 0.18 | −0.38 | 0.26 |
| Pingtung County | −0.65 | 0.70 ◎◎ | 0.09 | 0.13 | −0.16 | 0.15 |
| Taitung County | −0.54 | 0.20 | −0.53 * | 0.19 | −0.97 * | 0.59 ◎ |
| Hualien County | −1.49 | 0.30 | −0.85 * | 0.08 | −0.84 * | 0.09 |
| Keelung City | −1.51 | 0.47 ◎ | −0.14 | 0.02 | −0.05 | 0.11 |
| Hsinchu City | −1.37 | 0.33 | 0.47 | 0.03 | 0.77 | 0.02 |
| Chiayi City | 0.00 | 0.03 | 0.00 | 0.01 | 0.00 | 0.01 |
| Taipei City | 0.14 | 0.22 | 0.27 | 0.09 | −0.07 | 0.03 |
| Kaohsiung City | −2.05 * | 0.63 ◎◎ | −0.28 | 0.03 | −0.58 | 0.09 |
| New Taipei City | −0.66 | 0.48 ◎ | 0.00 | 0.08 | 0.13 | 0.11 |
| Taichung City | −1.13 | 0.55 ◎ | −0.31 | 0.18 | −0.08 | 0.20 |
| Tainan City | −1.55 | 0.59 ◎ | −0.40 * | 0.05 | −0.68 | 0.09 |
| Taoyuan City | 0.15 | 0.14 | −0.13 | 0.06 | 0.26 | 0.06 |
| Average | −1.80 | 0.53 ◎ | −0.31 | 0.05 | −0.69 | 0.11 |

Notes: 1. Level of distribution inequality estimated by score of "Median-Mean". *: smaller than average. 2. Level of distribution inequality estimated by Gini coefficient. ◎: 0.4~0.6, median inequality, ◎◎: > 0.6, high inequality.

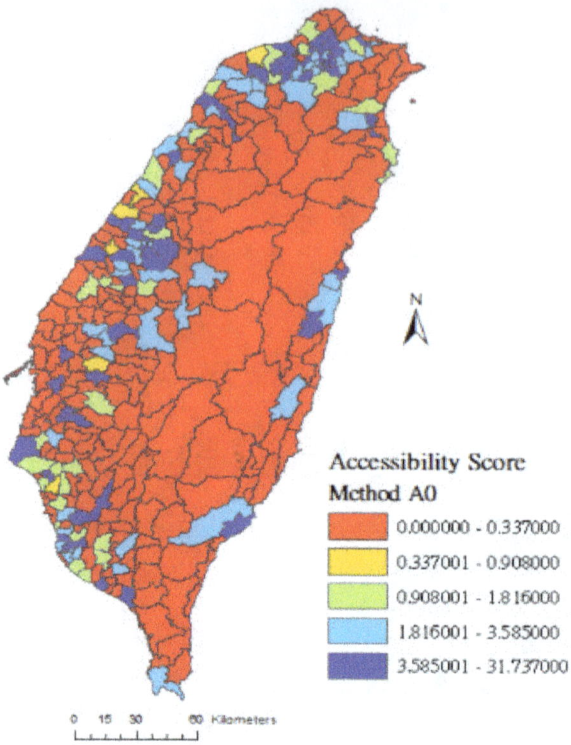


**Figure 1.** Accessibility score of rehabilitation physician service in Taiwan using method A0.

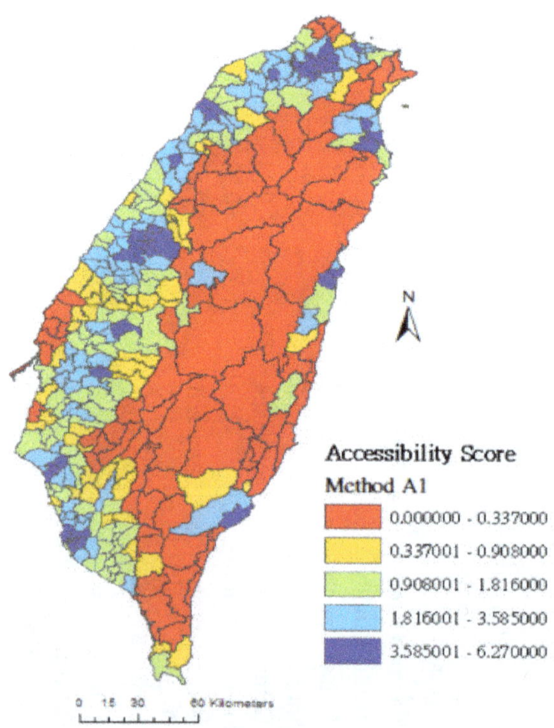


**Figure 2.** Accessibility score of rehabilitation physician service in Taiwan using method A1.

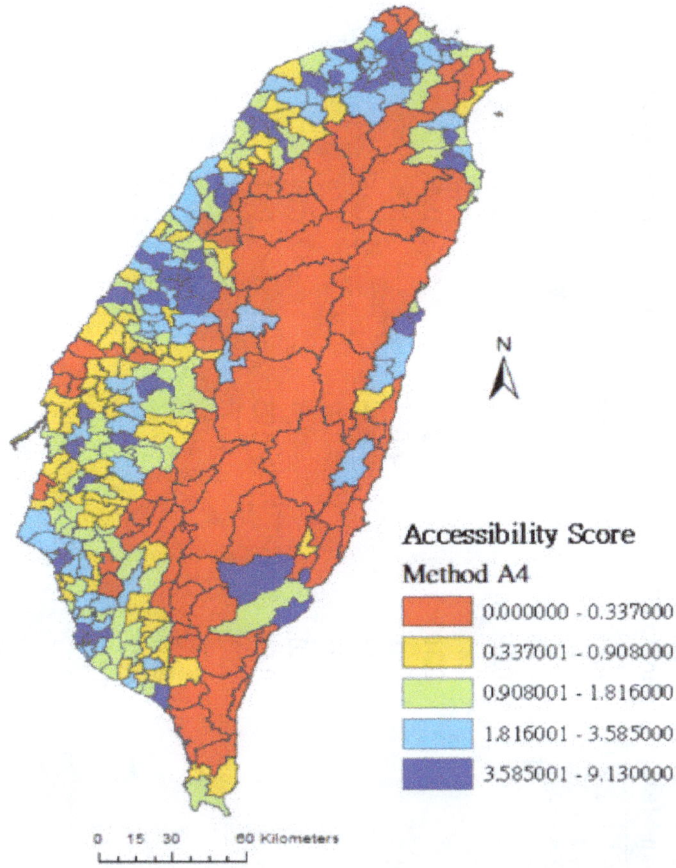


**Figure 3.** Accessibility score of rehabilitation physician service in Taiwan using method A4.

Based on the three-step floating catchment area method, Table 7 shows the results of assessments on inequality in resource accessibility with methods A2–A4. According to the calculation results of the three methods, 50% of the medical resources in towns had low accessibilities (negative value of "median value minus mean value") in Taitung County and Hualien County. When we compared the inequality in resource accessibilities of counties/cities with the Gini coefficient, the value of Taitung County was close to the critical value of high inequality.

Comparing Figures 3 5: Figure 3 was drawn by method A4 which considers both spatial factors (distance) and non-spatial factors (Google Rating score and resource capacity). In Figure 3, the number of dark-blue areas is increased in vast towns of the central and eastern administrative regions. Compared with method A2 (which only considers the selection probability of distance and resource capacity), and method A3 (which only considers the selection probability of distance and Google Rating score), the assessment result of method A4 may be more in line with people's decision-making in choosing rehabilitation medical resources.

With method A4 proposed in this study, an assessment of physiatrist resource allocation policies on the main island of Taiwan was carried out. The results of our study have important implications for rehabilitation physician services and elderly care policy in Taiwan. In the first stage, the improvement of resources in Taitung County should be prioritized. The next in line should be the three counties/cities (Taitung County, Changhua County and Hualien County) where 50% of the medical resources in towns had low accessibilities and scored lower than Taiwan's average. In the third stage, work should be carried out on the 12 counties/cities with low accessibilities in 50% of the medical resources in towns.

**Table 7.** Measures of geographic inequality of physical rehabilitation resources accessibility scores by methods A2–A4.

| Administrative District | Method A2 | | | | Method A3 | | | | Method A4 | | | |
|---|---|---|---|---|---|---|---|---|---|---|---|---|
| | Median-Mean | | Gini Coefficient | | Median-Mean | | Gini Coefficient | | Median-Mean | | Gini Coefficient | |
| | | | | | | | | | *Estimated by 10,000 * Capacity/People* | | | |
| Yilan County | −0.55 | | 0.12 | | −0.55 | | 0.18 | | −0.53 | | 0.11 | |
| Hsinchu County | −0.53 | | 0.16 | | −0.52 | | 0.16 | | −0.54 | | 0.16 | |
| Miaoli County | −0.35 | | 0.24 | | −0.35 | | 0.24 | | −0.36 | | 0.24 | |
| Changhua County | −0.73 | * | 0.28 | | −0.74 | | 0.26 | | −0.74 | * | 0.28 | |
| Nantou County | −0.42 | | 0.20 | | −0.38 | | 0.25 | | −0.40 | | 0.25 | |
| Yunlin County | −0.45 | | 0.29 | | −0.54 | | 0.31 | | −0.46 | | 0.29 | |
| Chiayi County | −0.39 | | 0.26 | | −0.44 | | 0.31 | | −0.38 | | 0.26 | |
| Pingtung County | −0.15 | | 0.15 | | −0.18 | | 0.17 | | −0.16 | | 0.15 | |
| Taitung County | −0.84 | * | 0.54 | ◎ | −0.85 | * | 0.55 | ◎ | −0.97 | * | 0.59 | ◎ |
| Hualien County | −0.84 | * | 0.09 | | −0.84 | * | 0.10 | | −0.84 | * | 0.09 | |
| Keelung City | 0.00 | | 0.11 | | −0.14 | | 0.14 | | −0.05 | | 0.11 | |
| Hsinchu City | 0.77 | | 0.02 | | 0.62 | | 0.03 | | 0.77 | | 0.02 | |
| Chiayi City | 0.00 | | 0.01 | | 0.00 | | 0.00 | | 0.00 | | 0.01 | |
| Taipei City | −0.08 | | 0.03 | | −0.17 | | 0.05 | | −0.07 | | 0.03 | |
| Kaohsiung City | −0.60 | | 0.09 | | −0.83 | * | 0.21 | | −0.58 | | 0.09 | |
| New Taipei City | 0.17 | | 0.12 | | 0.11 | | 0.22 | | 0.13 | | 0.11 | |
| Taichung City | −0.07 | | 0.19 | | −0.27 | | 0.23 | | −0.08 | | 0.20 | |
| Tainan City | −0.67 | | 0.09 | | −0.63 | | 0.10 | | −0.68 | | 0.09 | |
| Taoyuan City | 0.28 | | 0.07 | | 0.14 | | 0.07 | | 0.26 | | 0.06 | |
| **Average** | **−0.69** | | **0.10** | | **−0.75** | | **0.15** | | **−0.69** | | **0.11** | |

Notes: 1. Level of distribution inequality estimated by score of "Median-Mean". *: smaller than average. 2. Level of distribution inequality estimated by Gini coefficient. ◎: 0.4~0.6, median inequality.

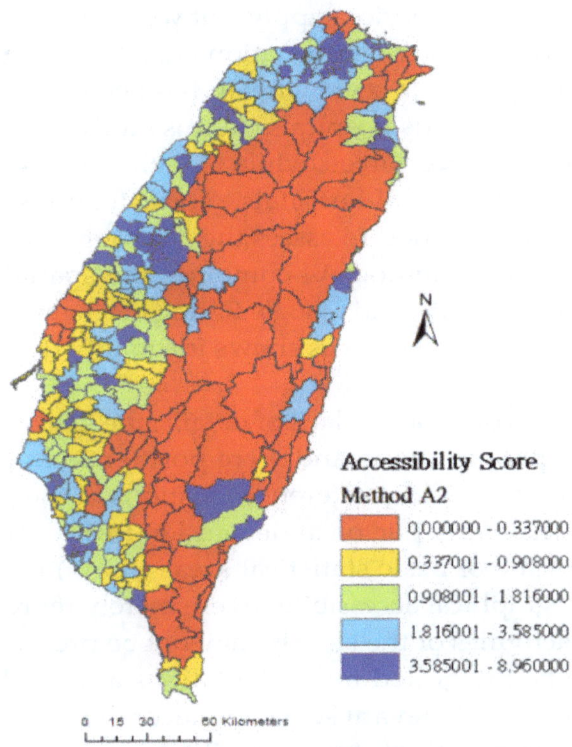


**Figure 4.** Accessibility score of rehabilitation physician service in Taiwan using method A2.

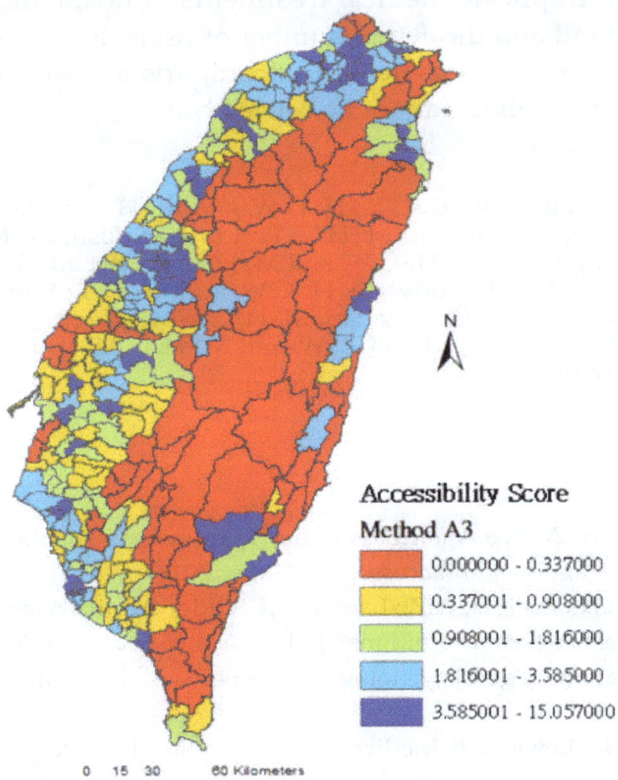


**Figure 5.** Accessibility score of rehabilitation physician service in Taiwan using method A3.

## 5. Conclusions

As the free market influences the medical environment, people have many choices of medical services and have access to duplicate medical treatments. Therefore, when we discuss the distributional fairness of physical rehabilitation resources, not only do we focus on the degree of coordination between the population at demand and service supply, but we also have to consider the distance factor when the patients travel to hospitals. In the information age, the rating mechanism of online service platforms, which allows ordinary people to review freely, has become a crucial source of references for people when they choose from numerous hospitals. In this study, various methods were utilized to assess the geographic accessibility of resources, while Google ratings were added as a choice factor for when people examine the credibility of hospitals. These features were integrated into method A4, which is an innovative research method to assess the appropriateness in the demand and supply of physical rehabilitation resources. Method A4 combines the spatial condition of travel distance, non-spatial conditions of hospital capacity, and Google Rating mechanism. This helps examine whether the allocation of physical rehabilitation resources shows inequality due to regional differences.

With restrictions on the access of data and lack of details, the limitations encountered in this study include the following: (1) The people at demand were positioned at the weighted center points of population in geometry. This only provides reference locations of the people at demand and cannot reflect the exact locations of each elderly person at demand in reality. The author suggests employing finer space scales such as the scale of basic statistical areas (BSAs) for better research in the future. (2) Activity areas regarding geographical accessibility were merely represented by route distances and the estimation of the range of activities of the elderly may not be precise. In the future, transportation time or different vehicles can be integrated into calculations and evaluations. (3) This study only examines geographical accessibility. Relevant social and economic conditions can be weighted and added to the calculation processes in the future to facilitate analysis combined with geographical accessibility. (4) The open data of the government do not disclose the number of users, statistics of service items, and details of duplicate medical treatments in hospitals. Therefore, concerning the differences in people at demand and the actual number of users, a cross-validation cannot be carried out in this study. At the same time, assessments and comparisons between the service loads and the actual service effectiveness of hospitals cannot be carried out.

**Author Contributions:** Conceptualization: C.-C.L., H.-C.W. and M.-H.T.; Data curation: C.-C.L., H.-C.W. and M.-H.T.; Formal analysis: C.-C.L., H.-C.W. and M.-H.T.; Funding acquisition: M.-H.T.; Investigation: C.-C.L., H.-C.W. and M.-H.T.; Methodology: C.-C.L., H.-C.W. and M.-H.T.; Project administration: H.-C.W. and M.-H.T.; Resources: C.-C.L., H.-C.W. and M.-H.T.; Software, H.-C.W. and M.-H.T.; Supervision: C.-C.L., H.-C.W. and M.-H.T.; Validation: H.-C.W. and M.-H.T.; Visualization: M.-H.T.; Writing—original draft preparation: C.-C.L. and H.-C.W.; Writing—review and editing: C.-C.L., H.-C.W. and M.-H.T. All authors have read and agreed to the published version of the manuscript.

## References

1. World Health Organization. Active Ageing: A Policy Framework. Available online: http://www.who.int/ageing/publications/active_ageing/en/ (accessed on 1 May 2018).
2. Phelan, E.A.; Anderson, L.A.; Lacroix, A.Z.; Larson, E.B. Older Adults' Views of "Successful Aging"—How Do They Compare with Researchers' Definitions? *J. Am. Geriatr. Soc.* **2004**, *52*, 211–216. [CrossRef] [PubMed]
3. Rusin, M. Stroke rehabilitation: A geropsychological perspective. *Arch. Phys. Med. Rehabil.* **1990**, *71*, 914–922. [PubMed]
4. Lippmann, H.I.; Fishman, L.; Lewis, C.B. Medical care for the elderly: A counterview. *Arch. Phys. Med. Rehabil.* **1991**, *72*, 430. [PubMed]
5. Strasser, D.C.; Solomon, D.H.; Burton, J.R. Geriatrics and physical medicine and rehabilitation: Common principles, complementary approaches, and 21st century demographics. *Arch. Phys. Med. Rehabil.* **2002**, *83*, 1323–1324. [CrossRef]
6. Momosaki, R.; Kakuda, W.; Kinoshita, S.; Yamada, N.; Abo, M. Clinical Effectiveness of Board-certificated Physiatrists on Functional Recovery in Elderly Stroke Patients During Convalescence: A Retrospective Cohort Study. *Int. J. Gerontol.* **2017**, *11*, 7–11. [CrossRef]

7. MInistry of Health and Welfare (Taiwan). National Health Insurrance Administration. Available online: http://www.nhi.gov.tw/english/index.aspx (accessed on 1 March 2020).
8. Franceschini, M.; Caso, V.; Zampolini, M.; Negrini, S.; Giustini, A. The Role of the Physiatrist in Stroke Rehabilitation: A European Survey. *Am. J. Phys. Med. Rehabil.* **2009**, *88*, 596–600. [CrossRef]
9. Lipner, R.S.; Hess, B.J.; Phillips, R.L.J. Specialty Board Certification in the United States: Issues and Evidence. *J. Contin. Educ. Health Prof. Autumn* **2013**, *33*, S20–S35. [CrossRef]
10. Sanchez, J.; Byfield, G.; Brown, T.T.; LaFavor, K.; Murphy, D.; Laud, P. Perceived Accessibility Versus Actual Physical Accessibility of Healthcare Facilities. *Rehabil. Nurs.* **2000**, *25*, 6–9. [CrossRef]
11. Burns, T.J.; Batavia, A.I.; Smith, Q.W.; DeJong, G. Primary health care needs of persons with physical disabilities: What are the research and service priorities? *Arch. Phys. Med. Rehabil.* **1990**, *71*, 138–143.
12. Veltman, A.; Stewart, D.E.; Tardif, G.S.; Branigan, M. Perceptions of primary healthcare services among people with physical disabilities. Part 1: Access issues. *Medscape Gen. Med.* **2001**, *3*, 18.
13. Pan, J.; Zhao, H.; Wang, X.; Shi, X. Assessing spatial access to public and private hospitals in Sichuan, China: The influence of the private sector on the healthcare geography in China. *Soc. Sci. Med.* **2016**, *170*, 35–45. [CrossRef] [PubMed]
14. Luo, W.; Wang, F. Spatial Accessibility to Primary Care and Physician Shortage Area Designation: A Case Study in Illinois with GIS Approaches. In *Geographic Information Systems and Health Applications*; Khan, O., Ed.; IDEA Group Publishing: London, UK, 2003; pp. 261–279.
15. Luo, W.; Wang, F. Measures of Spatial Accessibility to Health Care in a GIS Environment: Synthesis and a Case Study in the Chicago Region. *Environ. Plan. B* **2003**, *30*, 865–884. [CrossRef]
16. McLafferty, S.L. GIS and Health Care. *Annu. Rev. Public Health* **2003**, *24*, 25–42. [CrossRef] [PubMed]
17. Guagliardo, M.F. Spatial accessibility of primary care: Concepts, methods and challenges. *Int. J. Health Geogr.* **2004**, *3*, 3. [CrossRef] [PubMed]
18. Arcury, T.A.; Gesler, W.M.; Preisser, J.S.; Sherman, J.; Spencer, J.; Perin, J. The Effects of Geography and Spatial Behavior on Health Care Utilization among the Residents of a Rural Region. *Health Serv. Res.* **2005**, *40*, 135–156. [CrossRef] [PubMed]
19. Wang, F.; Luo, W. Assessing Spatial and Nonspatial Factors for Healthcare Access: Towards An Integrated Approach to Defining Health Professional Shortage Areas. *Health Place* **2005**, *11*, 131–146. [CrossRef]
20. Luo, W.; Qi, Y. An Enhanced Two-Step Floating Catchment Area (E2SFCA) Method for Measuring Spatial Accessibility to Primary Care Physicians. *Health Place* **2009**, *15*, 1100–1107. [CrossRef]
21. McGrail, M.R.; Humphreys, J.S. Measuring spatial accessibility to primary care in rural areas: Improving the effectiveness of the two-step floating catchment area method. *Appl. Geogr.* **2009**, *29*, 533–541. [CrossRef]
22. Wang, F. Measurement, optimization, and impact of health care accessibility: A methodological review. *Ann. Assoc. Am. Geogr.* **2012**, *102*, 1104–1112. [CrossRef]
23. Cabrera-Barona, P.; Blaschke, T.; Gaona, G. Deprivation, Healthcare Accessibility and Satisfaction: Geographical Context and Scale Implications. *Appl. Spat. Anal. Policy* **2017**. [CrossRef]
24. Wu, H.-C.; Tseng, M.-H. Evaluating Disparities in Elderly Community Care Resources: Using a Geographic Accessibility and Inequality Index. *Int. J. Environ. Res. Public Health* **2018**, *15*, 1353. [CrossRef]
25. Social and Family Affairs Adminstration Ministry of Health and Welfare. Assistive Benefits. Available online: https://newrepat.sfaa.gov.tw/home/repat-welfare (accessed on 1 June 2020).
26. Taiwan Academy of Physical Medicine and Rehabilitation. Associators. Available online: https://www.pmr.org.tw/associator/associator-all.asp (accessed on 1 March 2020).
27. Taiwan, M.O.I. Department of Household Resistration. Available online: http://www.ris.gov.tw/en/web/ris3-english/home (accessed on 1 June 2020).
28. Page, N.; Langford, M.; Higgs, G. An evaluation of alternative measures of accessibility for investigating potential 'deprivation amplification' in service provision. *Appl. Geogr.* **2018**, *95*, 19–33. [CrossRef]
29. Ministry of Transportation and Communications. GIS-T. Available online: https://gist.motc.gov.tw/gist_web (accessed on 1 March 2018).
30. Ruger, J.P.; Kim, H.J. Global health inequalities: An international comparison. *J. Epidemiol. Community Health* **2006**, *60*, 928–936. [CrossRef] [PubMed]
31. McGrail, M.R.; Humphreys, J.S. The Index of Rural Access: An Innovative Integrated Approach for Measuring Primary Care Access. *BMC Health Serv. Res.* **2009**, *9*, 124. [CrossRef] [PubMed]

32. Wan, N.; Zou, B.; Sternberg, T. A three-step floating catchment area method for analyzing spatial access to health services. *Int. J. Geogr. Inf. Sci.* **2012**, *26*, 1073–1089. [CrossRef]
33. Hearn, A. Structuring feeling: Web 2.0, online ranking and rating, and the digital'reputation'economy. *Ephemer. Theory Politics Organ.* **2010**, *10*, 421–438.
34. Dabholkar, P.A. Factors Influencing Consumer Choice of a "Rating Web Site": An Experimental Investigation of an Online Interactive Decision Aid. *J. Mark. Theory Pract.* **2006**, *14*, 259–273. [CrossRef]
35. Wang, S.; Zheng, Z.; Wu, Z.; Lyu, M.R.; Yang, F. Reputation measurement and malicious feedback rating prevention in web service recommendation systems. *IEEE Trans. Serv. Comput.* **2014**, *8*, 755–767. [CrossRef]
36. Ceriani, L.; Verme, P. The origins of the Gini index: Extracts from Variabilità e Mutabilità (1912) by Corrado Gini. *J. Econ. Inequal.* **2012**, *10*, 421–443. [CrossRef]
37. Yitzhaki, S. Relative Deprivation and the Gini Coefficient. *Q. J. Econ.* **1979**, *93*, 321–324. [CrossRef]
38. Tseng, M.-H.; Wu, H.-C. The geographic accessibility and inequality of community-based elderly learning resources: A remodeling assessment, 2009~2017. *Educ. Gerontol.* **2018**, *44*, 226–246. [CrossRef]
39. Cowell, F.A. *Measuring Inequality*, 3rd ed.; Oxford University Press: Oxford, UK, 2011.

# Links between Cognitive Status and Trace Element Levels in Hair for an Environmentally Exposed Population: A Case Study in the Surroundings of the Estarreja Industrial Area

**Marina M. S. Cabral Pinto [1,*], Paula Marinho-Reis [2], Agostinho Almeida [3], Edgar Pinto [3], Orquídia Neves [4], Manuela Inácio [1], Bianca Gerardo [5], Sandra Freitas [5], Mário R. Simões [5], Pedro A. Dinis [6], Luísa Diniz [3], Eduardo Ferreira da Silva [1] and Paula I. Moreira [7]**

[1] Geobiotec Research Centre, Department of Geosciences, University of Aveiro, 3810-193 Aveiro, Portugal; minacio@ua.pt (M.I.); eafsilva@ua.pt (E.F.d.S.)
[2] Instituto de Ciências da Terra, University of Minho, 4710–057 Braga, Portugal; pmarinho@dct.uminho.pt
[3] Laboratory of Applied Chemistry, Department of Chemical Sciences, Faculty of Pharmacy, LAQV/REQUIMTE, Porto University, 4050-313 Porto, Portugal; aalmeida@ff.up.pt (A.A.); ecp@ess.ipp.pt (E.P.); luisa2diniz@gmail.com (L.D.)
[4] CERENA, DECivil, Instituto Superior Técnico, University of Lisbon, 1049-001 Lisbon, Portugal; orquidia.neves@tecnico.ulisboa.pt
[5] Center for Research in Neuropsychology and Cognitive and Behavioral Intervention (CINEICC), University of Coimbra, 3030-548 Coimbra, Portugal; bianca.s.gerardo94@gmail.com (B.G.); sandrafreitas0209@gmail.com (S.F.); simoesmr@fpce.uc.pt (M.R.S.)
[6] MARE—Marine and Environmental Sciences Centre, Department of Earth Sciences, University of Coimbra, 3030-790 Coimbra, Portugal; pdinis@dct.uc.pt
[7] CNC—Center for Neuroscience and Cell Biology, University of Coimbra and Institute of Physiology, Faculty of Medicine, University of Coimbra, 3030-548 Coimbra, Portugal; pimoreira@fmed.uc.pt
* Correspondence: marinacp@ua.pt.

**Abstract:** In the present study, trace elements (TE) levels were evaluated in scalp hair along the continuum from healthy subjects (HS) to patients suffering from subjective memory concerns (SMC), and/or mild cognitive impairment (MCI), and those with already installed dementia (DEM) in order to: (i) assess the effects of environmental and lifestyle factors on TE concentrations and (ii) evaluate the analyzed elements as possible diagnostic biomarkers for the disease. The study involved 79 mainly permanent residents, >55 years old, from the city of Estarreja (northern Portugal), a former industrial area. The health status of the participants was assessed by means of a complete socio-demographic questionnaire and through cognitive screening tests, namely the Mini-Mental State Examination (MMSE). The test scores were categorized and used in the statistical analysis. Hair samples were collected and analyzed by inductively coupled plasma-mass spectrometry (ICP-MS) ICP-MS for selected TE. Dementia appears to be associated with higher age, the female gender, lower education level, and longer residence time in the study area. In addition, most of the participants diagnosed with dementia frequently consume home-grown foodstuffs, some irrigated with contaminated well water. The calculation of the TE enrichment factors of soil samples collected in kitchen gardens/small farms in the vicinity of the Estarreja Chemical Complex (ECC) reinforces the degree of Hg soil contamination in the area, due to anthropogenic sources that can be a source for the population Hg exposure route among others. Mercury levels in hair differed significantly between the four individual groups (HS, SMC, MCI, and DEM), increasing from healthy to dementia participants. Improved diagnostic results can be obtained using hair TE signatures coupled with MMSE scores. This strategy may prove useful for predictive diagnosis in population screening for cognitive impairment.

**Keywords:** exposure; trace elements; cognitive impairment; mercury

## 1. Introduction

The aging of human populations around the world is leading to an epidemic of Alzheimer's disease (AD), with the number of cases estimated to rise to nearly 106 million by 2050 [1]. Human AD is characterized by a progressive decline of cognitive function, with marked loss of memory and other cognitive functions, leading to a gradual loss of functionality and autonomy. It is the most frequent and fearsome form of dementia in the elderly, and its cure and prevention are among the primary challenges of modern medicine. Nowadays, it is accepted that the preclinical stage of AD can begin more than a decade before symptoms are evident, and therefore detection of preclinical stages is a critical factor to fight the disease. Mild cognitive impairment (MCI) is a prodromal stage of AD and affected individuals experience memory loss and/or other cognitive impairments greater than would be expected based on their age and level of education, but not enough to allow a diagnosis of dementia. Longitudinal studies show that MCI patients progress to overt dementia at a rate of 10–15% per year, compared with a rate of 1–2% in the control subjects [2]. This explains why MCI is now the focus of prediction studies and the target of clinical trials of new disease-modifying therapies. Research on significant memory impairment is very contradictory. On one hand, numerous studies have proven that subjective memory concerns (SMC) reflect objective cognitive impairment [3–7] and is associated with increased risk of MCI and dementia. For instance, a meta-analysis conducted by Mitchell and colleagues [8] shows that 6.6% of older individuals with SMC, but with no objective complaints, will convert to MCI per year, comparatively to 1% in those without SMC, and that individuals in this condition double their risk of dementia. Furthermore, there is clear evidence that SMC are a reliable predictor of conversion to dementia [9–12]. However, some other studies refute these findings by reporting a lack of value of SMC [13,14] and SMC severity [15] in predicting dementia. Additionally, a longitudinal study conducted with healthy elderly showed that SMC are not associated with significant changes in cognitive performances over time [16]. Also, a review conducted by Reid and MacLullich [17] reported inconsistency in the association between SMC and cognitive impairment.

Multiple factors have been reported as contributing to the etiology of sporadic (late-onset) AD including aging, genetics, head injury, and exposure to certain chemical compounds. Whilst the genetic component of sporadic AD risk has been increasingly recognized in recent years due to genome-wide association studies (GWAS) [18], the presence of the *APOE*ε4 allele being the strongest genetic risk factor, the role of environmental exposures and the mechanisms of their contribution to the pathogenesis of AD continues to be a subject of discussion. This is partly because of the extended time lapse between exposure and onset of the disease. However, while not all environmental contaminants and toxins have been tested in research studies in terms of their impact on the central nervous system (CNS), the risks of developing AD (and other neurodegenerative diseases, like Parkinson's disease) in elderly persons as a result of neurologic impairments caused by environmental toxins, is established [19]. Recent studies also support close gene-environment interactions [20].

Systemic human exposure to trace elements (TE) is a common circumstance worldwide, resulting from multiple exposure pathways including inhalation of ambient particulate matter, dermal absorption of trace elements from soil and dust, and ingestion of contaminated soil/dust (through hand-to-mouth movements or clearance of particulate matter from airways by swallowing), water and foodstuff, such as agricultural crops, meat and seafood. Toxicity of TE depends upon the absorbed dose, route, and duration of exposure.

Currently, no marker exists as an AD indicator in its early stages, only mainly in its prodromal stage, and the diagnosis of the disease is still based on clinical ground. Biomarkers capable of identifying the preclinical stage of AD have the potential to open a therapeutic window in which neurons would remain responsive to treatment. Furthermore, biomarkers capable of defining the at-risk state may drive novel therapeutic strategies to finally achieve a disease-modifying status of AD. Elemental profiling is an interesting approach for understanding neurodegenerative processes, considering that compelling evidence shows that element toxic effects might play a crucial role in the onset and progression of AD [21].

The use of hair for biomonitoring purposes offers the possibility of integrating a relatively long-term exposure in a single-specimen analysis and presents important practical aspects: non-invasive sample collection and easy sample transport and storage [22]. The concentration of trace elements in hair varies greatly according to the environmental exposure of the subject. On the other hand, hair is an inert tissue and trace elements are slowly incorporated into its structure. Their concentrations do not fluctuate over a short time scale, as in blood, and thus they can be used as a long-term diagnostic tool [23]. It has been reported that several characteristics such as age, sex, ethnicity, nutrition, and geographical location might affect some element concentrations in human hair [24–26]. It has been shown by clinical research that the levels of specific TE in hair, especially those with higher toxic potential, may present a strong correlation with specific pathological disorders. It should be noted that aging is associated with changes in metabolism and several nutrient imbalances [27] and the maintenance of an adequate TE status, in addition to macrominerals and vitamins, is particularly important in the elderly population, for maintenance of the physiological homeostasis and prevention of several age-associated diseases [28,29]. Many studies exist in elderly populations, focused on their health status regarding a specific area of gerontological epidemiology, such as mental diseases [30] and cognitive impairment [31].

The Estarreja Chemical Complex (ECC), located near Aveiro, in the center region of Portugal, has had intense industrial activity since the early 1950s, and high environmental levels of potentially toxic elements such as arsenic (As) and mercury (Hg) have been reported at this region [32,33]. Despite the significant number of studies focusing on determining the severity and the extension of the contamination [34–38] only a few have tried to assess potential health effects on the resident population [39–43]. Hence, the area displays interesting characteristics to study the potential effect of long-term exposures to chemicals in the environment.

In the present study, trace elements levels were evaluated in scalp hair along the continuum from healthy subjects (HS) to patients suffering from SMC and/or MCI and those with already installed AD in order to: (i) assess the effects of environmental and lifestyle factors on TE concentrations in the hair of the elderly and (ii) evaluate the analyzed elements as possible diagnostic biomarkers for the disease.

## 2. Methods

### *2.1. Participants*

Ethical approval for this study was obtained from the National Committee for Data Protection (n° 11726/2017). The study involved 79 participants, voluntarily recruited through convenience sampling, who met the criteria: (i) to have resided in the study area at least the last 5 previous years, in order to ensure minimum exposure time [44] and (ii) be over 55 years of age (age group of interest due to the likelihood of this population reporting subjective cognitive complaints and/or being involved in pathological aging processes, such as the dementia spectrum).

Additionally, the following exclusion criteria were considered: (a) absence of Portuguese language skills required for cognitive testing; (b) a current or past history of neurological disease (other than SCM, MCI or dementia), traumatic brain injury, or psychiatric disorder, including depression—depression at screening was assessed with the Geriatric Depression Scale (GDS), and participants with a GDS score ≥21 [45–47] were considered depressed and excluded from the study; (c) previous or current alcohol or other substance abuse; (d) severe visual or auditory impairment that would negatively affect the ability to satisfactorily complete tests or understand test instructions; or (V) current or prior use of antipsychotic medication.

The same expert neuropsychologist administered a battery of tests to all participants in a fixed order, which included the following instruments:

(1) Sociodemographic and clinical questionnaire: During a personal interview, demographic and clinical data were collected through an extensive sociodemographic questionnaire and an inventory of past habits, current clinical health status, and medical history. The following data were collected: age, marital status, weight, height, nationality, education level, the period of time working in agriculture, pesticide application methods and time of exposure, use of personal protective equipment, home-grown foodstuff consumption, irrigation water source, and drinking water source. Additionally, a full medical record was obtained during this interview, including information on 29 symptoms typically associated with toxic elements exposure or essential elements deficiency [48];

(2) The Mini-Mental State Examination (MMSE) [49–51] is a brief screening test for assessment of the global cognitive status. The MMSE score ranges from 0–30, with higher scores indicating better cognitive performance. In this study we considered the following categories: (i) 0–25 points: dementia; (ii) 26–28 points: mild cognitive impairment (MCI); and (iii) 29–30 points: normal cognitive functioning [51]. The MMSE is the most broadly used brief cognitive screening instrument in clinical, epidemiological, and research contexts. Despite the existence of other neuropsychological instruments with greater sensitivity in detecting cognitive decline at earlier stages (e.g., probably due to the lack of MMSE in including executive functioning assessment and the usage of rather simple tasks to assess short-term memory, working memory, attention, and concentration, language and visuospatial skills), the MMSE has been largely validated for different populations, thus representing a common reference in the communication between health professionals, including psychologists, neurologists, and psychiatrists [51];

(3) The Geriatric Depression Scale (GDS) [45–47] is a brief instrument to assess depressive symptoms in older adults, composed of 30 dichotomous questions that evaluate emotional and behavioral symptoms. The maximum score is 30 points, with higher scores indicating greater severity of depressive symptomatology. In this study, we considered the following categories: (i) 0–10 points: absence of depressive symptoms; (ii) 11–20 points: mild depressive symptoms; and (iii) 21–30 points: moderate to severe depressive symptoms.

### 2.2. *Study Groups*

The study groups were generated according to their education levels and their subjective memory complaints. The four groups were selected based on MMSE and memory scale complaints score ranges from 0–30. According to Kaup et al. [52] and O'Bryant et al. [53], the following categories were used in the statistical analysis described below: a) to lower education level [53], 0–23: dementia (DEM), 24–28: mild cognitive impairment (MCI), 29–30 (and reaching 4 on the scale of complaints): subjective memory complaints (SMC), and 29–30: healthy status (HS); b) to higher education level, 0–26: dementia, 27–28: mild cognitive impairment (MCI), 29–30 (and reaching 4 on the scale of complaints): SMC, 29–30: HS.

### 2.3. *Hair Samples and Analysis*

Human biomonitoring, defined as "the method for assessing human exposure to chemicals or their effects by measuring these chemicals, their metabolites or reaction products in human specimens", involves the measurement of biomarkers in different body fluids (e.g., blood, urine, and breast milk) or tissues (e.g., nails, and hair) [22].

Hair samples (ranging from 100 to 300 mg) were collected near the scalp from 79 inhabitants of the city of Estarreja. Only hair samples presenting their natural color and from individuals residing permanently in the study area were considered for analysis.

Following a procedure reported elsewhere [54], hair samples were duly washed to completely remove exogenous contamination without significantly altering the endogenous trace element content of the sample. After washing, samples were dried in a laboratory drying oven (Raypa, Spain) at 95 °C for 40 h (the time required to achieve a constant weight). Dried samples (~0.1 mg) were mineralized through a microwave-assisted acid digestion procedure with 1 ml of concentrated $HNO_3$ (>65% m/m;

TraceSELECT®, Fluka, France) and 0.5 ml of $H_2O_2$ (≥30% v/v; TraceSELECT®, Fluka, Seelze, Germany) in a Milestone (Sorisole, Italy) MLS 1200 Mega high performance microwave digestion unit, equipped with an HPR 1000/10 rotor. The flowing microwave oven program (W/min) was used: 250/1, 0/2, 250/5, 400/5, and 600/5. After cooling, sample solutions were made up to 8.5 mL with ultrapure water and stored in closed propylene tubes at 4 °C until analysis.

Ultrapure water (at 25 °C the resistivity value is 18.2 MΩ cm) produced in an Aarium®pro (Sartorius, Gottingen, Germany) water purification system was used throughout the work. All lab ware was duly decontaminated by 24 h immersion in a 10% HNO3 bath and thoroughly rinsing with ultrapure water. A sample blank was prepared in each digestion run (10 samples). Average blank levels were subtracted from the samples values. For analytical quality control, the certified reference material (CRM) ERM-DB001—human hair was used, using the same acid digested procedure.

Trace element concentrations were determined through inductively coupled plasma-mass spectrometry (ICP-MS) using a Thermo Fisher Scientific (Waltham, MA, USA) iCAP™ Q instrument, equipped with standard components and accessories: a MicroMist™ nebulizer (Glass Expansion, Port Melbourne, Australia), a Peltier-cooled baffled cyclonic spray chamber, a standard quartz torch, and a two-cone interface design (nickel sample and skimmer cones). High-purity argon (99.9997%; Gasin, Leça da Palmeira, Portugal) was used as the nebulizer and plasma gas. Before each analytical series, the ICP-MS instrument was tuned for maximum sensitivity and signal stability while keeping the formation of oxides and double-charged ions to a minimum. Commercially available multi-element standard solutions (Plasma CAL, SCP Science, Baie D'Urfé, Canada) were used to prepare calibration standards. The internal standard solution was prepared from an Isostandards Material (Madrid, Spain) commercial solution.

The limits of detection (LoD) were calculated as the concentration corresponding to three times the standard deviation of 10 replicate measurements of the blank solution (2% v/v $HNO_3$).

### *2.4. Soil Samples and Analysis*

Composite top layer (0–15 cm) soil samples were randomly collected from 26 kitchen gardens and/or small farms, in an area of approximately 20 $km^2$ in the vicinity of the Estarreja Chemical Complex. The sample sites were selected in order to provide a representative area of the agricultural soils throughout the community that surround this industrial area. All soil samples were air-dried and sieved at 2 mm. The analysis of soils was performed by ICP-MS in ACME certified laboratory after extraction with aqua regia, and according to the laboratory standards methods and quality assurance and quality control and protocols. The accuracy and precision were checked through the analysis of certified reference materials, blank spikes and duplicates (analytical splits) of randomly selected samples. The analytical precision was better than 10%.

#### Enrichment Factor

The environmental risk was evaluated calculating the enrichment factor (EFi), proposed by Buat-Menard and Chesselet [55], and calculated as follows:

$$\text{EFi} = (\text{CM}/\text{Cnor})\ \text{sample}/(\text{CM}/\text{Cnor})\ \text{background} \quad (1)$$

The enrichment factor is calculated for each element (i) and indicates the enrichment of the metal (CM: concentration of each metal) in a sample, relative to the concentration of that metal in the chosen geochemical background, after normalization (Cnor) to a conservative geogenic element, which is usually Al, Fe, Zr or Sc. The normalization points to differences caused by the sampling and by the variability of behaviors of the chemical elements in the surface geochemical processes [55]. The normalizing element used was Al, also used by Islam et al. [56–58]. Enrichment factors above 1.5 indicates anthropogenic influence in soil composition and it can be minor (2 < EF), moderate (2 < EF < 5), significant (5 < EF < 20), very strong (20 < EF < 40), and extreme (EF > 40) [59].

### 2.5. Statistical Techniques

Variables were examined for outliers and normal distribution by means of histograms, box plots, normal quantile-quantile plots, and the Shapiro–Wilk test. When normal distribution could not be accepted, variable transformations (square, square root, and logarithmic) were attempted. The base-10 logarithm of Al, Mn, Fe, and Pb levels, and the square root of Hg, Cu, and Zn concentrations helped to improve the distribution shape. Differences between groups were tested using the Kruskal–Wallis H test with post-hoc tests, with the results interpreted based on rank differences. A probability $\leq 0.05$ was assumed as significant in testing the null hypothesis of no differences across the considered clinical conditions. The one-way analysis of covariance was used to determine whether there were significant differences in element hair contents between the study groups, while "statistically controlling" the effects of the confounding variables (covariates) that were believed to affect the results. The assumption of equality of variance was assessed by means of Levene's test, which indicated heterogeneity of variances for hair Cu levels. Finally, post hoc pairwise multiple comparisons were performed using Bonferroni correction in order to detect significant differences between two specific groups.

## 3. Results and Discussion

### 3.1. The Study Population Cognitive Status

Clinical and demographic characteristics of the individuals recruited, divided by the four study groups, are reported in Table 1. The Kruskal–Wallis H test showed statistically significant differences for age ($\chi^2(3) = 11.626$, $p = 0.009$) between SMC and Dementia (DEM) subjects ($p = 0.013$), for the level of education ($\chi^2(3) = 26.988$, $p < 0.0001$) between HS and DEM ($p = 0.001$), as well as between SMC and DEM subjects ($p < 0.001$), and for the amount of time living in the city ($\chi^2(3) = 12.662$, $p = 0.005$) between HS and DEM subjects ($p = 0.012$). Table 1 shows that in this study, dementia appears to be associated with higher age, female gender, lower education level, and longer residence time in the study area, i.e., in the surrounding industrial zone. Several authors [51,60–64] have shown that sociodemographic variables have an important effect on cognitive-screening test performance, predominantly age and education level. Old age has been found to significantly increase the probability of obtaining lower scores, whereas the worst performance has been found among those with lower education levels and ceiling effects have been observed among highly educated individuals. The magnitude of the effect of education level is so strong that education is invariably considered a criterion for the establishment of normative data for the MMSE [65–67]. Some studies further suggest that high diet quality in terms of the consumption of vitamins, minerals, and trace elements can be expected when education levels are high [68,69]. However, the relationship between education and diet quality seems to be significant for deficient intake only. In our study, none of the participants were reported to have TE deficiency.

Previous studies regarding the effect of gender have proven to be more controversial; only a few have shown a significant association between this variable and cognitive-screening test performance [70–72]. However, a recent study report that female *APOE* ε4 carriers have faster rates of memory decline than their male counterparts among MCI individuals [73] corroborating the observations of Iwata and colleagues [74].

In Table 1 it is also noticeable that participants whose professional activity is associated with agriculture and fisheries seem the most vulnerable to dementia. In addition, most of the participants diagnosed with dementia frequently consumed local home-grown foodstuffs. Cabral Pinto et al. [41,43,44] reported a link between cognitive status and both agricultural activity and the consumption of crops cultivated in soils irrigated with groundwater from wells in the ECC surroundings. Regarding medical history, there was no clear relation with the cognitive levels, although it was found that cardiovascular diseases were relatively common in the participants with lower cognitive level, which is in accordance with previous observations of [75].

**Table 1.** Demographic, lifestyle habits, and clinical characteristics of the study groups. Healthy subjects (HS), patients suffering from subjective memory concerns (SMC) and/or mild cognitive impairment (MCI), and those with already installed dementia (DEM).

| | | HS | SMC | MCI | DEM |
|---|---|---|---|---|---|
| | | $n = 10$ | $n = 14$ | $n = 16$ | $n = 39$ |
| Age (mean ± SD) | | 74.0 ± 9.6 | 73.3 ± 7.2 | 78.4 ± 7.8 | 81.7 ± 9.0 |
| Gender (n; %) | Male | 2; 20% | 2; 14% | 6; 37% | 3; 8% |
| | Female | 8; 80% | 12; 86% | 10; 63% | 36; 92% |
| Level of Education (mean ± SD) | | 7.40 ± 5.10 ** | 4.43 ± 2.34 ** | 2.88 ± 1.54 | 1.91 ± 3.36 ** |
| Time of residence (mean ± SD) | | 53.20 ± 17.76 * | 58.36 ± 27.80 | 58.69 ± 26.64 | 70.82 ± 26.08 * |
| Profession | Housewife | 2; 20% | 2; 14% | 4; 25% | 8; 21% |
| | Agriculture/Fishery | 2; 20% | 3; 21 % | 4; 25% | 16; 41% |
| | Industry/Construction | - | 5; 36% | 4; 25% | 5; 13% |
| | Commerce/Services | 6; 60% | 4; 29 % | 4; 25% | 10; 26% |
| Medical History (*n*, %) | | | | | |
| Diabetes | | - | - | 2; 13% | 6; 15% |
| Dyslipidemia | | 1; 10% | 3; 21% | - | 1; 3% |
| Cardiovascular | | 3; 30% | 5; 36% | 8; 50% | 9; 25% |
| Respiratory | | 3; 30% | 1; 7% | 1; 6% | 2; 5% |
| Other | | 2; 20% | 3; 21% | 2; 13% | 5; 13% |
| Lifestyle factors (*n*; %) | | | | | |
| Supplements | | - | 5; 36% | 3; 19% | 7; 18% |
| Homegrown food | | 7; 70% | 11; 79% | 13; 81% | 26; 67% |
| Drinking Water | Bottled | 6; 60% | 11; 79% | 12; 75% | 22; 56% |
| | Bottled & tap water | 2; 20% | 1; 7% | - | 1; 3% |
| | Tap water | 2; 20% | 1; 7% | 4; 25% | 11; 28% |
| | Well or borehole | - | 1; 7% | - | - |

* ($p < 0.05$); ** ($p < 0.01$); age, level of education and time of residence are expressed in years; SD: standard deviation.

### *3.2. TE Levels in Hair and Population Cognitive Status Relations*

Biomonitoring is used as a means for assessing the impact of environmental chemical elements on living organisms [76]. Regardless of whether they are essential, nonessential, or highly toxic, human hair acts as an excretory tissue for all elements, which become incorporated into the hair matrix during its growth. In general, a target population's health and nutrition status regarding TE can be assessed by measuring the levels in hair samples [22], i.e., the determination of TE in hair is a way of indirectly testing for the body's overload or deficiency.

In this study, seven TE (Al, Mn, Fe, Cu, Zn, Hg, and Pb) were measured in the hair of Estarreja residents with or at risk of AD. Among them, two TE [Zn ($\chi^2$ (3) = 11.723, $p = 0.008$) and Hg ($\chi^2$ (3) = 17.772, $p < 0.001$)] showed statistically significant differences in their hair concentrations between the study groups (Table 2). Pairwise comparisons showed significant differences between SMC vs. DEM subjects for Zn and between HS vs. DEM subjects for Hg. Table 2 also presents reference interval values (P5–P95) for toxic TE (Al, Hg, and Pb) estimated in accordance with International Union of Pure and Applied Chemistry recommendations [77] and reference range values for essential TE (Cu, Fe, Mn, and Zn), reported from a review of Mikulewicz et al. [78]. In general, the mean levels of the analyzed TE found in this study were well within the range reported for non-exposed people, in all the groups (four different cognitive status), except for Hg (Table 2). The mean level of Pb in hair was above the reference range in the DEM group; the mean level of Mn in hair was above the reference range in the MCI and DEM groups, and the mean levels of Zn in the DEM group were also out of reference values for many participants, but there was no statistically significant difference ($p > 0.05$). The mean level of Hg in hair was above the reference range in the SMS, MCI, and DEM groups. The TE body burden in the study groups suggests a potential long-term environmental exposure, which is evidenced by the significantly higher ($p < 0.05$) hair Hg content compared to non-exposed people.

**Table 2.** Trace elements levels (μg/g) in hair samples of the study participants, reported as mean ± standard deviation (SD) and (range) according to the cognitive status (HS = healthy subjects; SMC = subjective memory complaints; MCI = mild cognitive impairment; DEM = Dementia). Also present are hair reference value ranges for non-exposed people reported from Skalny et al. [77] and Mikulewicz et al. [78].

| Element | HS | SMC | MCI | DEM | Kruskal-Wallis | | Pairwise Comparisons [(a)] ($p$-Value) | | | | | | Hair Reference Values * |
|---|---|---|---|---|---|---|---|---|---|---|---|---|---|
| μg/g | $n = 10$ | $n = 14$ | $n = 16$ | $n = 39$ | H(3) | $p$-Value | DEM vs. HS | DEM vs. SMC | DEM vs. MCI | HS vs. SMC | HS vs. MCI | SMC vs. MCI | |
| Al | 6.67 ± 10.69 (0.23–35.96) | 4.18 ± 3.36 (0.16–11.70) | 3.80 ± 3.95 (0.78–14.02) | 6.26 ± 8.76 (0.51–49.08) | 1.222 | 0.748 | N/A | N/A | N/A | N/A | N/A | N/A | 2.91–11.63 |
| Mn | 0.19 ± 0.17 (0.04–0.54) | 0.18 ± 0.24 (0.05–0.87) | 0.85±1.38 (0.03–5.46) | 1.39 ± 2.83 (0.03–16.10) | 7.089 | 0.069 | N/A | N/A | N/A | N/A | N/A | N/A | 0.002–0.91 |
| Fe | 16.37 ± 12.24 (0.59–43.77) | 10.79 ± 7.75 (0.59–27.78) | 12.67 ± 9.65 (3.60–40.47) | 18.47 ± 40.24 (3.41–259.26) | 2.266 | 0.519 | N/A | N/A | N/A | N/A | N/A | N/A | 3.66–36.8 |
| Cu | 10.57 ± 10.24 (1.62–37.10) | 10.77 ± 5.82 (0.56–22.20) | 9.85 ± 2.54 (3.83–13.39) | 17.99 ± 36.71 (2.66–237.40) | 1.581 | 0.664 | N/A | N/A | N/A | N/A | N/A | N/A | 7.2–82.7 |
| Zn | 140.29 ± 65.66 (10.70–234.86) | 118.17 ± 58.33 (3.99–229.00) | 151.69 ± 28.64 (103.00–200.38) | 258.10 ± 519.33 (63.25–3396.26) | 11.723 | 0.008 ** | n.s. | 0.006 ** | n.s. | n.s. | n.s. | n.s. | 30–327 |
| Hg | 0.88 ± 0.92 (0.12–3.24) | 1.48 ± 1.40 (0.11–5.38) | 1.63 ± 1.18 (0.63–5.13) | 4.43 ± 13.86 (0.06–88.46) | 17.772 | <0.001 ** | 0.001** | n.s. | n.s. | n.s. | n.s. | n.s. | 0.17–1.19 |
| Pb | 0.29 ± 0.39 (0.048–1.33) | 0.46 ± 0.56 (0.001–2.20) | 0.33 ± 0.42 (0.03–1.58) | 1.02 ± 2.11 (0.03–12.75) | 8.839 | 0.077 | N/A | N/A | N/A | N/A | N/A | N/A | 0.19–1.39 |

[a] Post hoc analysis using Bonferroni method; N/A: not applicable; n.s.: not significant; * ($p < 0.05$); ** ($p < 0.01$).

It has been demonstrated that several factors have the potential to influence TE concentrations in human hair (e.g., Hartmann and Kist [23], and references therein). Our study groups presented significant differences in age and education level (Table 1), suggesting that they are potential confounding variables. The effect of factors such as sex, age, education level, etc., on hair TE concentration has been shown by many authors (e.g., [26,79–83]). Therefore, an analysis of covariance was performed considering several variables such as age, gender, level of education, use of minerals-containing food supplements, time period of residence in Estarreja, home-grown foodstuff consumption, and type of drinking water as covariates. Three outliers were excluded from this covariance analysis, which is highly influenced by the presence of extreme values in the dataset. The homogeneity of regression slopes was tested and no interaction existed between the covariates and the independent variable. However, the hair Cu levels failed Levene's test, and it was not possible to assume the equality of variance. These results could indicate that the significant differences in hair Zn levels between SMC and DEM subjects could result from one or more of the confounding factors, such as the time period of residence or the use/exposure of water from wells containing high levels of Zn, as reported in Cabral Pinto et al. [41]. Nevertheless, available data on the exposure of the Estarreja population through soil/dust, and home-grown foodstuffs consumption [44] did not support an excessive Zn exposure. Zinc is an essential element involved in many metabolic functions and is crucial for a healthy body status [84]; however, excess Zn can be harmful and cause toxicity [85]. Excessive ingestion of Zn can suppress Cu and Fe gastrointestinal absorption [84,86]. Toxicity associated with excessive exposure to Zn is not well known. Situations in which toxicity has been observed include inhalation of zinc fumes, deliberate ingestion, exposure to contaminated food and/or drinking water and epidemiological causes [85,87]. Occupational and environmental (chronic) exposure to specific levels of Zn has led it to be suggested as a possible cause of cognitive dysfunction and dementia ([88] and references therein).

As previous reported, only Hg differed significantly between the four groups. Pairwise multiple comparisons showed that Hg was significantly higher in DEM compared to HS (Table 3). While none of the covariates significantly predicted ($p > 0.050$) the hair Hg content, MMSE was a good predictor of the dependent variables ($p = 0.005$). Hence, the results indicate that none of the tested covariates seems to influence the concentration of Hg in the hair of the elderly. This element has been identified as one of the most toxic nonradioactive materials known to man [89]. Although it is a naturally occurring element, anthropogenic Hg is now a major worldwide concern and is an international priority pollutant as it is persistent, bioaccumulative, and toxic even at very low levels to humans and aquatic/terrestrial ecosystems. Sensory disturbances (hypoesthesia), lack of coordination of voluntary movements (ataxia), impairment of hearing, concentric constriction of the visual field, and slurred speech (dysarthria) are some of the signs of Hg poisoning [90]. It was recently shown that circulatory levels of Hg are significantly higher in AD patients [91] and it is known that Hg favors misfolding and aggregation of amyloid beta (A$\beta$) protein, a neurotoxic protein present in AD patients brain [92], which supports the idea that Hg is a risk factor for this neurodegenerative disease.

The risk of mercury toxicity depends very much on the form of Hg (elemental, organic, and inorganic) and route of exposure. Due to the health risks of excessive Hg exposure, the FAO/WHO Joint Expert Committee on Food Additives established a "Provisional Tolerable Weekly Intake" (PTWI) of 4 $\mu g\ kg^{-1}$ body weight per week for inorganic Hg and 1.6 $\mu g\ kg^{-1}$ bw per week for MeHg [93,94]. The United States Environmental Protection Agency [76] presented a lower value for the intake of MeHg, setting a reference dose (RfD) of 0.1 $\mu g\ kg^{-1}$ bw per day [95]. The PTWI and RfD correspond to a hair Hg concentration of 2.2 and 1.0 $\mu g\ g^{-1}$, respectively [93–95]. WHO, through the analysis of neurotoxicological data, considered the Hg concentration of 50 $\mu g\ g^{-1}$ in the human hair as the "no observed adverse effect level" value for MeHg [95]. During hair growth (~1 cm per month), circulating Hg passes from the bloodstream to the hair follicle and is incorporated into the hair shaft, where it becomes stable and is carried along its length as the hair grows, providing an accumulation pattern and the history of exposure [96–99]. Mercury concentration in scalp hair is used to assess blood Hg concentrations during hair growth [76], and the methylmercury (MeHg) exposure level can also be estimated from hair Hg levels, since approximately 80% of hair Hg is MeHg [100].

**Table 3.** Results of the analysis of covariance (ANCOVA) performed on the hair trace elements data.

| Element | ANCOVA [a] F(3, 68) | *p-Value* | *partial* $\eta^2$ | Pairwise Comparisons [b] (*p*-Value) DEM*vs*HS | DEM*vs*SMC | DEM*vs*MCI | HS*vs*SMC | HS*vs*MCI | SMC*vs*MCI |
|---|---|---|---|---|---|---|---|---|---|
| Al | 0.597 | 0.620 | 0.03 | N/A | N/A | N/A | N/A | N/A | N/A |
| Mn | 1.13 | 0.344 | 0.05 | N/A | N/A | N/A | N/A | N/A | N/A |
| Fe | 0.982 | 0.407 | 0.05 | N/A | N/A | N/A | N/A | N/A | N/A |
| Cu | 0.455 | 0.715 | 0.22 | N/A | N/A | N/A | N/A | N/A | N/A |
| Zn | 2.477 | 0.07 | 0.11 | N/A | N/A | N/A | N/A | N/A | N/A |
| Hg | 4.411 | 0.007 * | 0.18 | 0.005 * | n.s. | n.s. | n.s. | n.s. | n.s. |
| Pb | 2.757 | 0.500 | 0.02 | N/A | N/A | N/A | N/A | N/A | N/A |

[a] Covariates: age, education level, mineral-containing food supplements use, residence time in the study area, home-grown foodstuff consumption and type of drinking water. [b] Post hoc analysis using Bonferroni method; N/A: not applicable; n.s.: not significant; * ($p < 0.01$).

### 3.3. *Trace Elements Hair Versus Risk of AD*

TE are present at very low concentrations in tissues, and even small variations in these concentrations could be harmful or a sign of disease [23]. The levels of a biomarker may change over time after the onset of AD, namely from the early to the middle and the advanced stages. TE profiling is an interesting approach for understanding neurodegenerative processes, considering that compelling evidence shows that TE toxic effects might play a crucial role in the onset and progression of AD. Due to its more stable nature, the analysis of these elements in human hair is potentially more reliable than in blood [23]. Figure 1 shows a trend of an increase of Hg hair mean values between the four groups, increasing from HS to SMS, SMS to MCI (less evident), and from MCI to DEM groups. However, for the other toxic TE (Al and Pb) this signature is not followed. Lead in hair, for example shows higher mean concentrations in SMS and DEM groups. On the other hand, aluminum seems to decrease with the increase of cognitive decline groups.

Essential TE in hair (Mn, Fe, Cu, and Zn) showed a different behavior between the four cognitive statuses. Manganese and Zn in hair tended to increase from the HS to DEM group (Figure 1). Cabral Pinto et al. [42] found the highest contents of Zn and Mn in fingernails associated with the group of dements. Many studies have investigated the association between TE levels in hair and the risk of AD, but the results have also been ambiguous. Koseoglu et al. [101] found that AD patients presented significantly different concentrations of Al, Pb, Fe, Mn, Hg, and Cu in hair compared to control individuals. Vance et al. [102] compared the results of the hair analysis of 63 AD patients and 117 controls and also found that Zn levels were higher in AD patients and that no significant difference existed for the Fe concentrations, similar to our results. Koc et al. [101], in a study involving 45 patients with AD and 33 controls, found that AD patients had significantly higher hair Cu and Mn levels, but no significant differences for Fe, in accordance with our results. In contrast to our results, these authors found significantly lower Zn levels in AD patients compared to control participants; however they found no significant difference between the hair Zn levels of the two groups. Koc and co-authors [103] also found that some TE levels were changed in patients with AD. The small number of participants is a limitation of the study, which might not be enough to reach significant values in some cases.

The discovery of biomarkers that could confer high confidence to a presymptomatic AD diagnosis would be a great step forward to study the etiology of the disease, to identify the risk factors and to ultimately discover effective treatments [23]. According to the previous authors, studies such as the one presented here have the interesting potential to find reliable biomarkers in noninvasive samples for AD and MCI. However, many more studies must be conducted to allow the extraction of much more accurate information that may eventually lead to a panel of analyses that produce high-confidence AD diagnoses, especially in the early stages. The findings of our results suggest that Hg could contribute to generating a distinctive signature during the progression of dementia, and monitoring them in the elderly might help to detect preclinical stages of AD.

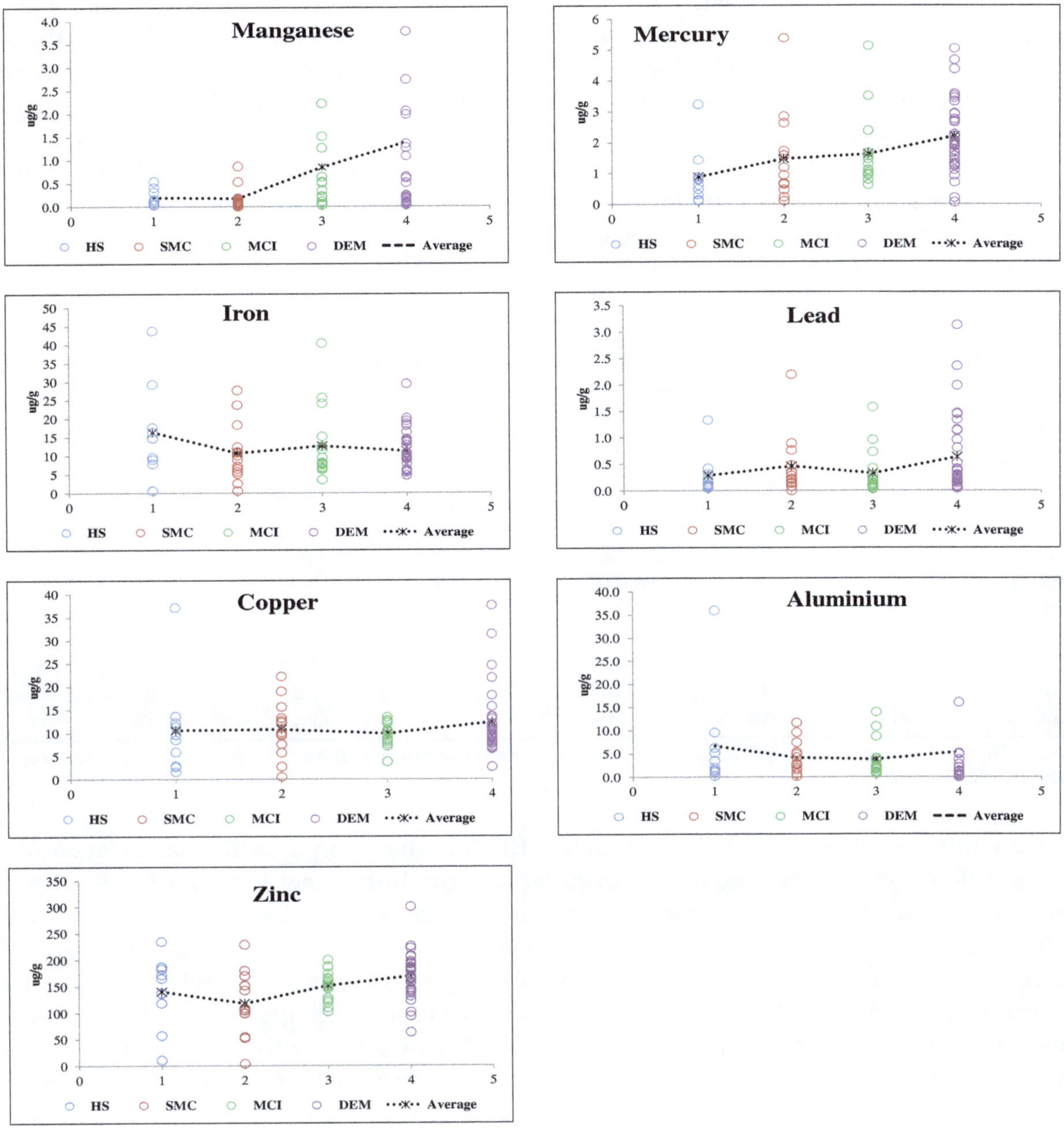


**Figure 1.** Trace elements hair levels in the different study groups.

Plagia et al. [104] also found a variable behavior in the TE serum of subjects with or at risk of AD relatively to a control group. Paglia et al. [104] found a decreased signature for essential TE (Zn, Mn, Fe) in the serum of subjects with or at risk of AD relative to a control group. The results of a meta-analysis of Li et al. [105] provided rigorous statistical support for the association of the serum levels of TE and the risk of AD, suggesting a positive relationship between the serum Cu levels and AD risk, and a negative relationship between the serum Zn levels and AD risk. The differences between our observations and those from other authors may be simply related to the nature of the biological specimen: hair vs. serum.

### *3.4. Metal Population Environmental Exposure*

Table 4 summarizes (a) the quality of the soil samples collected in the ECC surroundings, used by the population for agricultural proposes; (b) the soil background values (BG), presented by Inácio [106], calculated by the average of Cambisoil and Podzol toplayers representative soils of the study area;

(c) and the Canadian Environmental Quality Guidelines for agricultural use [107]. Manganese, iron, and aluminum have no international guidelines for any use, but their mean values in the sampled agricultural soils are impoverished relative to the mean local background values. The content of Cu, Hg, Pb, and Zn of the agricultural soil were higher than the background values, particularly Hg, Cu, and Pb (100%, 46%, and 42% of the samples). In general, the mean values of metals in soils are below Canadian limits, except for the Hg mean value which exceeded the Canadian guideline for agricultural uses. However, in some soil sampling locations, maximum Cu concentrations were also well above the Canadian guidelines for agricultural proposes.

**Table 4.** Descriptive statistics for metal contents (mg $kg^{-1}$) in Estarreja Chemical Complex (ECC) agricultural soils, soil top layer local background (BG) values from Inácio [104] and (CEQG) Canadian Environmental Quality Guidelines (mg $kg^{-1}$) for agricultural land use [107].

| | Al | Cu | Hg | Mn | Fe | Pb | Zn |
|---|---|---|---|---|---|---|---|
| Minimum | 0.31 | 3 | 0.03 | 36 | 0.26 | 6 | 16 |
| Mean | 0.60 | 33 | **1.5** | 120 | 0.66 | 33 | 84 |
| Median | 0.56 | 23 | 0.15 | 110 | 0.62 | 23 | 67 |
| Maximum | 1.15 | 103 | 14 | 255 | 1.15 | 109 | 199 |
| SD | 0.21 | 27 | 4 | 56 | 0.2 | 23 | 44 |
| Cambisoil | 2.14 | 2 | 0.05 | 251 | 2.39 | 20 | 55 |
| Podzol | 0.13 | 12 | 0.05 | 154 | 0.17 | 6 | 4 |
| BG | 1.14 | 7 | 0.05 | 203 | 1.28 | 13 | 30 |
| CEQG | NA | 62 | 0.16 | NA | NA | 45 | 290 |
| % > BG | 0 | **96** | **100** | **12** | 0 | **84** | **96** |
| % > CEQG | NA | **15** | **35** | NA | NA | **23** | 0 |

Note: In bold are the trace elements (TE) values relatively higher than the respective guidelines; SD: standard deviation.

As previous reported, the enrichment factor (EF) for studied metals [55] has been employed for calculating differences between the metals originating from human activities and those from natural sources and it can also be used to assess and explain the contamination of metals in soils [108]. To access the degree of soil contamination through this factor, the local background values were used as it is easier to isolate anthropogenic factors from the geogenic ones, because when using generalized values, spurious enrichment due to natural local concentrations of elements may appear [109–112].

The calculated EF values are presented in Figure 2. Considering the average EF value for each element, the enrichment decreases in the order: Hg > Cu > Zn > Pb > Mn > Fe = Al. The observed EF values indicate that 26% of the sampled agricultural soils are very strong and extremely contaminated in Hg and 65% of the samples reach significant contamination; 58%, 42%, and 27% of the sampled agricultural soils reach significant contaminant in Cu, Zn, and Pb, respectively. The soil enrichment on these elements is mainly due to the former industrial activity and agriculture practices that were greatly lowering soil quality. The degradation of soil quality by metals/metalloids pollution causes environmental risks, leads to groundwater pollution, is harmful for human health and lowers human life quality.

Biomonitoring and cognitive status relations of residents' hair in the ECC surrounding area studied in this work highlight that Hg has features which distinguish it from remaining studied chemical elements and could be the result of long-term environmental exposure. The causes of total resident exposure to Hg may be due to the use of contaminated water for cooking and showering, inhalation, ingestion, and dermal contact of soil/dust, consumption of Hg-contaminated home-grown foodstuffs, and even fish.

The assessment of health risks for ECC surrounding residents done by Cabral Pinto et al. [44] showed that the elements of greatest concern were As and Hg, regarding both carcinogenic and non-carcinogenic risk. Reis [112] detected that total Hg concentrations in well water samples ranged between 26 and 846 ng $L^{-1}$, and all samples presented concentrations below the maximum level allowable for drinking water as defined in Portuguese law (1.0 μg $L^{-1}$). Even at low concentrations, water from these wells that was used for irrigation may be a problem due to the Hg bioaccumulation and biomagnification capacity associated with high toxicity. The highest concentration was detected near the S. Filipe effluent (10 m away). However, Cabral Pinto et al. [43] reported a mean/max total Hg concentration of 23.6/659 μg $L^{-1}$ (in 2006), 60/473 μg $L^{-1}$ (in 2010) and 1.2/4 μg $L^{-1}$ in groundwater (in 2013). In this sampling campaign, these Hg levels exceeded both the limit for Portuguese drinking water and the groundwater intervention value of the Dutch legislation [113] for soil remediation (0.3 μg $L^{-1}$). When this groundwater was considered for ingestion and dermal contact exposure, Hg concentrations such as those reported for some locations may constitute a non-cancer health risk.

According to Cabral Pinto et al. [44], fresh cabbage leaves (*Brassica Oleracea* L.), one of the home-grown foodstuffs consumed by the residents, yielded a total Hg concentration of 0.012 mg/kg in 4% of the studied samples, which is slightly above the limit of 0.01 mg/kg for fresh vegetable consumption, proposed in 2005 by the Ministry of Health of the People's Republic of China [114]. Reis [112] also observed a Hg concentration of 0.01 to 0.42 mg/kg in different fish captured in the Ria de Aveiro (site of effluent discharges for many years, among others, from the Estarreja chlor-soda industrial plant). These Hg concentrations did not exceed the limits defined in Commission Regulation (EC) No. 466/2001 of 8 March (0.5 mg kg $^{-1}$ fresh weight) or official journal of European Communities [115] recommendation (1.0 mg $kg^{-1}$).

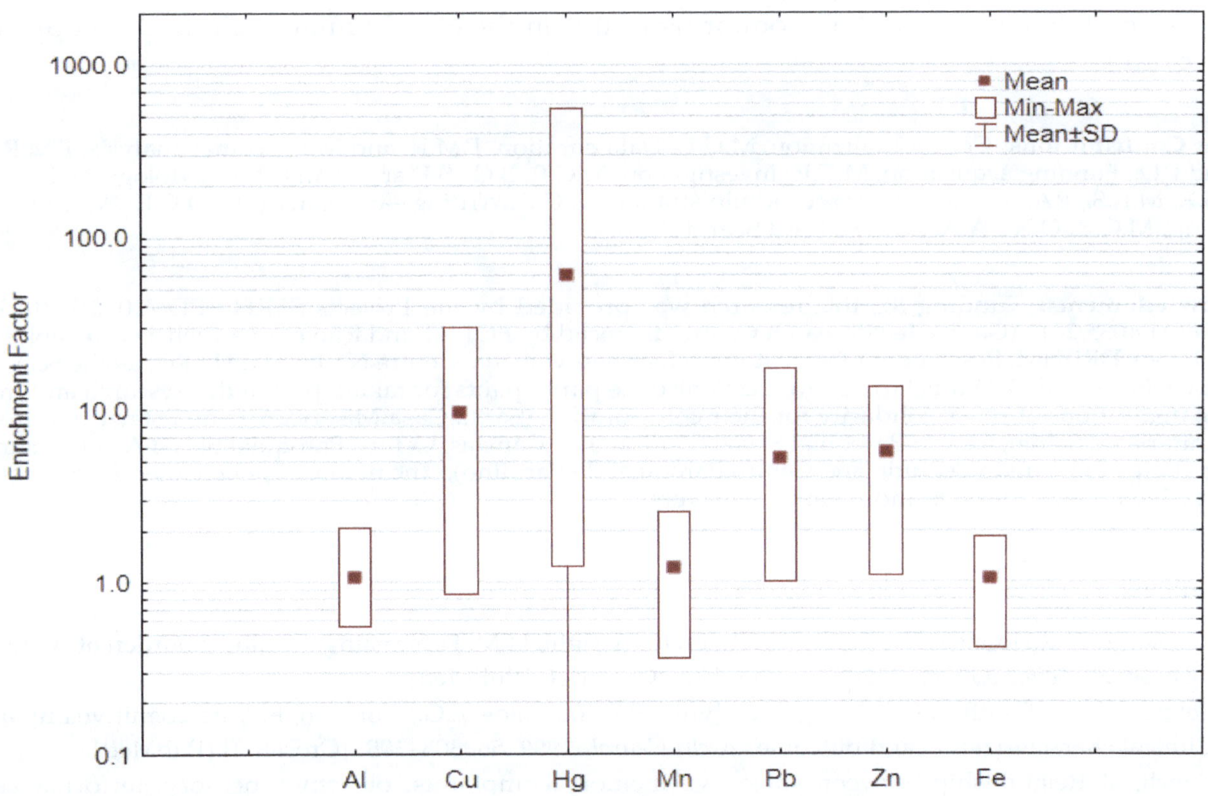


**Figure 2.** Enrichment factors for the studied elements in sampled agricultural soils from ECC surroundings.

## 4. Conclusions

The inter-disciplinary approach applied in this study was successful in identifying links between different datasets. The following conclusions can be drawn:

- Participants whose professional activity was associated with agriculture and fisheries were shown to be the most vulnerable to dementia.

- Participants diagnosed with dementia frequently consume home-grown foodstuffs, some of them probably irrigated with contaminated well water.
- Biomonitoring and the analysis of the cognitive status of the residents surrounding ECC suggest that Hg levels in hair differed significantly between the four cognitive groups (healthy, subjective memory complaint, mild cognitive impairment, and dementia), increasing from healthy to dementia participants.
- Mercury mean levels in soil samples were above Canadian guidelines for soil for agricultural uses and the enrichment factor calculation values highlighted that 26% of the studied soils reached the "extremely contaminated" class for Hg, with 65% of the soils reaching the "significant contamination" class.
- Improved diagnostic results can be obtained using hair TE signatures coupled with MMSE scores. This strategy may prove useful for predictive diagnosis in populations screening of cognitive impairment.

The discovery of biomarkers that could confer high confidence to presymptomatic AD diagnosis would be a great step forward to study the etiology of the disease, to identify the risk factors, and to ultimately discover effective treatments. Supplementary studies must be conducted to achieve an adequate panel of analyses capable of providing high-confidence AD diagnoses, especially in the early stages. Our research suggest that Hg could contribute to generate a distinctive signature during the progression of dementia, and the monitoring of this in the elderly might help to detect preclinical stages of AD.

**Author Contributions:** Conceptualization, M.C.P.; Data curation, P.M.R. and M.I.; Formal analysis, P.M.R., A.A., E.P. and P.D.; Funding acquisition, M.C.P.; Investigation, M.C.P., B.G., S.F. and P.A.D.; Methodology, M.C.P., P.M.R., B.G., S.F., M.R.S., P.A.D. and L.D.; Project administration, M.C.P.; Writing—original draft, M.C.P.; Writing – review & editing, M.C.P., O.N., A.A., M.R.S., E.F.d.S. and P.I.M.

**Acknowledgments:** Funding for this research was provided by the Projects SFRH/BPD/71030/2010, Project UID/GEO/04035/2019 (GeoBioTec Research Centre) financed by FCT – Fundação para a Ciência e Tecnologia and by the Labex DRIIHM, Réseau des Observatoires Hommes-Millieux–Centre National de la Recherche Scientifique (ROHM–CNRS) and OHMI-Estarreja. We thank also the participants for taking part in this research and the local private institutions of social solidarity for their collaboration (Santa Casa Misericórdia de Estarreja, Associação Humanitária de Salreu, Centro Paroquial Social São Tomé de Canelas, Centro Paroquial Social Avanca, Fundação Cónego Filipe Figueiredo Beduído and Centro Paroquial de Pardilhó). The manuscript benefited from careful and constructive reviews by three anonymous reviewers.

## References

1. Brookmeyer, R.; Johnson, E.; Ziegler-Graham, K.; Arrighi, H.M. Forecasting the global burden of Alzheimer's disease. *Alzheimers Dement.* **2007**, *3*, 186–191. [CrossRef] [PubMed]
2. Petersen, R.C.; Smith, G.E.; Waring, S.C.; Ivnik, R.J.; Tangalos, E.G.; Kokmen, E. Mild cognitive impairment: Clinical characterization and outcome. *Arch. Neurol.* **1999**, *56*, 303–308. [CrossRef] [PubMed]
3. Zandi, T. Relationship between subjective memory complaints, objective memory performance, and depression among older adults. *Am. J. Alzheimers Dis.* **2004**, *19*, 353–360. [CrossRef] [PubMed]
4. Youn, J.C.; Kim, K.W.; Lee, D.Y.; Jhoo, J.H.; Lee, S.B.; Park, J.H.; Woo, J.I. Development of the subjective memory complaints questionnaire. *Dement. Geriatr. Cogn. Disord.* **2009**, *27*, 310–317. [CrossRef] [PubMed]
5. Benito-León, J.; Mitchell, A.J.; Vega, S.; Bermejo-Pareja, F. A population-based study of cognitive function in older people with subjective memory complaints. *J. Alzheimers Dis.* **2010**, *22*, 159–170. [CrossRef]
6. Genziani, M.; Stewart, R.; Béjot, Y.; Amieva, H.; Artero, S.; Ritchie, K. Subjective memory impairment, objective cognitive functioning and social activity in French older people: Findings from the Three Cities study. *Geriatr. Gerontol. Int.* **2013**, *13*, 139–145. [CrossRef]

7. Vaskivuo, L.; Hokkanen, L.; Hänninen, T.; Antikainen, R.; Bäckman, L.; Laatikainen, T.; Ngandu, T. Associations between Prospective and Retrospective Subjective Memory Complaints and Neuropsychological Performance in Older Adults: The Finger Study. *J. Int. Neuropsychol. Soc.* **2018**, *24*, 1099–1109. [CrossRef]
8. Mitchell, A.J.; Beaumont, H.; Ferguson, D.; Yadegarfar, M.; Stubbs, B. Risk of dementia and mild cognitive impairment in older people with subjective memory complaints: Meta–analysis. *Acta Psychiatr. Scand.* **2014**, *130*, 439–451. [CrossRef]
9. Geerlings, M.I.; Jonker, C.; Bouter, L.M.; Adèr, H.J.; Schmand, B. Association between memory complaints and incident Alzheimer's disease in elderly people with normal baseline cognition. *Am. J. Psychiatry* **1999**, *156*, 531–537.
10. Jorm, A.F.; Christensen, H.; Korten, A.E.; Jacomb, P.A.; Henderson, A.S. Memory complaints as a precursor of memory impairment in older people: A longitudinal analysis over 7–8 years. *Psychol. Med.* **2001**, *31*, 441–449. [CrossRef]
11. Wang, L.; Van Belle, G.; Crane, P.K.; Kukull, W.A.; Bowen, J.D.; McCormick, W.C.; Larson, E.B. Subjective memory deterioration and future dementia in people aged 65 and older. *J. Am. Geriatr. Soc.* **2004**, *52*, 2045–2051. [CrossRef] [PubMed]
12. Waldorff, F.B.; Siersma, V.; Vogel, A.; Waldemar, G. Subjective memory complaints in general practice predicts future dementia: A 4-year follow-up study. *Int. J. Geriatr. Psychiatry* **2012**, *27*, 1180–1188. [CrossRef] [PubMed]
13. Amieva, H.; Letenneur, L.; Dartigues, J.F.; Rouch-Leroyer, I.; Sourgen, C.; D'Alchée-Birée, F.; Fabrigoule, C. Annual rate and predictors of conversion to dementia in subjects presenting mild cognitive impairment criteria defined according to a population-based study. *Dement. Geriatr. Cogn. Disord.* **2004**, *18*, 87–93. [CrossRef] [PubMed]
14. Devier, D.J.; Villemarette-Pittman, N.; Brown, P.; Pelton, G.; Stern, Y.; Sano, M.; Devanand, D.P. Predictive utility of type and duration of symptoms at initial presentation in patients with mild cognitive impairment. *Dement. Geriatr. Cogn. Disord.* **2010**, *30*, 238–244. [CrossRef] [PubMed]
15. Silva, D.; Guerreiro, M.; Faria, C.; Maroco, J.; Schmand, B.A.; Mendonça, A.D. Significance of subjective memory complaints in the clinical setting. *J. Geriatr. Psychiatry Neurol.* **2014**, *27*, 259–265. [CrossRef] [PubMed]
16. Mol, M.E.; Van Boxtel, M.P.; Willems, D.; Jolles, J. Do subjective memory complaints predict cognitive dysfunction over time? A six-year follow-up of the Maastricht Aging Study. *Int. J. Geriatr. Psychiatry J. Child Psychol. Psychiatry* **2006**, *21*, 432–441. [CrossRef]
17. Reid, L.M.; MacLullich, A.M. Subjective memory complaints and cognitive impairment in older people. *Dement. Geriatr. Cogn. Disord.* **2006**, *22*, 471–485. [CrossRef]
18. Dourlen, P.; Kilinc, D.; Malmanche, N.; Chapuis, J.; Lambert, J.C. The new genetic landscape of Alzheimer's disease: From amyloid cascade to genetically driven synaptic failure hypothesis? *Acta Neuropathol.* **2019**, *138*, 221–226. [CrossRef]
19. Yegambaram, M.; Manivannan, B.; Beach, G.T.; Halden, U.R. Role of environmental contaminants in the etiology of Alzheimer's disease: A review. *Curr. Alzheimer Res.* **2015**, *12*, 116–146. [CrossRef]
20. Eid, A.; Mhatre, I.; Richardson, J.R. Gene-environment interactions in Alzheimer's disease: A potential path to precision medicine. *Clin. Pharm.* **2019**, *199*, 173–187. [CrossRef]
21. Paglia, G.; D'Alessandro, A.; Rolfsson, Ó.; Sigurjónsson, Ó.E.; Bordbar, A.; Palsson, S.; Palsson, B.O. Biomarkers defining the metabolic age of red blood cells during cold storage. *Blood* **2016**, *128*, 43–50. [CrossRef] [PubMed]
22. WHO 2015 Human Biomonitoring: Facts and Figures. Available online: http://www.euro.who.int/__data/assets/pdf_file/0020/276311/Human--biomonitoring-facts-figures-en.pdf (accessed on 1 September 2018).
23. Hartmann, S.; Ledur Kist, T.B. A review of biomarkers of Alzheimer's disease in noninvasive samples. *Biomark. Med.* **2018**, *12*, 677–690. [CrossRef] [PubMed]
24. Wilhelm, M.; Hafner, D.; Lombeck, I.; Ohnesorge, F.K. Monitoring of cadmium, copper, lead and zinc status in young-children using toenails—Comparison with scalp hair. *Sci. Total Environ.* **1991**, *103*, 199–207. [CrossRef]
25. Savarino, L.; Granchi, D.; Ciapetti, G.; Cenni, E.; Ravaglia, G.; Forti, P. Serum concentrations of zinc and selenium in elderly people: Results in healthy nonagenarians/ centenarians. *Exp. Gerontol.* **2001**, *36*, 327–339. [CrossRef]

26. Hao, Z.; Li, Y.; Liu, Y.; Li, H.; Wang, W.; Yu, J. Hair elements and healthy aging: A cross-sectional study in Hainan Island, China. *Environ. Geochem. Health* **2016**, *38*, 723–735. [CrossRef]
27. Ahluwalia, N.; Gordon, M.A.; Handte, G.; Mahlon, M.; Li, N.Q.; Beard, J.L. Iron status and stores decline with age in Lewis rats. *J. Nutr.* **2000**, *130*, 2378–2383. [CrossRef]
28. Pike, J.; Chandra, R.K. Effect of vitamin and traceelement supplementation on immune indexes in healthy elderly. *Int. J. Vitam. Nutr. Res.* **1995**, *65*, 117–121.
29. Schmidt, K. Vitamins, minerals, and trace-element in elderly people. *Zentralblatt für Hygiene und Umweltmedizin* **1991**, *191*, 327–332.
30. Selim, A.J.; Fincke, G.; Berlowitz, D.R.; Miller, D.R.; Qian, S.X.; Lee, A. Comprehensive health status assessment of centenarians: Results from the 1999 large health survey of veteran enrollees. *J. Gerontol. A Biol. Sci. Med. Sci.* **2005**, *60*, 515–519. [CrossRef]
31. Andersen-Ranberg, K.; Vasegaard, L.; Jeune, B. Dementia is not inevitable: A population-based study of Danish centenarians. *J. Gerontol. B Psychol. Sci. Soc. Sci.* **2001**, *56*, 152–159. [CrossRef]
32. Pereira, M.E.; Lillebø, A.I.; Pato, P.; Válega, M.; Coelho, J.P.; Lopes, C. Mercury pollution in Ria de Aveiro (Portugal): A review of the system assessment. *Environ. Monit. Assess.* **2009**, *155*, 39–49. [CrossRef] [PubMed]
33. Cachada, A.; Pereira, M.E.; Ferreira da Silva, E.; Duarte, A.C. Sources of potentially toxic elements and organic pollutants in an urban area subjected to an industrial impact. *Environ. Monit. Assess.* **2012**, *184*, 15–32. [CrossRef] [PubMed]
34. Inácio, M.; Neves, O.; Pereira, V.; da Silva, E.F. Levels of selected potential harmful elements (PHEs) in soils and vegetables used in diet of the population living in the surroundings of the Estarreja Chemical Complex (Portugal). *J. Appl. Geochem.* **2014**, *44*, 38–44. [CrossRef]
35. Patinha, C.; Reis, A.P.; Dias, A.C.; Abduljelil, A.A.; Noack, Y.; Robert, S.; Cave, M.; da Silva, E.F. The mobility and human oral bioaccessibility of Zn and Pb in urban dusts of Estarreja (N Portugal). *Environ. Geochem. Health* **2015**, *37*, 115–131. [CrossRef] [PubMed]
36. Leitão, T.B.E. Metodologia Para A Reabilitação De Aquíferos Poluídos. Ph.D. Thesis, Faculdade de Ciências da Universidade de Lisboa, Lisboa, Portugal, 1996.
37. Van der Weijden, C.; Pacheco, F.A.L. Hydrogeochemistry in the Vouga River basin (central Portugal): Pollution and chemical weathering. *J. Appl. Geochem.* **2006**, *21*, 580–613. [CrossRef]
38. Ordens, C.M. Estudo Da Contaminacão Do Aquífero Superior Na Região De Estarreja. Master's Thesis, Coimbra University, Coimbra, Portugal, 2007. Unpublished. Available online: http://www.lneg.pt/download/3268/carlos_ordens.pdf (accessed on 11 March 2018).
39. Reis, A.P.; Costa, S.; Santos, I.; Patinha, C.; Noack, Y.; Wragg, J. Investigating relationships between biomarkers of exposure and environmental copper and manganese levels in house dusts from a Portuguese industrial city. *Environ. Geochem. Health* **2015**, *37*, 725–744. [CrossRef]
40. Plumejeaud, S.; Reis, A.P.; Tassistro, V.; Patinha, C.; Noack, Y.; Orsière, T. Potentially harmful elements in house dust from Estarreja, Portugal: Characterization and genotoxicity of the bioaccessible fraction. *Environ. Geochem. Health* **2018**, *40*, 127–144. [CrossRef]
41. Cabral Pinto, M.M.S.; Marinho-Reis, A.P.; Almeida, A.; Ordens, C.M.; Silva, M.M.; Freitas, S.; Simões, M.; Dinis, P.; Moreira, P.; de Melo, T.C.; et al. Human predisposition to cognitive impairment and its relation with environmental exposure to potentially toxic elements. *Environ. Geochem. Health* **2018**, *40*, 1767–1784. [CrossRef]
42. Cabral Pinto, M.M.S.; Marinho-Reis, A.P.; Almeida, A.; Freitas, S.; Simões, M.R.; Diniz, M.L.; da Silva, E.F.; Moreira, P.I. Fingernail trace element content in environmentally exposed individuals and its influence on their cognitive status in ageing. *Expo. Health* **2018**, *11*, 181–194. [CrossRef]
43. Cabral Pinto, M.M.S.; Ordens, C.M.; de Melo, M.T.C.; Inácio, M.; Almeida, A.; Pinto, E.; da Silva, E.A.F. An Inter-disciplinary Approach to Evaluate Human Health Risks Due to Long-Term Exposure to Contaminated Groundwater Near a Chemical Complex. *Expo. Health* **2019**, 1–16. [CrossRef]
44. Cabral-Pinto, M.M.S.; Inácio, M.; Neves, O.; Almeida, A.A.; Pinto, E.; Oliveiros, B.; Ferreira da Silva, E.A.F. Human Health Risk Assessment Due to Agricultural Activities and Crop Consumption in the Surroundings of an Industrial Area. *Expo. Health* **2019**, 1–12. [CrossRef]
45. Yesavage, J.A.; Brink, T.L.; Rose, T.L.; Lum, O.; Huang, V.; Adey, M.; Leirer, V.O. Development and validation of a geriatric depression screening scale: A preliminary report. *J. Psychiatr. Res.* **1983**, *17*, 37–49. [CrossRef]

46. Pocinho, M.T.S.; Farate, C.; Dias, C.A.; Lee, T.T.; Yesavage, J.A. Clinical and psychometric validation of the Geriatric Depression Scale (GDS) for Portuguese Elders. *Clin. Gerontol.* **2009**, *32*, 223–236. [CrossRef]
47. Simões, M.R.; Prieto, G.; Pinho, M.S.; Firmino, H. Geriatric Depression Scale (GDS-30). In *Escalas e Testes na Demência (3a. edição) (Scales and tests in dementia)*, 3rd ed.; Simões, M.R., Isabel Santana e Grupo de Estudos de Envelhecimento Cerebral e Demência, Eds.; Novartis: Lisboa, Portugal, 2015; pp. 128–133.
48. Kuiper, N.; Rowell, C.; Nriagu, J.; Shomar, B. What do the trace metal contents of urine and toenail samples from Qatar's farm workers bioindicate? *Environ. Res.* **2014**, *131*, 86–94. [CrossRef] [PubMed]
49. Folstein, M.; Folstein, S.; McHugh, P. Mini-mental state: A practical method for grading the cognitive state of patients for the clinician. *J. Psychiatr. Res.* **1975**, *12*, 189–198. [CrossRef]
50. Freitas, S.; Simões, M.R.; Alves, L.; Santana, I. Montreal Cognitive Assessment (MoCA): Validation study for mild cognitive impairment and Alzheimer's disease. *Alzheimer Dis. Assoc. Disord.* **2013**, *27*, 37–43. [CrossRef] [PubMed]
51. Freitas, S.; Simões, M.R.; Alves, L.; Santana, I. The relevance of sociodemographic and health variables on MMSE normative data. *Appl. Neuropsychol. Adult* **2015**, *22*, 311–319. [CrossRef]
52. Kaup, A.R.; Nettiksimmons, J.; LeBlanc, E.S.; Yaffe, K. Memory complaints and risk of cognitive impairment after nearly 2 decades among older women. *Neurology* **2015**, *85*, 1852–1858. [CrossRef]
53. O'Bryant, S.E.; Humphreys, J.D.; Smith, G.E.; Ivnik, R.J.; GraffRadford, N.R.; Petersen, R.C.; Lucas, J.A. Detecting dementia with the Mini-Mental State Examination (MMSE) in highly educated individuals. *Arch. Neurol.* **2008**, *65*, 963–967. [CrossRef]
54. Bass, D.A.; Hickok, D.; Quig, D.; Urek, K. Trace element analysis in hair: Factors determining accuracy, precision, and reliability. *Altern. Med. Rev.* **2001**, *6*, 472–481.
55. Buat-Menard, P.; Chesselet, R. Variable influence of the atmospheric flux on the trace metal chemistry of oceanic suspended matter. *Earth Planet. Sci. Lett.* **1979**, *42*, 399–411. [CrossRef]
56. Islam, S.; Ahmed, K.; Masunaga, S. Potential ecological risk of hazardous elements in different land-use urban soils of Bangladesh. *Sci. Total Environ.* **2015**, *512*, 94–102. [CrossRef] [PubMed]
57. Islam, M.N.; Park, J.H. Immobilization and reduction of bioavailability of lead in shooting range soil through hydrothermal treatment. *J. Environ. Manag.* **2017**, *191*, 172–178. [CrossRef] [PubMed]
58. Cabral Pinto, M.M.S.; Silva, M.M.; da Silva, E.A.F.; Dinis, P.A.; Rocha, F. Transfer processes of potentially toxic elements (PTE) from rocks to soils and the origin of PTE in soils: A case study on the island of Santiago (Cape Verde). *J. Geochem. Explor.* **2017**, *183*, 140–151. [CrossRef]
59. Sutherland, R.A. Depth variation in copper, lead and zinc concentrations and mass enrichment ratios in soils of an urban watershed. *J. Environ. Qual.* **2000**, *29*, 1414–1422. [CrossRef]
60. Anderson, T.M.; Sachdev, P.S.; Brodaty, H.; Trollor, J.; Andrews, G. Effects of sociodemographic and health variables on MiniMental State Exam scores in older Australians. *Am. J. Geriatr. Psychiatry* **2007**, *15*, 467–476. [CrossRef]
61. Bravo, G.; Hébert, R. Age and education specific reference values for the Mini-Mental and Modified Mini-Mental State Examination derived from a non-demented elderly population. *Int. J. Geriatr. Psychiatry* **1997**, *12*, 1008–1018. [CrossRef]
62. Gallacher, J.E.; Elwood, P.C.; Hopkinson, C.; Rabbitt, P.M.; Stollery, B.T.; Sweetnam, P.M.; Huppert, F.A. Cognitive function in the Caerphilly study: Associations with age, social class, education and mood. *Eur. J. Epidemiol.* **1999**, *15*, 161–169. [CrossRef]
63. Matallana, D.; Santacruz, C.; Cano, C.; Reyes, P.; Samper-Ternent, R.; Markides, K.S.; Reyes-Ortiz, C.A. The relationship between educational level and Mini-Mental State Examination domains among older Mexican Americans. *J. Geriatr. Psychiatry Neurol.* **2011**, *24*, 9–18. [CrossRef]
64. Moraes, C.; Pinto, J.A.; Lopes, M.A.; Litvoc, J.; Bottino, C.M. Impact of sociodemographic and health variables on Mini-Mental State Examination in a community-based sample of older people. *Eur. Arch. Psychiatry Clin. Neurosci.* **2010**, *260*, 535–542. [CrossRef]
65. Han, C.; Jo, S.A.; Jo, I.; Kim, E.; Park, M.H.; Kang, Y. An adaptation of the Korean Mini-Mental State Examination (K-MMSE) in elderly Koreans: Demographic influence and population-based norms (the AGE Study). *Arch Gerontol. Geriatr.* **2008**, *47*, 302–310. [CrossRef] [PubMed]
66. Mathuranath, P.S.; Cherian, J.P.; Mathew, R.; George, A.; Alexander, A.; Sarma, S.P. Mini Mental State Examination and the Addenbrooke's Cognitive Examination: Effect of education and norms for a multicultural population. *Neurol. India* **2007**, *55*, 106–110. [CrossRef] [PubMed]

67. Measso, G.; Cavarzeran, F.; Zappalà, G.; Lebowitz, B.D.; Crook, T.H.; Pirozzollo, F.J.; Grigoletto, F. The Mini-Mental State Examination: Normative study of an Italian random sample. *Dev. Neuropsychol.* **1993**, *9*, 77–85. [CrossRef]
68. Thiele, S.; Mensink, G.B.; Beitz, R. Determinants of diet quality. *Public Health Nutr.* **2004**, *7*, 29–37. [CrossRef] [PubMed]
69. Li, J.; Powdthavee, N. Does more education lead to better health habits? *Evidence from the school reforms in Australia. Soc. Sci. Med.* **2015**, *127*, 83–91. [PubMed]
70. Mías, C.D.; Sassi, M.; Masih, M.E.; Querejeta, A.; Krawchik, R. Deterioro cognitivo leve: Estúdio de prevalência y factores sociodemográficos en la ciudad de Córdoba, Argentina [Mild cognitive impairment: A prevalence and sociodemographic factors study in the city of Córdoba, Argentina]. *Rev. Neurol.* **2007**, *44*, 733–738. [PubMed]
71. Ribeiro, P.C.; Oliveira, B.H.; Cupertino, A.P.; Neri, A.L.; Yassuda, M.S. Desempenho de idosos na bateria cognitiva CERAD: Relações com variáveis sociodemográficas e saúde percebida [Performance of the elderly in the CERAD Cognitive Battery: Relations with socio-demographic variables and perceived health]. *Psicologia, Reflexão e Crítica* **2010**, *23*, 102–109. [CrossRef]
72. Scazufca, M.; Almeida, O.P.; Vallada, H.P.; Tasse, W.A.; Menezes, P.R. Limitations of the Mini-Mental State Examination for screening dementia in a community with low socioeconomic status. *Eur. Arch. Psychiatry Clin. Neurosci.* **2009**, *259*, 8–15. [CrossRef]
73. Wang, X.; Zhou, W.; Ye, T.; Lin, X.; Zhang, J. Alzheimer's Disease Neuroimaging Initiative. Sex Difference in the Association of APOE4 with Memory Decline in Mild Cognitive Impairment. *J. Alzheimer's Dis.* **2019**, *69*, 1161–1169. [CrossRef]
74. Iwata, A.; Iwatsubo, T.; Ihara, R.; Suzuki, K.; Matsuyama, Y.; Tomita, N.; Ikeuchi, T. Effects of sex, educational background, and chronic kidney disease grading on longitudinal cognitive and functional decline in patients in the Japanese Alzheimer's Disease Neuroimaging Initiative study. *Alzheimer's Dement. Transl. Res. Clin. Interv.* **2018**, *4*, 765–774. [CrossRef]
75. Gorelick, P.B.; Scuteri, A.; Black, S.E.; DeCarli, C.; Greenberg, S.M.; Iadecola, C.; Petersen, R.C. Vascular contributions to cognitive impairment and dementia: A statement for healthcare professionals from the American Heart Association/American Stroke Association. *Stroke* **2011**, *42*, 2672–2713. [CrossRef] [PubMed]
76. USEPA. *Mercury Study Report to Congress*; Office of Air Quality Planning and Standards and Office of Research and Development: Washington, DC, USA; USEPA: Washington, DC, USA, 1997.
77. Skalny, A.V.; Skalnaya, M.G.; Tinkov, A.A.; Serebryansky, E.P.; Demidov, V.A.; Lobanova, Y.N.; Skalnaya, O.A. Hair concentration of essential trace elements in adult non-exposed Russian population. *Environ. Monit. Assess.* **2015**, *187*, 677–688. [CrossRef] [PubMed]
78. Mikulewicz, M.; Chojnacka, K.; Gedrange, T.; Górecki, H. Reference values of elements in human hair: A systematic review. *Environ. Toxicol. Pharmacol.* **2013**, *36*, 1077–1086. [CrossRef] [PubMed]
79. Rodushkin, I.; Axelsson, M.D. Application of double focusing sector field ICP-MS for multielemental characterization of human hair and nails. Part II. A study of the inhabitants of northern Sweden. *Sci. Total Environ.* **2000**, *262*, 21–36. [CrossRef]
80. Kazi, T.G.; Memon, A.R.; Afridi, H.I.; Jamali, M.K.; Arain, M.B.; Jalbani, N.; Sarfraz, R.A. Determination of cadmium in whole blood and scalp hair samples of Pakistani male lung cancer patients by electrothermal atomic absorption spectrometer. *Sci. Total Environ.* **2008**, *389*, 270–276. [CrossRef] [PubMed]
81. Brulle, R.J.; Pellow, D.N. Environmental justice: Human health and environmental inequalities. *Annu. Rev. Publ. Health.* **2006**, *27*, 103–124. [CrossRef] [PubMed]
82. Ginter, E.; Simko, V. Women live longer than men. *Bratisl. Lek. Listy* **2013**, *114*, 45–49. [CrossRef]
83. Luy, M.; Gast, K. Do women live longer or do men die earlier? Reflections on the causes of sex differences in life expectancy. *Gerontology* **2014**, *60*, 143–153. [CrossRef]
84. Serdar, M.A.; Akin, B.S.; Razi, C.; Akin, O.; Tokgoz, S.; Kenar, L.; Aykut, O. The correlation between smoking status of family members and concentrations of toxic trace elements in the hair of children. *Biol. Trace Elem. Res.* **2012**, *148*, 11–17. [CrossRef]
85. Zatta, P.; Lucchini, R.; Van Rensburg, S.J.; Taylor, A. The role of metals in neurodegenerative processes: Aluminum, manganese, and zinc. *Brain Res. Bull.* **2003**, *62*, 15–28. [CrossRef]
86. Loef, M.; Von Stillfried, N.; Walach, H. Zinc diet and Alzheimer's disease: A systematic review. *Nutr. Neurosci.* **2012**, *15*, 2–12. [CrossRef] [PubMed]

87. Prodan, C.I.; Holland, N.R. CNS demyelination from zinc toxicity? *Neurology* **2000**, *54*, 1705–1706. [CrossRef] [PubMed]
88. Genuis, S.J.; Kelln, K.L. Toxicant exposure and bioaccumulation: A common and potentially reversible cause of cognitive dysfunction and dementia. *Behav. Neurol.* **2015**, *2015*, 620143. [CrossRef] [PubMed]
89. Zahir, F.; Rizwi, S.J.; Haq, S.K.; Khan, R.H. Low dose mercury toxicity and human health. *Environ. Toxicol. Chem.* **2005**, *20*, 351–360. [CrossRef] [PubMed]
90. Waldron, I.; Johnston, S. Why do women live longer than men? *J. Hum. Stresss* **1976**, *2*, 19–30. [CrossRef] [PubMed]
91. Ninomiya, T.; Ohmori, H.; Hashimoto, K.; Tsuruta, K.; Ekino, S. Expansion of methylmercury poisoning outside minamata: An epidemiological study on chronic methylmercury poisoninig outside of Minamata. *Environ. Res.* **1995**, *70*, 47–50. [CrossRef]
92. Xu, L.; Zhang, W.; Liu, X.; Zhang, C.; Wang, P.; Zhao, X. Circulatory Levels of Toxic Metals (Aluminum, Cadmium, Mercury, Lead) in Patients with Alzheimer's Disease: A Quantitative Meta-Analysis and Systematic Review. *J. Alzheimers Dis.* **2018**, *62*, 361–372. [CrossRef]
93. Meleleo, D.; Notarachille, G.; Mangini, V.; Arnesano, F. Concentration-dependent effects of mercury and lead on Aβ42: Possible implications for Alzheimer's disease. *Eur. Biophys. J.* **2019**, *48*, 173–187. [CrossRef]
94. Food and Agriculture Organization World Health of the United Nations. In Proceedings of the Joint FAO/WHO Expert Committee on Food Additives Sixty-Third Meeting, Geneva, Switzerland, 8–17 June 2004. Available online: http://www.fao.org/3/a-at878e.pdf (accessed on 1 April 2019).
95. JECFA. Evaluation of Certain Contaminants in Food. In *The Seventy-Second Report of Joint FAO/WHO Expert Committee on Food Additives*; WHO technical report series; JECFA: Geneva, Switzerland, 2011; Volume 959, pp. 1–115.
96. WHO. *Environmental Health Criteria 101—Methylmercury*; World Health Organization: Geneva, Switzerland, 1990.
97. Agusa, T.; Kunito, T.; Sudaryanto, A.; Monirith, I.; Kan-Atireklap, S.; Iwata, H.; Tanabe, S. Exposure assessment for trace elements from consumption of marine fish in Southeast Asia. *Environ. Pollut.* **2007**, *145*, 766–777. [CrossRef]
98. Freire, C.; Ramos, R.; Lopez-Espinosa, M.J.; Díez, S.; Vioque, J.; Ballester, F.; Fernández, M.F. Hair mercury levels, fish consumption, and cognitive development in preschool children from Granada, Spain. *Environ. Res.* **2010**, *110*, 96–104. [CrossRef]
99. Díez, S.; Esbrí, J.M.; Tobias, A.; Higueras, P.; Martínez-Coronado, A. Determinants of exposure to mercury in hair from inhabitants of the largest mercury mine in the world. *Chemosphere* **2011**, *84*, 571–577. [CrossRef] [PubMed]
100. Dolbec, J.; Mergler, D.; Larribe, F.; Roulet, M.; Lebel, J.; Lucotte, M. Sequential analysis of hair mercury levels in relation to fish diet of an Amazonian population, Brazil. *Sci. Total Environ.* **2001**, *271*, 87–97. [CrossRef]
101. Koseoglu, E.; Koseoglu, R.; Kendirci, M.; Saraymen, R.; Saraymen, B. Trace metal concentrations in hair and nails from Alzheimer's disease patients: Relations with clinical severity. *J. Trace Elem. Med. Biol.* **2017**, *39*, 124–128. [CrossRef] [PubMed]
102. Vance, D.E.; Ehmann, W.D.; Markesbery, W.R. Trace element imbalances in hair and nails of Alzheimer's disease patients. *Neurotoxicology* **1998**, *9*, 197–208.
103. Koc, E.R.; Ilhan, A.; Ayturk, Z.; Acar, B.; Gürler, M.; Altuntaş, A.; Karapirli, M.; Bodur, A.S. A comparison of hair and serum trace elements in patients with Alzheimer disease and healthy. *Turk. J. Med. Sci.* **2015**, *45*, 1034–1039. [CrossRef]
104. Paglia, G.; Miedico, O.; Cristofano, A.; Vitale, M.; Angiolillo, A.; Chiaravalle, A.E.; DiCostanzo, A. Distinctive pattern of serum elements during the progression of Alzheimer's disease. *Sci. Rep.* **2016**, *6*, 22679. [CrossRef]
105. Li, Y.H.; Zou, X.Y.; Lv, J.M.; Yang, L.S.; Li, H.R.; Wang, W.Y. Trace elements in fingernails of healthy Chinese centenarians. *Biol. Trace Elem. Res.* **2012**, *145*, 158–165. [CrossRef]
106. Inácio, M.M.S. Dados Geoquímicos De Base De Solos De Portugal Continental, Utilizando Amostragem De Baixa Densidade. Ph.D. Thesis, University of Aveiro, Aveiro, Portugal, 2004. Available online: http://hdl.handle.net/10773/18827 (accessed on 20 February 2019).
107. Canadian Guideline: Minister of the Environment (Canada). Soil, Ground Water and Sediment Standards for Use under Part XV.1 of the Environmental Protection Act. Available online: http://www.mah.gov.on.ca/AssetFactory.aspx?did=8993 (accessed on 15 March 2018).

108. Thiombane, M.; Di Bonito, M.; Albanese, S.; Zuzolo, D.; Lima, A.; De Vivo, B. Geogenic versus anthropogenic behaviour and geochemical footprint of Al, Na, K and P in the Campania region (Southern Italy) soils through compositional data analysis and enrichment factor. *Geoderma* **2019**, *335*, 12–26. [CrossRef]
109. Reimann, C.; Garrett, R.G. Geochemical background—concept and reality. *Sci. Total Environ.* **2005**, *350*, 12–27. [CrossRef]
110. Cabral Pinto, M.M.S.; da Silva, E.F.; Silva, M.; Melo-Gonçalves, P.; Candeias, C. Environmental risk assessment based on high-resolution spatial maps of potentially toxic elements sampled on stream sediments of Santiago, Cape Verde. *Geosciences* **2014**, *4*, 297–315. [CrossRef]
111. Cabral Pinto, M.M.S.; da Silva, E.F.; Silva, M.; Melo-Gonçalves, P. Heavy metals of Santiago Island (Cape Verde) top soils: Estimated background value maps and environmental risk assessment. *J. Afr. Earth Sci.* **2015**, *101*, 162–176. [CrossRef]
112. Reis, A.T. Impact of Mercury on Human Health: Aveiro, A Study Case. Master's Thesis, University of Aveiro, Aveiro, Portugal, 2008; p. 112.
113. Dutch Soil Remediation Circular. Esdat Environmental Database Managment Software. 2009. Available online: www.esdat.net (accessed on 10 October 2019).
114. Ministry of Health of the People's Republic of China. The maximum levels of contaminants in foods (GB 2762-2005). Maximum Levels of Contaminants in Foods_Beijing_China—Peoples Republic of_12-11-2014.pdf. 2005. Available online: https://apps.fas.usda.gov/newgainapi/api/report/downloadreportbyfilename?filename=Maximum%20Levels%20of%20Contaminants%20in%20Foods%20_Beijing_China%20-%20Peoples%20Republic%20of_12-11-2014.pdf (accessed on 11 April 2019).
115. Jornal Oficial das Comunidades Europeias, L77/1 de 16/3/2001. p. 1. Available online: https://www.europarl.europa.eu/RegData/PDF/r1049_pt.pdf (accessed on 6 March 2019).

8

# Lithium in Portuguese Bottled Natural Mineral Waters—Potential for Health Benefits?

**Maria Orquídia Neves [1,*], José Marques [1] and Hans G.M. Eggenkamp [2]**

1 Department of Civil Engineering, Architecture and Georesources, CERENA (Centro de Recursos Naturais e Ambiente), Instituto Superior Técnico, University of Lisbon, 1049-001 Lisbon, Portugal; jose.marques@tecnico.ulisboa.pt

2 Department of Petrology and Mineral Resources, Eberhard-Karls University of Tübingen, Schnarrenbergstraße 94-96, 72076 Tübingen, Germany; hans@eggenkamp.info

* Correspondence: orquidia.neves@tecnico.ulisboa.pt

**Abstract:** There is increasing epidemiologic and experimental evidence that lithium (Li) exhibits significant health benefits, even at concentrations lower than the therapeutic oral doses prescribed as treatment for mental disorders. The aim of this study is to determine the content of Li in 18 brands of bottled natural mineral waters that are available on the Portuguese market and from which the sources are found within the Portuguese territory, to provide data for Li intake from drinking water. Analyses of Li were performed by inductively coupled plasma-mass spectrometry. The results indicate highly different Li concentrations in natural mineral waters: one group with low Li concentrations (up to 11 μg Li/L) and a second group with Li concentrations higher than 100 μg/L. The highest Li concentrations (>1500 μg Li/L) were observed in the highly mineralized Na-$HCO_3$ type waters that are naturally carbonated (>250 mg/L free $CO_2$). As a highly bioavailable source for Li dietary intake these natural mineral waters have potential for Li health benefits but should be consumed in a controlled manner due to its Na and $F^-$ contents. The consumption of as little as 0.25 L/day of Vidago natural mineral water (2220 μg Li/L), can contribute up to 50% of the proposed daily requirement of 1 mg Li/day for an adult (70 kg body weight). In future, Li epidemiological studies that concern the potential Li effect or health benefits from Li in drinking water should consider not only the Li intake from tap water but also intake from natural mineral water that is consumed in order to adjust the Li intake of the subjects.

**Keywords:** lithium intake; natural mineral water; health benefits; public health

## 1. Introduction

One of the challenges of the present century is the improvement of human health and to prevent the spreading of diseases. This specifically also applies to mental disorders that occur in all regions and cultures of the world. The most prevalent of these being depression and anxiety, which are estimated to affect nearly one in ten people on the planet. At its worst, depression can lead to suicide [1].

Lithium is the gold standard treatment for several psycho-neurological diseases (e.g., as bipolar disorders). The relationship between Li and health has been shown over time, since the time of the Roman Empire but the clinical history of lithium only started in the mid-19th century when it was used to treat gout, that proved to be ineffective. Its use in the treatment of psycho-neurological began in 1948, by John Cade [2] in Australia. It is administered essentially as carbonate ($Li_2 CO_3$) and at therapeutic doses within the limits of 600 to 1200 mg/day (113–226 mg Li/day [3]). Due to the toxicity of Li there is a rather narrow therapeutic window (between 0.6 and 1.2 mmol/L blood serum) for Li medication, which must be continuously monitored. Lithium is known to interact with neurotransmitters and receptors in the human brain, increasing serotonin levels and reducing brain production of norepinephrine. The mechanisms under which Li acts neurologically have yet to be fully understood, although several hypotheses exist [4].

In the human body the average (total) Li quantity is approximately 7 mg [5] and it is found in various organs and tissues. Schrauzer [6] reported that Li appears to have an important role in fetal development, considering the relatively high Li content within embryos during the early pregnancy. Post-mortem human studies revealed that the cerebellum, cerebrum, and the kidneys retain more Li than other organs [6]. Although, Li was not yet officially recognized as an essential element and no recommended dietary allowance was proposed, in 2002 Schrauzer [6] indicated for a 70-kg adult a provisional daily requirement of 1 mg Li/day (14.3 μg/kg body weigh).

All the essential elements and those considered beneficial must be provided by the diet or nutritional supplements.

Environmental Li exposure and population diet intake can vary greatly from region to region. The available data of daily Li intake point low to average doses in Belgium (8.6 μg/day) [7], the United Kingdom (16 μg/day) [8], France (48.2 μg/day) [9], Hanoi (Vietnam) (36 μg/day) [10] and New Zealand (20–29 μg/day) [11]. Depending on different Li content in food and beverages and to different ingestion habits, the intake could be significantly higher as in Canary Islands (3.6 mg/day [12]) and vary over a wide range [6]. Based on literature data, some grains and vegetables are the primary sources of Li (0.5–3.4 mg Li/kg) as compared to dairy products (0.5 mg/kg) and meat (0.012 mg/kg) [13]. To meet the nutritional demand for Li, Goldstein and Mascitelli [14] suggested that cereal grain products should be fortified with Li or it be added to dietary supplements. Mleczek et al. [15] also investigate mushrooms Li fortification, as food or alternative medicine in various cultures but further studies are necessary to investigate the safety implications of these Li-enriched food items.

Not only solid food, which is the major source of mineral nutrients in the human diet but also drinking water can contribute with variable amounts to the total intake. The role of non-alcoholic beverages was reported in a French diet survey where it was observed that important contributions to Li intake were water (35% for adults), followed by coffee (17%) and other hot beverages (14%) [9]. Lithiated beverages were common in the beginning of the twentieth century, as they were believed to mediate health benefits. One of the most popular soft drinks in the world was launched in 1929; the "*Lithiated Lemon Soda*" that was supplemented with 5 mg Li (as Li citrate/L) until 1948 [16], when it was banned by the government. It was believed to cure alcohol-induced hangover symptoms, make people more energetic and give lust for life and on the top of that shinier hair and brighter eyes [17]. In fact, it is still on the market but since 1936 its name changed to 7UP. In 1949, John Cade discovered that higher Li concentrations were toxic. Nowadays, according Seidel et al. [16] 7UP only contains 1.4 μg Li/L.

In recent years, there have been ecological studies on aggregate data that suggest that long-term intake of low Li concentrations, such as occurring in public drinking water (tap water), may also promote mental health benefits for the general population. This research found that higher concentrations of Li in the tap water are associated with lower suicide mortality rates. These results were observed in Texas (1–160 μg Li/L [18]), Japan (1–60 μg Li/L [19]), Austria (<3–1300 μg Li/L [20]), Greece (0.1–121 μg Li/L [21]) and Lithuania (0.5–35 μg Li/L [22]). This inverse association was found with or without adjustment for additional confounding factors such as the socioeconomic factors that are closely related to suicide. However, in the east of England where Li concentrations in tap water are between 0.1 and 21 μg/L [23], in Italy (0.11–60.8 μg Li/L [24]), Denmark (0.6–30.7 μg Li/L [25]) and in Portugal (<1–191 μg Li/L [26]) the association that high Li concentration in drinking water may protect against suicide was not well supported.

Lithium has also been considered as a possible therapeutic agent for treating chronic neurodegenerative diseases such as Alzheimer's, Parkinson's, and Huntington's diseases [27]. A Li dose of 300 μg/day has been reported to stabilize cognitive impairment in patients with Alzheimer's disease although the underlying molecular mechanisms have not yet been fully understood [28].

Furthermore, there is experimental evidences that Li may have positive effects on bone health [29] and muscle function [30].

From these findings it has been suggested that Li should be added to public drinking water supplies to improve the mental health of the general population, although this would be premature and raises ethical concerns [31] and further research on this subject is necessary.

Lithium doses used for mental health treatment are considerably higher than those obtained from daily exposure to Li in tap water. This raises questions regarding whether (an increased) daily intake of Li from tap water can reduce the risk of suicide or otherwise be beneficial towards the mental health of the population. As reported above, tap water is not the only liquid dietary source of Li and the earlier discussed studies did not take it into account.

Several studies have shown that various bottled waters are rich in Li; the highest values reported (9860 and 5450 μg Li/L) were from bottled waters from Slovakia [32] and Armenia (Hankavan-Lithia: 5.45 mg Li/L). Mineral waters such as Vichy Catalan (1.3 mg Li/L) and Evian (6.6 μg Li/L) were initially also promoted as Li waters based on their Li content [33].

Bottled water plays a more and more important role in daily life. The worldwide bottled water consumption is characterized by a significant growth over the last decade. Its consumption is still increasing [34], especially in developed countries, even though tap water quality is good and several orders of magnitudes less expensive than bottled water. Although all bottled waters might look the same, in fact each natural mineral or spring water has its own distinctive taste, a unique set of properties and a specific chemical composition at the source from where it is extracted, that reflects the geological characteristics of the region and water-rock interactions occurring at depth. Moreover, the defining characteristics of naturally sourced waters are reflected in their protected origin status and are guaranteed by strict European Union (EU) legislation governing the extraction and packaging of the product [35].

European and national legislation distinguish three categories of waters: natural mineral water, spring water and drinking water.

The EU has laid down specific rules for natural mineral and spring waters, which clearly set them apart from drinking water, more commonly known as tap water [36]. Lithium is one of the elements for which no potable water standards are defined in Europe. In Australia Li is listed as a pollutant that causes environmental harm and it is limited to 2.5 mg/L for general irrigation and to a limit of 0.075 g/L for the irrigation of citrus cultures, respectively [37].

Bottled water in the EU is predominantly made up of the natural mineral water category [35]. The bottling and commercialization of natural mineral waters first began in Europe in the mid-16th century, with the mineral waters from Spa in Belgium, Vichy in France, Ferrarelle in Italy, and Apollinaris in Germany.

In 2016, natural mineral water accounted for 83% of EU bottled water retail, with spring water accounting for 14% [34].

In Portugal, like in other countries, natural mineral and spring waters have always aroused great interest due to their exceptional quality, diversity, and health-friendly effects. In 2018, the Portuguese per capita consumption of bottled waters was 134 L/year, which is the 7th highest European Union average consumption (EU average is 119 L/year) [35].

Each natural mineral water in Europe must receive official recognition from the State's competent authority. In the list of natural mineral waters brands recognized by the EU 27 Member States, Portugal accounts for 22 brands [38].

According to Portuguese Legislation [39] natural mineral waters (the subject of this study) are bacteriologically pure waters, of underground circulation, with stable physico-chemical characteristics at the source within the range of natural fluctuations and which may result in possible therapeutic properties or favorable health effects. Spring waters are also natural waters of underground circulation, bacteriologically pure, which do not have the characteristics necessary for qualification as natural mineral waters, provided that at the source they are suitable for drinking.

The present study aims to quantify the Li concentration in Portuguese marketed bottled natural mineral waters to identify if they may represent a significant source of dietary Li intake. Due to its biological impact, it becomes more and more important to understand the Li content in drinking water obtained from different sources, particularly in bottled natural mineral waters, especially in big cities. These data will be very helpful for evaluating the future Li intake via drinking water and diet of Portuguese epidemiological studies related with Li health effects on the population or at an individual level, also contributing to the development of a Li food data base.

## 2. Materials and Methods

### *2.1. Sample Collection*

The eighteen brands of bottled natural mineral waters characterized in this study were selected from the list of the bottled natural mineral waters recognized by Portugal, updated in September 2019 [39]. On this list, the brands Monchique and Chic are, according their label, the same mineral water but commercialized in different packages. Also, the mineral waters Pedras Salgadas and Pedras Levíssima only differ in their $CO_2$ content [40].

All studied natural mineral waters were purchased in supermarkets and local shops. Regarding the type of packaging, nine of the natural mineral waters were bottled in polyethylene terephthalate (PET) and ten in glass bottles. It is not expected that the content of Li and other trace element under study in these mineral waters can be affected by bottled material leaching. According some experiments [32] this becomes problematic for Sb in PET bottles and for Pb, Cr and Ce in glass bottles at acid pH but not for Li.

These natural mineral waters have their catchment area on the Portugal mainland. Only the brand Magnificat issues at the volcanic island of São Miguel, Azores Archipelago (Figure 1).

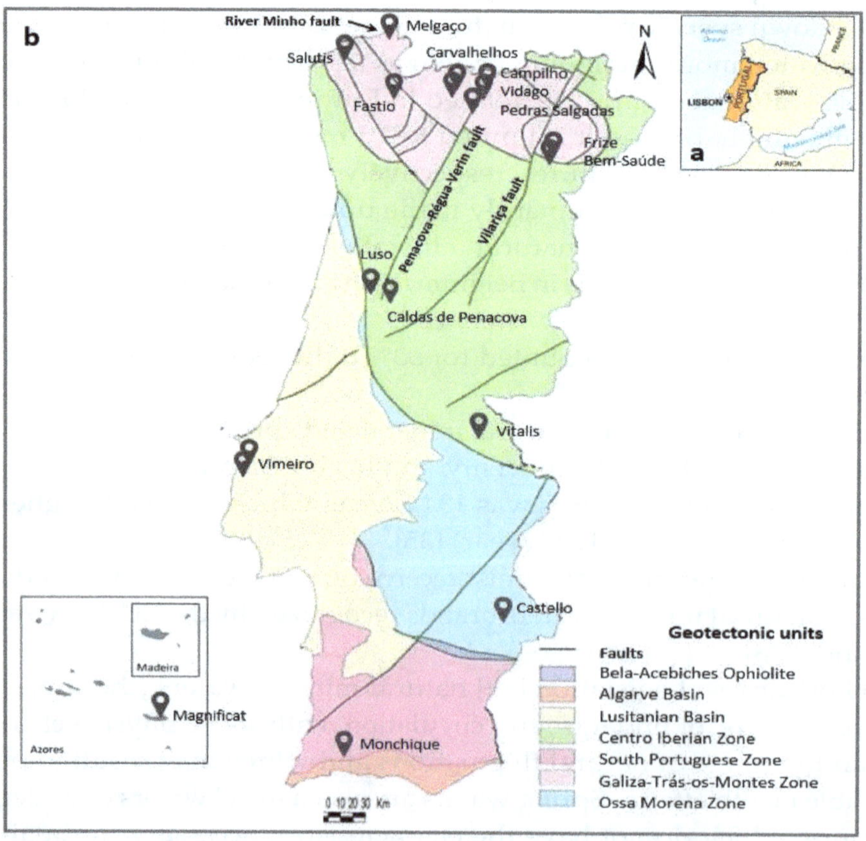


**Figure 1.** (**a**) Portugal geographic location; (**b**) Distribution of the catchment location of the studied Portuguese bottled natural mineral waters (adapted from Reference [41]).

### 2.2. *Sample and Data Analysis*

The pH value of the samples was measured with a glass electrode connected to a WTW pH 325-meter, previously calibrated against buffer solutions at pH 4.0, pH 7.0 and 10.0 (Merck), with an accuracy of the pH measurement of about ±0.05 pH units. Electric conductivity corrected to a temperature of 25 °C was measured using a WTW Cond 330i probe (WTW, Weilheim, Germany), previously calibrated with a 0.01 M KCl standard solution (WTW).

The anions, fluoride, chloride, nitrate, and sulphate were analyzed in non-acidified samples at the Laboratory of Mineralogy and Petrology of the Instituto Superior Técnico (LAMPIST, Lisboa, Portugal) by ion chromatography. A Thermo Scientific™ Dionex™ ICS-900 (Dionex, Sunnyvale, CA, USA) with auto-sampler equipped with a conductivity detector, an IonPac AS22 column and a self-regenerating suppressor using a sodium carbonate–sodium bicarbonate eluent was used for the analyses. Standard solutions (Merck) and ultrapure deionized water (resistivity 18.2 MΩ.cm at 25 °C) produced in a Direct-Q®3 water purification system (Merck Millipore) were used to prepare calibration standards. A multi ion anion IC standard solution (Alfa Aesar Specure) was used every batch of samples as reference and recalibration was performed if the average of triplicate measurements deviated by more than 10%.

The alkalinity was determined by volumetric titration on unfiltered and unacidified samples (50 mL) with a 0.02 N HCL solution, using an automatic Metrohm titrator (titration end point pH 8.3 to determine OH- and/or $CO_2^{3-}$, followed to titration end point pH 4.5 to determine $HCO_3^-$ concentration, according to Standard Method 2.320B [42]).

The dry residue (DR) content,was obtained from evaporation and drying the water sample at 180 °C.

The cations were analyzed at Activation Laboratories, Ltd. (Actlabs) (Vancouver, BC, Canada), an accredited Laboratory, by inductively coupled plasma-mass spectrometry using a Thermo iCAP Q, after the samples had been acidified with concentrated $HNO_3$ (≥65%) to a pH < 2. The Detection and Quantification Limits (LOD/LOQ) in µg/L are as follow: Li (0.02/0.06), Na (2/7), K (1/5), Ca (20/70), Mg (2/6), Mn (0.1/0.3), Rb (0.002/0.006) and Cs (0.001/0.003). Quality controls were achieved according to the Laboratory standards methods and quality assurance and protocols. The standards NIST 1643e and SLRS was used by Actlabs to check the validity and reproducibility of the results.

All samples were analyzed without filtration to represent the water consumed as it is in the bottle.

The charge-balance errors, based on the percentage difference between the total positive charge and the total negative charge (mEq), was below 10% for each sample.

Pearson's correlation coefficient and linear regression with a confidence interval of 95% were calculated with TIBCO®Data Science—Statistica® (Palo Alto, CA, USA) software (version 13.5.017) and Piper diagram projections with RockWorks17 software.

## 3. Results and Discussion

The natural mineral waters discussed in this study are groundwaters abstracted from boreholes and bottled directly at the source. Their distribution across the Portuguese mainland is uneven, with a greater concentration in northern part of the country (Figure 1), mainly caused by the (i) geomorphologic and climatic conditions (higher mountains, colder climate, more rainfall/recharge), (ii) structural characteristics such as the prevalence of deep faults responsible for meteoric waters infiltration at deep and natural mineral waters up flow to the surface as springs and (iii) geological signatures (more fractured and permeable rocks promoting water-rock interaction at depth and developing different water geochemical characteristics (e.g., Reference [43]).

The main characteristics of the studied Portuguese bottled natural mineral waters are presented in Table 1.

**Table 1.** Characteristics of the studied Portuguese bottled natural mineral waters.

| | Mineral Water Brand | Type | pH | EC µS/cm | DR mg/L | $F^-$ µg/L | $Cl^-$ mg/L | $HCO_3^-$ mg/L | $NO_3^-$ mg/L | $SO_4^{2-}$ mg/L | $Na^+$ mg/L | $K^+$ mg/L | $Ca^{2+}$ mg/L | $Mg^{2+}$ mg/L | $Mn^{2+}$ µg/L | $Li^+$ µg/L | $Rb^+$ µg/L | $Cs^+$ µg/L | Water Type | Place of Exploitation |
|---|---|---|---|---|---|---|---|---|---|---|---|---|---|---|---|---|---|---|---|---|
| 1 | Salutis | S | 5.2 | 49 | 32 | 16 | 7.8 | 1.2 | 3.8 | 1.8 | 4.3 | 0.7 | 0.8 | 0.57 | 10 | 1 | 2.4 | 0.04 | Na-Cl | Ferreira—Paredes de Coura |
| 2 | Fastio | S | 6.0 | 35 | 141 | 19 | 3.8 | 8.5 | 2.3 | 0.8 | 4.3 | 0.6 | 1.2 | 0.47 | 1 | <1 | 1.7 | 0.11 | $Na\text{-}HCO_3$ | Chamoim—Terras de Bouro |
| 3 | Monchique | S | 9.4 | 419 | 107 | 1058 | 31.7 | 126.9 | nd | 47.8 | 76.9 | 2.0 | 1.3 | 0.05 | <1 | 1 | 5.1 | 0.03 | $Na\text{-}HCO_3$ | Caldas de Monchique—Monchique |
| 4 | Vimeiro Lisa | S | 7.1 | 83 | 75 | <10 | 8.4 | 21.4 | 0.4 | 3.6 | 9.2 | 0.3 | 5.6 | 1.46 | <1 | 1 | 0.3 | 0.02 | $Na\text{-}HCO_3$ | Maceira—Torres |
| 5 | Vitalis | S | 5.7 | 50 | 71 | 28 | 6.3 | 3.7 | 2.2 | 2.7 | 5.2 | 2.1 | 0.9 | 0.55 | 8 | 1 | 14.5 | 0.34 | Na-Cl | Castelo de Vide |
| 6 | Caldas de Penacova | S | 5.5 | 48 | 41 | <10 | 7.8 | 4.3 | 1.8 | 1.3 | 6.1 | 0.3 | 0.7 | 1.10 | 4 | 2 | 0.9 | 0.19 | Na-Cl | Penacova |
| 7 | Magnificat | NC | 5.0 | 166 | 234 | 533 | 18.0 | 81.0 | 13.6 | 4.2 | 21.9 | 9.7 | 8.4 | 5.04 | 142 | 3 | 33.9 | 0.15 | $Na\text{-}HCO_3$ | Serra do Trigo—Açores |
| 8 | Castello | AC | 5.4 | 793 | 472 | 123 | 46.2 | 361.7 | 18.3 | 18.0 | 34.3 | 0.9 | 94.8 | 26.20 | <1 | 7 | 0.4 | 0.04 | $Ca\text{-}HCO_3$ | Pisões -Moura |
| 9 | Luso | S | 5.6 | 58 | 56 | 31 | 7.3 | 12.2 | 1.5 | 1.4 | 7.4 | 0.8 | 0.8 | 1.81 | 3 | 7 | 3.5 | 0.39 | Na-Cl | Luso—Mealhada |
| 10 | Vimeiro | AC | 5.7 | 1050 | 2291 | 221 | 176.3 | 425.8 | 7.8 | 79.4 | 144.0 | 3.7 | 112 | 27.1 | <1 | 11 | 3.4 | 0.19 | $Na/Ca\text{-}HCO_3$ | Maceira—Torres |
| 11 | Carvalhelhos | S | 7.0 | 248 | 453 | 963 | 2.6 | 141.5 | 0.2 | 6.9 | 55.0 | 1.4 | 5.6 | 0.70 | 1 | 173 | 16.3 | 30.90 | $Na\text{-}HCO_3$ | Carvalhelhos -Boticas |
| 12 | Carvalhelhos | AC | 5.3 | 189 | 208 | 498 | 2.7 | 124.4 | 1.1 | 7.6 | 52.3 | 1.4 | 5.9 | 0.62 | <1 | 177 | 17.9 | 32.60 | $Na\text{-}HCO_3$ | Carvalhelhos -Boticas |
| 13 | Melgaço | NC | 5.7 | 840 | 439 | 657 | 10.6 | 691.7 | 1.2 | 7.5 | 87.5 | 3.6 | 145 | 3.01 | 275 | 600 | 19.5 | 6.73 | $Ca\text{-}HCO_3$ | Quinta do Peso—Melgaço |
| 14 | Campilho | AC | 5.9 | 1892 | 1289 | 4131 | 15.6 | 1288.3 | 0.9 | 8.7 | 428.0 | 27.1 | 37.4 | 10.06 | 1 | 1590 | 215.0 | 222.00 | $Na\text{-}HCO_3$ | Vidago—Chaves |
| 15 | Frize | NC | 6.5 | 2300 | 2336 | 1440 | 100.2 | 1941.0 | 1.6 | nd | 630.0 | 41.1 | 75.6 | 25.05 | 53 | 1760 | 335.0 | 331.00 | $Na\text{-}HCO_3$ | Sampaio—Vila Flor |
| 16 | Pedras Salgadas | NC | 6.1 | 2660 | 1825 | 1265 | 22.9 | 1897.1 | 0.2 | 7.5 | 594.0 | 34.4 | 95.6 | 24.70 | 213 | 1800 | 238.0 | 49.00 | $Na\text{-}HCO_3$ | Pedras Salgadas—Vila Pouca de Aguiar |
| 17 | Bem-Saúde | NC | 6.0 | 2310 | 1600 | 2100 | 90.0 | 1596.0 | 20.3 | 7.2 | 510.0 | 46.0 | 84.0 | 21.00 | 100 | 2000 | na | na | $Na\text{-}HCO_3$ | Sampaio -Vila Flor |
| 18 | Vidago | NC | 6.0 | 1554 | 1797 | 25 | 25.5 | 1869.0 | 0.8 | 7.7 | 624.0 | 53.8 | 73.7 | 14.40 | 20 | 2210 | 415.0 | 253.00 | $Na\text{-}HCO_3$ | Vidago—Chaves |

Notes: S—still natural mineral water; NC—naturally carbonated natural mineral water; AC—artificially carbonated natural mineral water; na—not available; nd—not detected.

The natural mineral waters are commercialized with or without $CO_2$ gas (carbonated or still water, respectively). In some of the sources, the dissolved $CO_2$ can be present due to natural geological processes. If bottled as such it must be labelled as "naturally carbonated natural mineral water" (Table 1: NC waters). For example, in the case of Vidago and Pedras Salgadas waters, following [44] and references therein, their $\delta^{13}C_{CO2}$ values vary between −7.2 and −5.1‰ vs. V-PDB and $CO_2/^3He$ ratios range from $1 \times 10^8$ to $1 \times 10^9$, indicating a deep (upper mantle) source for the $CO_2$. It may also be possible to capture the natural source of $CO_2$ and re-inject it into the water prior to bottling or added it artificially, being, in the second case, described as "artificially carbonated natural mineral water" (Table 1: AC waters). If the waters are subjected to gasification processes, this must be indicated on the label.

Considering the major ions present (expressed as percentage of the total mEq/L), the natural mineral waters under study are mainly of the Na-$HCO_3$ type, follow by the Na-Cl type (Salutis, Vitalis, Caldas de Penacova and Luso) and Ca-$HCO_3$ type (Castello and Melgaço) (Table 1 and Figure 2).

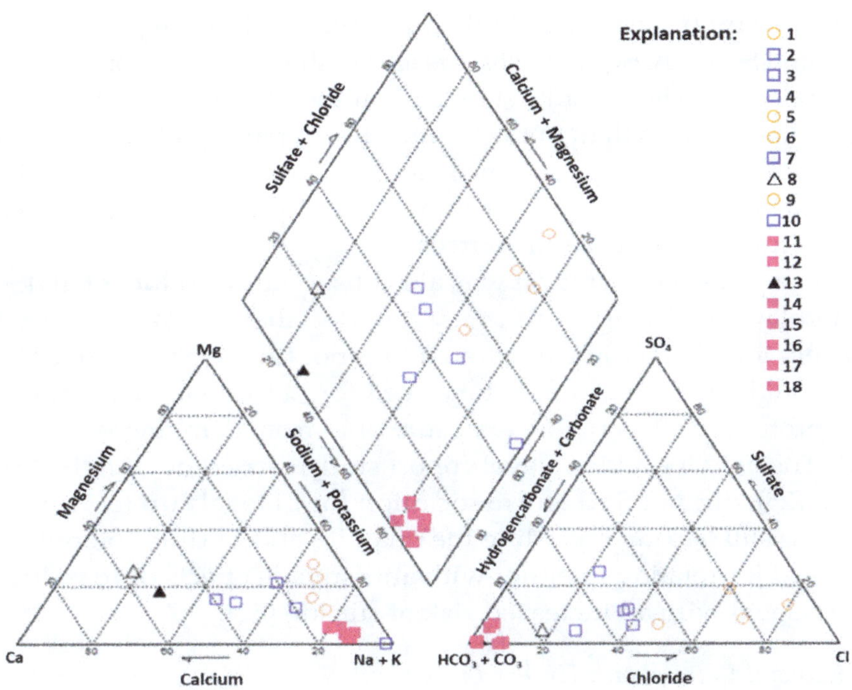


**Figure 2.** Piper Diagram showing the chemical composition of the studied Portuguese natural mineral waters (water type legend: Ca-$HCO_3$ (triangles); Na-Cl (circles), Na-$HCO_3$ (squares); low and high Li content (open and closed symbols, respectively); sample identification (number) as indicated in Table 1).

The natural mineral waters studied present a large range of Li concentrations. It ranges from less than 1 to 2210 μg/L and two groups can be recognized from the dataset: one group with low Li content (up to 11 μg/L) that represents 55.5% of the natural mineral water samples and a second group with higher Li contents (173 to 2210 μg/L).

### 3.1. *Bottled Natural Mineral Waters with Low Li Content*

The group with low Li samples ascribed to natural mineral waters with Li ranging between <1 and 11 μg/L) are also waters with very low (DR < 50 mg/L) or low dissolved salts (50 < DR < 500 mg/L). An exception is Vimeiro, with 2291 mg/L, due to water circulation through evaporite (with halite and gypsum) and carbonate rocks that occur at the contact of diapiric structures [45]. The Li concentrations measured in this group compare to Li concentrations observed in public drinking waters from 54 Portuguese municipalities [26]. According to Neves et al. [46], 75% of the water samples studied by Oliveira et al. [26] show concentrations below 10 μg Li/L, with a median of 4 μg Li/L. This is also lower than the median of 14.9 μg/L detected for Li, in 1785 samples of bottled waters collected in 38 European countries and analyzed for the European Groundwater Geochemistry Atlas [32]. In comparison to Li data available from German beverages such as wine (11.6 ± 1.97 μg/L), beer (8.5 ± 0.77 μg/L), soft and energy drinks (10.2 ± 2.95 μg/L) as reported by Reference [16], these Portuguese natural mineral waters can be considered Li-poor food items. The contribution of these natural mineral waters to the dietary Li supply will not be significantly different from the contribution of tap water.

Little is known on the effects of dietary Li on the Li status in the human body that is estimated either from the concentration in blood (plasma or serum) or from urinary Li excretion. Like it is for sodium, Li homeostasis is adaptively regulated by the kidney and Li is mainly reabsorbed in the proximal tubule. Under normal conditions, approximately 80% of Li is reabsorbed by renal tubes [47]. Excretion of Li occurs within 24 h after its oral intake and is facilitated by the kidneys. A small extent (2–3%) it is also excreted with feces and sweat [48].

Considering that the amount of Li taken by drinking water or food is probably reflected in serum or urinary Li levels, it will be necessary measure such levels, particularly in individuals that do not receive Li therapy.

In a study performed by Bochud et al. [49] both serum and urinary lithium concentrations were measured in Belgians and South Africa participants and in the tap water consumed by them (10 μg Li/L and 0.21 μg Li/L, respectively). Their results showed that the 24-h urinary lithium excretion was higher and more dispersed in the Belgians than in the South African participants (8.2 ± 5.6 and 3.1 ± 4.1 μmol per 24 h) but serum lithium levels were almost identical (0.31 ± 0.16 and 0.32 ± 0.21 μmol Li/L). These observations suggested that serum lithium is tightly regulated even when there are large variations in Li dietary intake from natural sources.

No increase in serum Li concentration was also observed by Seidel et al. [50] in the group that received low Li mineral water (1.7 μg/L) and they reported that the 24-h urinary Li excretion exceed the total uptake. At very low dietary intake, filtered Li is not fully reabsorbed [51] and Seidel et al. [50] suggested that there could be a minimum dietary need for Li to ensure a positive Li balance.

Concerning the protective effect of low exposure of Li from drinking water, future epidemiologic studies are required. The only individual-level cohort study carried out, on the Danish population for the period 1991–2012 [25], did not find any association for Li levels up to 31 μg/L in drinking water. However, information is still lacking regarding the quantity and/or duration of low Li exposure that is necessary to be achieved for relating Li either with anti-suicide effects or to reduced aggressivity and impulsivity, both associated with an increased risk of suicide [52].

### *3.2. Bottled Natural Mineral Waters with Higher Li Content*

The group of the natural mineral waters with higher Li contents (173 to 2210 μg/L) are mostly of the Na-$HCO_3$ water type, except in the case of Melgaço that is of the Ca-$HCO_3$ type (ascribed to granodioritic rocks [53]). The highest Li content is measured in Vidago (Table 1). These waters are mainly exploited from Hercynian granitic rocks in the north of Portugal, from within the Minho and Trás-os-Montes regions, in the Geotectonic unit Galiza-Trás-os-Montes Zone (Figure 1). The catchment areas of these Li-rich waters is well correlated with regional fault systems, such as the "Penacova-Régua-Verin Fault"(Campilho, Vidago and Pedras Salgadas), the "Vilariça Fault" (Frize and Bem-Saúde) and the "River Minho Fault" (Melgaço) (Figure 1), since they normally provide the best conditions for the rising of fluids from deep crustal zones [43–45,53,54]. With exception of Campilho, all these Li-rich natural mineral waters are naturally carbonated, with free $CO_2$ contents above 250 mg/L, identified on the bottle label as a "gasocarbonic" water. Carbonated waters were preferred by the consumers, as in addition to the slightly acidic taste, it stimulate the papillae tastes, favors digestion and especially if they are sodium carbonated water, they help to neutralize the acidity of the stomach [55].

Natural mineral waters with Li contents above 1500 μg/L present also high dissolved solids (DR > 1000 mg/L) and they are rich in sodium (Na > 400 mg/L), potassium (K > 27 mg/L) and magnesium (Mg > 10 mg/L). A good correlation was observed between Li and Na ($r = 0.966$, $p < 0.05$) and between Li and K ($r = 0.976$ $p < 0.05$), as also reported by Reference [50]. As their Na contents are higher than

200 mg/L they are classified as "water with sodium" [36] and so a regular consumption of these waters is not recommended for individuals that are on a low sodium diet.

Other elements that stand out in this Li rich group of bottled natural mineral waters, are fluorine ($F^-$), rubidium (Rb) and cesium (Cs) (Table 1). Fluorine ranges from 1265 μg/L (Pedras Salgadas) to 4131 μg/L (Campilho), which according to Reference [36] can be also classified as "water with fluorine" ($F^- > 1$ mg/L). In this group the Vidago natural mineral water with only 25 μg $F^-$/L, is the exception. According to Calado and Almeida [56], this anomalous $F^-$ content, that does not result from the dissolution of fluorite, as it was supposed but has a deep genesis, related to the circulation of mineralizing fluids meso and/or infra-crustal origin. These fluids will be related to the lifting crustal phenomena (uplift) that mainly affect the north and center of the country [56].

The European Union (EU) Directive [57] defines a maximum admissible concentration of 5 mg $F^-$/L in natural mineral waters and requires that $F^-$ concentrations above 1.5 mg/L are indicated on the label, following the general EU Directive [58] for drinking water. Excessive $F^-$ (>1.5 mg/L) incurs a risk of possible dental fluorosis, especially when the water is drunk regularly by children below the age of 7 and should be avoided also by adults.

Rubidium and Cs are also elements that are reported, together with Li, as a characteristic for some mineral waters related with water circulation through Hercynian granites. The maps of Rb (maximum 673 μg/L) and Cs (maximum 415 μg/L) in European bottled water [32] also shows up flow sites in northern Portugal and France (Massif Central) related to young granitic intrusions and complex type pegmatites of the LTC (Li-Ca-Ta) family.

It should be noted here that Li in water is present in aqueous solution in the form of hydrated $Li^+$ ions. Being in solution, it may assimilate in the human body more easily as compared to solid food or with the salts as used in medication. If the beneficial effect of Li could be achieved at safer lower doses, increasing its dietary intake would offer an approach to the prevention of the incidence of mental disorders and a reduction in suicide attempts, aggressive behavior and conducted disorder as reported [18,59–61]. This way, it may be possible to reduce the amounts necessary to be administered and could also reduce side effects.

A recent study with healthy male volunteers [50] indicates that Li derived from medium to high Li mineral water (171 and 1724 μg/L, respectively), is highly bioavailable. The consumption of the mineral water with higher Li concentrations resulted in a peak serum content of up to 10–12 μmol Li/L which did not return to baseline levels within 24 h. Also, the total urinary excretion of Li was positively associated with Li uptake via mineral water. The data suggested that some minerals waters are an important and bioavailable Li source for human intake. If confirmed, these findings have public health relevance and emphasize the need for more data on Li concentrations in drinking water, as bottled natural mineral water, and their intake in a daily basis.

Among the Portuguese bottled natural mineral waters, Campilho, Frize, Pedras Salgadas, Bem-Saúde and Vidago waters, are the ones that can contribute more intensively to a significant Li absorption or even to reach the provisional daily adult intake (1 mg Li/day).

A consumption of 0.5 L of each of these waters may provide between 0.75 and 1.1 mg Li/day, assuring adequate Li intakes, especially people that are at risk of Li nutritional deficiencies. So, as a source of bioavailable Li, its amount and frequency of ingestion cannot be ignored in ecologic/epidemiologic Portuguese studies evaluating relations with intake of natural doses of Li via drinking water or diet and mental health benefits.

It must be realized that the studied Li mineral waters are also rich in other elements as $F^-$ and Na, which can limit recommendations to be consumed regularly as source of Li for all individuals. For example, Vidago natural mineral water, due its lower $F^-$ content could be indicated as one that can be used for this purpose but attention should be put to its $Na^+$ concentration, as excessive Na in the human diet can harm the kidneys and acerbate high blood pressure that is associated with hypertension and coronary diseases in some individuals [62]. On the other hand, a higher Na content in the water can also modify the Li absorption process, as an increased of Na intake may increase the excretion of Li [48].

Considering a consumption of 0.5 L, Vidago natural mineral water may provide up to 315 mg Na/day, which is approximately 15.7% of the Daily Value (DV: 2 g Na/day) recommended by World Health Organization (WHO), for consumption. As a general guide, 5% DV or less of Na per serving is considered low and 20% DV or more is considered high [63].

However, Vidago natural mineral water also contains higher amounts of bicarbonate ion ($HCO_3^-$) instead of chloride ($Cl^-$) as the anion associated with the $Na^+$ cation. This is relevant because it is established that the effect of sodium in blood pressure depends on the corresponding anion; the blood pressure effect of sodium bicarbonate is much lower than that of equivalent amounts of sodium chloride [64]. A crossover, non-blinded study that evaluated 17 individuals ingesting 0.5 L/day of Pedras Salgadas and Vitalis natural mineral water (on Table 1) for 7 weeks, shows no effect on blood pressure values on normotensive individuals [64]. Another study conducted by Schorr et al. [65] also found that the ingestion of $HCO_3^-$ rich water (1.5 L/day) had hypotensive effects in and elderly population. However, this study was not replicate with hypertensive individuals, more prone to salt sensitivity. So, further research is necessary to improve knowledge on human body interactions with these anions and the benefits of these Li-rich natural mineral waters for our mental health.

Vidago natural mineral water is usually sold and consumed in 0.25 L bottles. The regular daily consumption of this natural mineral water volume can also provide a continuous supply of 550 μg Li per drink (half of the provisional Li daily intake) and Na with lesser health concerns, as it supplies 160 mg Na (8% DV) per serving. In that condition this mineral natural water will have potential to be regarded as an available natural nutritional Li supplement for suggested health benefits.

## 4. Conclusions

Depending on the chemical composition, natural mineral waters may significantly contribute to the recommended daily intake of minerals and provide us with a natural source of healthy hydration.

The health effects of Li in drinking water, both tap waters and bottled waters is not fully understood yet but there are indications that natural moderately high Li contents may be beneficial to the mental health situation of the population. In the present study the Li content from a set of bottled natural mineral waters from Portugal was evaluated. Based in the Li content two sets of natural mineral waters could be recognized. A set with low Li concentrations (up to 11 μg/L) that will not have any different effects on the dietary Li supply as compared to the contribution of tap water to the dietary contribution of Li and a second set with higher Li concentrations (173 to 2210 μg/L). The natural mineral water with higher Li contents (>1500 μg/L) is highly mineralized, mostly $Na$-$HCO_3$ type waters and naturally carbonated ($CO_2$-rich waters with > 250 mg/L free $CO_2$). These Li-rich natural mineral waters can be a source of bioavailable Li and its consumption cannot be ignored in studies evaluating the intake of natural doses of Li via drinking water or diet. It thus is important to take into consideration for studies towards Li health effects or benefits for the population or at an individual level. Among the studied bottled natural mineral waters, the consumption of 0.25 L/day of Vidago natural mineral water can contribute significantly to reach the proposed provisional Li daily intake.

It should be noted that the dose of Li ingested through bottled natural mineral water is significantly less than the recommended doses for therapeutic purposes and for that reason these waters can be also regarded as a natural nutritional supplement.

Lithium's interaction in human biochemistry is complex and should be a subject for continuous research to demonstrate the possible clinical effects of natural low-dose Li intake on mental health of the public.

**Author Contributions:** Conceptualization, M.O.N. and H.G.M.E.; methodology, M.O.N. and H.G.M.E.; validation, M.O.N. and H.G.M.E.; formal analysis of the literature, M.O.N.; investigation, M.O.N. and H.G.M.E.; resources,

J.M.; writing original draft, M.O.N.; review and extend the original draft, M.O.N., J.M. and H.G.M.E.; visualization, M.O.N. and J.M.; funding acquisition, J.M. All authors have read and agreed to the published version of the manuscript.

**Acknowledgments:** The authors would like to thank Antunes da Silva (SuperBock group) for helpful discussions during the development of the study. The authors are also grateful to Sumol + Compal (Companhia Portuguesa de Conservas Alimentares, SA) for the availability of data of the bottled natural mineral water "Bem-Saúde." An early draft of the manuscript was critically read by four anonymous reviewers and we gratefully acknowledge their contributions.

## References

1. World Health Statistics 2016: Monitoring Health for the SDGs. Chapter 6. pp. 29–42. Available online: https://www.who.int/gho/publications/world_health_statistics/2016/en/ (accessed on 12 May 2020).
2. Cade, J.F. Lithium salts in the treatment of psychotic excitement. *Med. J. Aust.* **1949**, *2*, 349–352. [CrossRef] [PubMed]
3. Young, W. Review of lithium effects on brain and blood. *Cell Trans.* **2009**, *18*, 951–975. [CrossRef] [PubMed]
4. Oruch, R.; Elderbi, M.A.; Khattab, H.A.; Pryme, I.F.; Lund, A. Lithium: A review of pharmacology, clinical uses, and toxicity. *Eur. J. Pharmacol.* **2014**, *740*, 464–473. [CrossRef] [PubMed]
5. Emsley, J. *The Elements*, 2nd ed.; Clarendon Press: Oxford, UK, 1991; p. 107.
6. Schrauzer, G.N. Lithium: Occurrence, dietary intakes, nutritional essentiality. *J. Am. Coll. Nutr.* **2002**, *21*, 14–21. [CrossRef]
7. Cauwenbergh, R.V.; Hendrix, P.; Robberecht, H.; Deelstra, H. Daily dietary lithium intake in Belgium using duplicate portion sampling. *Eur. Food Res. Technol.* **1991**, *208*, 153–155. [CrossRef]
8. Ysart, G.; Miller, P.; Crews, H.; Robb, P.; Baxter, M.; De L'Argy, C.; Lofthouse, S.; Sargent, C.; Harrison, N. Dietary exposure estimates of 30 elements from the UK Total Diet Study. *Food Addit. Contam.* **1999**, *16*, 391–403. [CrossRef]
9. ANSES. Second French Total Diet Study (FTDS2) Report 1. Inorganic Contaminants, Minerals, Persistent Organic Pollutants, Mycotoxins and Phytoestrogens. 2011, p. 60. Available online: https://www.anses.fr/en/system/files/PASER2006sa0361Ra1EN.pdf (accessed on 12 May 2020).
10. Marcussen, H.; Jensen, B.H.; Petersen, A.; Holm, P.E. Dietary exposure to essential and potentially toxic elements for the population of Hanoi, Vietnam. *Asia Pac. J. Clin. Nutr.* **2013**, *22*, 300–311.
11. Pearson, A.J.; Ashmore, E. Risk assessment of antimony, barium, beryllium, boron, bromine, lithium, nickel, strontium, thallium and uranium concentrations in the New Zealand diet. *Food Addit. Contam. Part A* **2020**, *37*, 451–464. [CrossRef]
12. González-Weller, D.; Rubio, C.; Gutiérrez, A.J.; González, G.L.; Mesa, J.M.C.; Gironés, C.R.; Ojeda, A.B.; Hardisson, A. Dietary intake of barium, bismuth, chromium, lithium, and strontium in a Spanish population (Canary Islands, Spain). *Food Chem. Toxicol.* **2013**, *62*, 856–868. [CrossRef]
13. Weiner, M.I. Overview of lithium toxicity. In *Lithium in Biology and Medicine*; Schrauzer, G.N., Klippel, K.F., Eds.; VCH Verlag: Weinheim, Germany, 1991; pp. 83–99.
14. Goldstein, M.R.; Mascitelli, L. Is violence in part a lithium deficiency? *Med. Hypotheses* **2016**, *89*, 40–42. [CrossRef]
15. Mleczek, M.; Siwulski, M.; Rzymski, P.; Budzyńska, S.; Gąsecka, M.; Kalač, P.; Niedzielski, P. Cultivation of mushrooms for production of food biofortified with lithium. *Eur. Food Res. Technol.* **2017**, *243*, 1097–1104. [CrossRef]
16. Seidel, U.; Jans, K.; Hommen, N.; Ipharraguerre, I.R.; Lüersen, K.; Birringer, M.; Rimbach, G. Lithium Content of 160 Beverages and Its Impact on Lithium Status in Drosophila melanogaster. *Foods* **2020**, *9*, 795. [CrossRef] [PubMed]
17. El-Mallakh, R.S.; Roberts, R.J. Lithiated Lemon-Lime Sodas. *Am. J. Psychiatr.* **2007**, *164*, 1662. [CrossRef] [PubMed]
18. Schrauzer, G.N.; Shrestha, K.P. Lithium in drinking water and the incidences of crimes, suicides, and arrests related to drug addictions. *Biol. Trace Elem. Res.* **1990**, *25*, 105–113. [CrossRef] [PubMed]

19. Ohgami, H.; Terao, T.; Shiotsuki, I.; Ishii, N.; Iwata, N. Lithium levels in drinking water and risk of suicide. *Br. J. Psychiatr.* **2009**, *194*, 464–465. [CrossRef] [PubMed]
20. Kapusta, N.D.; Mossaheb, N.; Etzersdorfer, E.; Hlavin, G.; Thau, K.; Willeit, M.; Praschak-Rieder, N.; Praschak-Rieder, N.; Sonneck, G.; Leithner-Dziubas, K. Lithium in drinking water and suicide mortality. *Br. J. Psychiatr.* **2011**, *198*, 346–350. [CrossRef]
21. Giotakos, O.; Nisianakis, P.; Tsouvelas, G.; Giakalou, V. Lithium in the public water supply and suicide mortality in Greece. *Biol. Trace Elem. Res. Biolog.* **2013**, *156*, 376–379. [CrossRef]
22. Liaugaudaite, V.; Mickuviene, N.; Raskauskiene, N.; Naginiene, R. Lithium levels in the public drinking water supply and risk of suicide: A pilot study. *J. Trace Elem. Med. Biol.* **2017**, *43*, 197–201. [CrossRef]
23. Kabacs, N.; Memom, A.; Obinwa, T.; Stochl, J.; Perez, J. Lithium in drinking water and suicide rates across the East of England. *Br. J. Psychiatr.* **2011**, *198*, 406–407. [CrossRef]
24. Pompili, M.; Vichi, M.; Dinelli, E.; Pycha, R.; Valera, P.; Albanese, S.; Lima, A.; De Vivo, B.; Cicchella, D.; Fiorillo, A.; et al. Relationships of local lithium concentrations in drinking water to regional suicide rates in Italy. *World J. Biol. Psychiatr.* **2015**, *16*, 567–574. [CrossRef]
25. Knudsen, N.N.; Schullehner, J.; Hansen, B.; Jørgensen, L.F.; Kristiansen, S.M.; Voutchkova, D.D.; Gerds, T.A.; Andersen, P.K.; Bihrmann, K.; Grønbæk, M.; et al. Lithium in Drinking Water and Incidence of Suicide: A Nationwide Individual-Level Cohort Study with 22 Years of Follow-Up. *Int. J. Environ. Res. Public Health* **2017**, *14*, 627. [CrossRef] [PubMed]
26. Oliveira, P.; Zagalo, J.; Madeira, N.; Neves, O. Lithium in Public Drinking Water and Suicide Mortality in Portugal: Initial Approach. *Acta Med. Port.* **2019**, *32*, 47–52. [CrossRef] [PubMed]
27. Marmol, F. Lithium: Bipolar disorder and neurodegenerative diseases. Possible cellular mechanisms of the therapeutic effects of lithium. *Prog. NeuroPsychopharmacol. Biol. Psychiatry* **2008**, *32*, 1761–1771. [CrossRef]
28. Andrade Nunes, M.; Viel, T.A.; Buck, H.S. Microdose lithium treatment stabilized cognitive impairment in patients with Alzheimer's disease. *Curr. Alzheimer Res.* **2013**, *10*, 104–107.
29. Liu, B.; Wu, Q.; Zhang, S.; Del Rosario, A. Lithium Use and Risk of Fracture: A Systematic Review and Meta-Analysis of Observational Studies. *Osteoporos. Int.* **2018**, *30*, 257–266. [CrossRef] [PubMed]
30. Jung, S.; Koh, J.; Kim, S.; Kim, K.E. Effect of Lithium on the Mechanism of Glucose Transport in Skeletal Muscles. *J. Nutr. Sci. Vitaminol.* **2017**, *63*, 365–371. [CrossRef] [PubMed]
31. Ng, J.; Manne Sjöstrand, M.; Eyal, N. Adding Lithium to Drinking Water for Suicide Prevention—The Ethics. *Public Health Ethics* **2019**, *12*, 274–286. [CrossRef]
32. Reimann, C.; Birke, M. *Geochemistry of European Bottled Water*; Borntraeger Science Publishers: Stuttgart, Germany, 2010; pp. 46–159.
33. Schafer, U. Evaluation of beneficial and adverse effects on plants and animals following lithium deficiency and supplementation, and on humans following lithium treatment of mood disorders. *Trace Elem. Electrolytes* **2012**, *29*, 91–112. [CrossRef]
34. STATISTA. Per Capita Consumption of Bottled Water in Europe in 2017, by Country. Available online: https://www.statista.com/statistics/455422/bottled-water-consumption-in-europe-per-capita/ (accessed on 15 May 2020).
35. EFBW. Natural Waters: The Natural Choice for Hydration. Available online: https://www.efbw.org/ (accessed on 15 May 2020).
36. Council Directive 2009/54/EC/18-6-2009 on the exploitation and marketing of natural mineral waters. *Off. J. EU* **2009**, *L164*, 45–58.
37. ANZ Guidelines. ANZECC-ARMCANZ-2000-Guidelines-Vol1. Available online: https://www.waterquality.gov.au/media/57 (accessed on 20 May 2020).
38. Publications Office of the EU. List of Natural Mineral Waters Recognized by Member States. Available online: https://op.europa.eu/en/publication-detail/-/publication/b9453284-4083-11e3-b4f5-01aa75ed71a1/language-en/format-PDF/source-140676132 (accessed on 15 May 2020).
39. Lei 54/2015. Bases do regime jurídico da revelação e do aproveitamento dos recursos geológicos existentes no território nacional, incluindo os localizados no espaço marítimo nacional. *DRE* **2015**, *54*, 4296–4308.
40. Super Bock Group. Águas. Água das Pedras. *Ficha Técnica.* Available online: https://www.superbockgroup.com/produto/aguas/ (accessed on 15 May 2020).

41. Lourenço, C.; Pascoal, R. O Estudo Metagenómico das Águas Minerais Naturais tendo em vista o Reconhecimento Científico das Vocações Terapêuticas. *Bol. Minas (Ed. Espec. Term.)* **2018–2019**, *53*, 39–53.
42. Greenberg, A.; Clesceri, L.S.; Andrew, D.; Eaton, A.D. *Standard Methods for the Examination of Water and Wastewater*, 18th ed.; American Public Health Association, American Water Works association, Water Environment Federations: Washington, DC, USA, 1992; pp. 25–27.
43. Marques, J.M.; Carreira, P.M.; Neves, O.; Espinha Marques, J.; Teixeira, J. Revision of the hydrogeological conceptual models of two Portuguese thermomineral water systems: Similarities and differences. *Sustain. Water Resour. Manag.* **2019**, *5*, 117–133. [CrossRef]
44. Marques, J.M.; Carreira, P.M.; Aires-Barros, L.A.; Monteiro Santos, F.A.; Antunes da Silva, M.; Represas, P. Assessment of Chaves Low-Temperature $CO_2$-Rich Geothermal System (N-Portugal) Using an Interdisciplinary Geosciences Approach. *Geofluids* **2019**, *2019*, 1379093. [CrossRef]
45. Lourenço, C.; Ribeiro, L.; Cruz, J. Classification of natural mineral and spring bottled waters of Portugal using Principal Component Analysis. *J. Geoch. Explor.* **2010**, *107*, 362–372. [CrossRef]
46. Neves, O.; Machete, I.; Marques, J.M.; da Silva, J.A.L.; Simões do Couto, F. Lítio em águas engarrafadas e de abastecimento público portuguesas. *Com. Geol.* **2015**, *102*, 103–106.
47. Rybakowski, J.; Drogowska, J.; Abramowicz, M.; Chłopocka-Woźniak, M.; Czekalski, S. The effect of long-term lithium treatment on kidney function. *Psychiatr. Pol.* **2012**, *246*, 627–636.
48. Szklarska, D.; Rzymski, P. Is Lithium a Micronutrient? From Biological Activity and Epidemiological Observation to Food Fortification. *Biol. Trace Elem. Res.* **2019**, *189*, 18–27. [CrossRef]
49. Bochud, M.; Staessen, J.A.; Woodiwiss, A.; Norton, G.; Maillard, M.; Burnier, M. Context dependency of serum and urinary lithium: Implications for measurement of proximal sodium reabsorption. *Hypertension* **2007**, *49*, e34. [CrossRef]
50. Seidel, U.; Baumhof, E.; Hägele, F.A.; Bosy-Westphal, A.; Birringer, M.; Rimbach, G. Lithium-Rich Mineral Water is a Highly Bioavailable Lithium Source for Human Consumption. *Mol. Nutr. Food Res.* **2019**, *63*, e1900039. [CrossRef]
51. Uwai, Y.; Arima, R.; Takatsu, C.; Furuta, R.; Kawasaki, T.; Nabekura, T. Sodium-Phosphate Cotransporter Mediates Reabsorption of Lithium in Rat Kidney. *Pharmacol. Res.* **2014**, *87*, 94–98. [CrossRef]
52. Kovacsics, C.E.; Gottesman, I.I.; Gould, T.D. Lithium's Antisuicidal Efficacy: Elucidation of Neurobiological Targets Using Endophenotype Strategies. *Annu. Rev. Pharmacol. Toxicol.* **2009**, *49*, 175–198. [CrossRef] [PubMed]
53. Carreira, P.M.; Marques, J.M.; Carvalho, M.R.; Nunes, D.; Antunes da Silva, M. Carbon isotopes and geochemical processes in $CO_2$-rich cold mineral water, N-Portugal. *Environ. Earth Sci.* **2014**, *71*, 2941–2953. [CrossRef]
54. Lourenço, M.C. Modelação Estatística das Águas Gasocarbónicas de Vidago e Pedras Salgadas. Master's Thesis, Instituto Superior Técnico, Lisbon, Portugal, April 2000.
55. APIAM. Águas Minerais Naturais e Águas de Nascente—Livro Branco. Available online: http://www.apiam.pt/publicacoes/Livro-Branco-APIAM/-/47/23/175 (accessed on 18 May 2020).
56. Calado, C.; Almeida, C. Geoquímica do Flúor em Águas Minerais da Zona Centro Ibérica. *Memórias* **1993**, *3*, 319–323.
57. Council Directive 2003/40/EC/16-5-2003/ establishing the list, concentration limits and labelling requirements for the constituents of natural mineral waters and the conditions for using ozone-enriched air for the treatment of natural mineral waters and spring waters. *Off. J EU* **2003**, *L126*, 34–39.
58. Council Directive 98/83/EC/3-11-1998/on the quality of water intended for human consumption. *Off. J EU* **1998**, *L330*, 32–54.
59. Campbell, M.; Adams, P.B.; Small, A.M.; Kafantaris, V.; Silva, R.R.; Shell, J.; Perry, R.; Overall, J.E. Lithium in hospitalized aggressive children with conduct disorder: A double-blind and placebo-controlled study. *J. Am. Acad. Child Adolesc. Psychiatr.* **1995**, *34*, 445–453. [CrossRef]
60. Cipriani, A.; Pretty, H.; Hawton, K.; Geddes, J. Lithium in the prevention of suicidal behavior and all-cause mortality in patients with mood disorders: A systematic review of randomized trials. *Am. J. Psychiatr.* **2005**, *162*, 1805–1819. [CrossRef]

61. Jones, R.; Arlidge, J.; Gillham, R.; Reagu, S.; van den Bree, M.; Taylor, P. Efficacy of mood stabilisers in the treatment of impulsive or repetitive aggression: Systematic review and meta-analysis. *Br. J. Psychiatr.* **2011**, *198*, 93–98. [CrossRef]
62. Rodriguez, C.J.; Bibbins-Domingo, K.; Jin, Z.; Daviglus, M.L.; Goff, D.C., Jr.; Jacobs, D.R. Association of sodium and potassium intake with left ventricular mass: Coronary artery risk development in young adults. *Hypertension* **2011**, *58*, 410–416. [CrossRef]
63. U.S. Food & Drug Administration. Sodium in Your Diet—Use the Nutrition Facts Label and Reduce Your Intake. Available online: https://www.fda.gov/food/nutrition-education-resources-materials/sodium-your-diet (accessed on 18 September 2020).
64. Santos, A.; Martins, M.J.; Severo, M.; Guimarães, J.; Azevedo, I. Ingestão de água mineral natural gasocarbónica hipersalina e pressão arterial. *Rev. Port. Cardiol.* **2010**, *29*, 159–172.
65. Schorr, U.; Distler, A.; Sharma, A.M. Effect of sodium chloride and sodium bicarbonate-rich mineral water on blood pressure and metabolic parameters in elderly normotensive individuals: A randomized double-blind crossover trial. *J. Hypertens.* **1996**, *14*, 131–135. [PubMed]

9

# Lead Toxicity: Health Hazards, Influence on Food Chain and Sustainable Remediation Approaches

**Amit Kumar [1], Amit Kumar [2,*], Cabral-Pinto M.M.S. [3], Ashish K. Chaturvedi [4], Aftab A. Shabnam [2], Gangavarapu Subrahmanyam [2], Raju Mondal [5], Dipak Kumar Gupta [6], Sandeep K. Malyan [7], Smita S. Kumar [8], Shakeel A. Khan [9] and Krishna K. Yadav [10]**

[1] School of Hydrology and Water Resources, Nanjing University of Information Science and Technology, Nanjing 210044, China; amitkdah@nuist.edu.cn
[2] Central Muga Eri Research and Training Institute, Central Silk Board, Jorhat, Assam 785000, India; aftab.csb@gov.in (A.A.S.); subrahmanyamg.csb@gov.in (G.S.)
[3] Geobiotec Research Centre, Department of Geosciences, University of Aveiro, 3810-193 Aveiro, Portugal; marinacp@ua.pt
[4] Water Management (Agriculture) Division, Centre for Water Resources Development and Management, Kozhikode, Kerala 673571, India; ashispc@cwrdm.org
[5] Central Sericultural Germplasm Resources Centre (CSGRC), Central Silk Board, Ministry of Textiles, Thally Road, Hosur, Tamil Nadu 635109, India; rmtarapur@gmail.com
[6] ICAR-Central Arid Zone Research Institute Regional Research Station Pali Marwar, Rajasthan 342003, India; deepak.gupta@icar.gov.in
[7] Institute of Soil, Water and Environmental Sciences, Agricultural Research Organization (ARO), The Volcani Center, Rishon LeZion 7505101, Israel; sanm@volcani.agri.gov.il
[8] Department of Environment science, J.C. Bose University of Science & Technology, YMCA, NH-2, Sector-6, Mathura Road, Faridabad, Haryana 121006, India; smita@jcboseust.ac.in
[9] Centre for Environment Science and Climate Resilient Agriculture, ICAR-Indian Agricultural Research Institute, New Delhi 110012, India; shakeel.khan@icar.gov.in
[10] Institute of Environment and Development Studies, Bundelkhand University, Kanpur Road, Jhansi 284128, India; envirokrishna@gmail.com
* Correspondence: amitkumar.csb@gov.in

**Abstract:** Lead (Pb) toxicity has been a subject of interest for environmental scientists due to its toxic effect on plants, animals, and humans. An increase in several Pb related industrial activities and use of Pb containing products such as agrochemicals, oil and paint, mining, etc. can lead to Pb contamination in the environment and thereby, can enter the food chain. Being one of the most toxic heavy metals, Pb ingestion via the food chain has proven to be a potential health hazard for plants and humans. The current review aims to summarize the research updates on Pb toxicity and its effects on plants, soil, and human health. Relevant literature from the past 20 years encompassing comprehensive details on Pb toxicity has been considered with key issues such as i) Pb bioavailability in soil, ii) Pb biomagnification, and iii) Pb- remediation, which has been addressed in detail through physical, chemical, and biological lenses. In the review, among different Pb-remediation approaches, we have highlighted certain advanced approaches such as microbial assisted phytoremediation which could possibly minimize the Pb load from the resources in a sustainable manner and would be a viable option to ensure a safe food production system.

**Keywords:** lead toxicity; lead contamination; health hazards; remediation

## 1. Introduction

Lead (Pb) is a highly noxious, non-disintegrative heavy metal with a bluish-gray color, an atomic number of 82, molecular weight 207.2, density 11.34 g/cm$^3$, and a melting point of 621.43 °F. It can be easily shaped, molded, and used to form alloys through mixing with other metals. It can exist in both organic as well as inorganic form. The inorganic Pb dominantly occurs in dust, soil, old paint, and other different user products, while organic Pb (Tetra-ethyl Pb) is predominantly found in leaded gasoline. Both of these forms of Pb are toxic, however organic Pb-complexes are excessively toxic to biological systems compared to inorganic Pb [1]. Pb is the second most toxic metal after Arsenic (As), comprises 0.002% of Earth's crust [2,3], and its natural level remains to be below 50 mg kg$^{-1}$ [4]. Although earlier literature did not focus on the biological importance of Pb, recent findings suggest that traces of Pb (~29 ng/g diet) is important for enzyme activities and cellular systems, especially during cell development, hematopoiesis, and reproduction [5].

In general, Pb salts/oxides through atmospheric dust, automobile exhaust, paint, polluted food, and water are the key pathways for human exposure. The food canning industry is also an important source of Pb intake due to its leaching ability into canned foods. Currently, humans are exposed to Pb through dust particles from soil transmitted into homes and/or drinking water. Lead is considered carcinogenic (Group 2B) to humans [6]. Humans are impacted by Pb primarily through ingestion as 20–70% of ingested Pb is absorbed by the human body. Children have a high absorption capacity of Pb [7,8].

Enhanced Pb concentration in blood affects behavior, cognitive performance, postnatal growth, delays puberty, and reduces hearing capacity in infants and children. In adults, Pb causes cardiovascular, central nervous system, kidney, and fertility problems. During pregnancy, Pb can also hamper fetal growth in the early stage [1,3]. The Commission Regulation E.C., No 1881/2006, documented the Pb concentration (0.3 mg kg$^{-1}$) thresholds for different agriculture products such as leafy vegetables and fresh aromatic herbs [9]. Pb sources, their inclusion in soil, Pb bioavilability to plants, soil role for Pb transfer to plants, plant toxicity and accumulation mechanism, Pb effect on plants and humans, and different remediation technologies are basically covered in the present review. The main objective of this review is to summarise the research updates on Pb toxicity, bioavailability, and its imposed toxic effects on plants and on human health, including recently tested/recommended remediation options.

## 2. Methodological Approach for Selecting and Reviewing the Literature in a Meaningful Way for Targeting Specific Objectives

### *2.1. Collection, Compilation, and Identification of Relevant Literature for the Study*

The criteria for selection of recent literature for targeting up-to-date information on the topic was done through search string/keywords such as "Lead", "sources of Lead", "Lead toxicity", "bioaccumulation of Pb in food and human", "toxic forms of Pb", "Pb tolerance in human and plants", "health effect of Pb toxicity", and "Pb remediation". The extensive search of existing literature on the specific keywords was performed to collect the data from Scopus, Science Direct and Google scholar, MDPI, and other academic university websites. Three important criteria were considered for addressing relevant updated information (i) peer-reviewed, (ii) highly cited (i–10), and (iii) articles appearing in journals with a minimum impact factor (>1.0, *Thomson Reuters*) (iv) few articles except the above listed criteria based on recent/specific information was also included. The selection criteria/rules were adopted and modified from Sandin and Peters [10].

### *2.2. Extraction of Data and Data Representation*

All the available relevant literature was studied carefully based on the key objectives of the present review. Omission of work was based on literature that was published before 2000, was without quantitative results, non-English, and/or was general/duplicate/similar in nature, which did not fit the questions of this review. Later, the results from all representative literature published from the year 2000 onwards were extracted and represented in tabular form.

## 3. Sources of Pb Contamination in Soil, Crops, and Water Resources

Pb contamination in air, soil, and water resources has been associated with natural causes, such as geochemical weathering, sea spray emissions, volcanic activity, and remobilization of sediment, soil, and water from mining areas [11–13]. Table 1 represents the various sources of lead contamination in agricultural soils, crops, and water in different countries/regions of the world. It is evidenced in Table 1 that the anthropogenic products and processes (such as industrial, oil-processing activities, agrochemicals, paint, smelting, mining, refining, informal recycling of lead, cosmetics, peeling window and door frames, jewelry, toys, ceramics, pottery, plumbing materials and alloys, water from old pipes, vinyl mini-blinds, stained glass, lead-glazed dishes, firearms with lead bullets, batteries, radiators for cars and trucks, and some colors of ink) are considered to be major sources of Pb contamination in the environment [14–21].

Pb is available in soil/sediments as a free metal ion, is associated with inorganic molecules (e.g., $HCO_3^-$, $CO_3^{2-}$, $SO_4^{2-}$, and $Cl^-$), and can also exist as organic ligands (e.g., amino acid, fulvic acid, and humic acid). Pb can also be adsorbed onto particle surfaces such as biological material, oxides of iron, clay particles, and organic matter [22,23]. In general, a higher concentration of anthropogenic Pb accumulates on the soil surface and can decrease with depth [24]. Pb has a high affinity with organic and colloidal materials and, thereby, is readily available for plant uptake [25].

**Table 1.** Table of Pb contamination in agricultural soils, crops, and water in different countries [18].

| Sources | Contaminati-on | Plant Species | Region | References |
|---|---|---|---|---|
| Wastewater of Shitalakhya river | Soil and vegetables | *Amaranthus lividus*, *Basella alba*, *Cucurbita moschata*, *Spinacia oleracea*, and *Trichosanthes cucumerina* | Bangladesh | [26] |
| Wastewater treatment plant | Soil, water, and crops | *Eruca sativa*, *Madia sativa*, *Malus sylvestris*, *Triticum æstivum*, *Triticum turgidum*, *Urtica dioica*, and *Vicia faba* | Morocco | [27] |
| Mine affected area | Soil and vegetable | *Amaranthus dubius*, *Ipomoea aquatic*, *Ipomoea batatas*, *Phaseolus vulgaris*, *Piper nigrum*, *Solanum lycopersicum*, and *Solanum melongena* | China | [28] |
| Sewage water | Soil and crop | *Oryza sativa* | Iran | [29] |
| Agricultural/Urbanisation activities | Water and sediments | *Lemna minor* | India | [30] |
| Urbanization | Soil, water, and vegetables | *Brassica oleracea*, *Momordica charantia*, *Phaseolus vulgaris*, *Raphanus raphanistrum*, *Solanum lycopersicum*, and *Triticum aestivum* | China | [31] |
| Anthropogenic activities | Soil and vegetables | *Cucurbita maxima*, *Lagenaria siceraria*, *Solanum melongena*, and *Spinacia oleracea* | Pakistan | [32] |
| Glass industry | Soil and agricultural crops | *Brassica juncea*, *Hordeum vulgare*, and *Triticum aestivum* | India | [33] |

## 4. Pb Bioavailability in Soil and Its Influencing Factors

Lead bioavailability in soil is strongly controlled by its species, especially free-Pb ions concentration [22,34]. Plants absorb lead in dissolved form via the soil solution [25]. Moreover, the concentration of the free lead ion in soils depends on its physical process (e.g., adsorption/desorption) [23].

The behavior of lead species (bioavailability, mobility, and solubility) in soil is controlled by complex interactions of different biogeochemical factors [25]. These factors are redox conditions [35], pH [23,36], cation-exchange capacity [23], soil mineralogy, biological and microbial conditions [2], lead quantity [26,37,38], inorganic and organic legend concentration [22,34,39], competing cation concentration [40,41], and the type of plant species involved [37]. The behavior (uptake rate) of lead species in soil and plants is influenced by either biogeochemical factor independently or in combination with geochemical factors. The effects of some factors on Pb bioavailability are summarized below:

### *4.1. Soil pH*

Soil pH is the most important factor that controls Pb availability to plants. Soil pH dictates Pb availability in soil as a negative correlation between Pb solubility and soil pH is noticed [42]. In acidic soil (pH < 7), Pb exists as aqueous $Pb(H_2O_6)^{+2}$, while in alkaline soil (pH > 7), Pb forms aqueous complexes with $OH^-$ (hydroxyl ions). Specific adsorption of Pb is directly proportional to soil pH [16,43]. At a low soil pH (3–5), adsorption is the dominant process, whereas at a high pH (6–7), precipitation is the dominant process [16,44].

### *4.2. Soil Redox Potential*

Redox potential controls Pb dynamics in soil. The solubility of Pb is inversely proportional to soil redox potential (i.e., Pb solubility increases along with a decrease in soil redox potential). Generally, heavy metals dissolve easily in waterlogged soils. Pb was dissolved by acetic acid in highly impeded drainage soil (1.9 $\mu g\ g^{-1}$) as compared to freely drained soils (0.1 $\mu g\ g^{-1}$) in a region of slate bedrock [16].

### *4.3. Soil Texture*

Soil texture significantly affects Pb solubility. In Clay soils, heavy metal ions are adsorbed through ion exchange and specific adsorption mechanisms [45]. Pb adsorption also varies between types of clay minerals [16]. For example, the affinity between iolite and Pb is ~32 times higher than montmorillonite [46]. Mao et al. [47] observed low Pb adsorption on montmorillonite due to competition between Ca and Pb for cation exchange sites on clay.

### *4.4. Soil Minerals*

Soil minerals such as Mn and Fe affect Pb solubility in soils. Mn oxides have a high affinity towards Pb, thus they significantly decrease Pb uptake by plants grown in Pb contaminated soil [2,16,48–50]. O'Reilly and Hochella [50] emphasized that microbial activity is responsible for Pb mobilization from oxides and carbonate. Tao et al. [51] reported that earthworms could enhance Pb availability to plants.

### *4.5. Nutrients, Organic Carbon, and $O_2$*

These are the essential factors for microbial growth and metabolism and are directly involved in the degradation of contaminants. Some of the bioactive nutritional elements include carbon (e.g., backbone of all organic compounds), nitrogen (e.g., cellular protein and cell wall component synthesis), phosphorus (e.g., cell membrane, ATP, and nucleic acid), sulfur (e.g., amino acid synthesis), calcium (signaling transport), and magnesium (e.g., enzymatic activities functioning) [52,53] etc. Zhao et al. [54] concluded that soil physical properties such as permeability and fracturing could also affect Pb dynamics in soils. Li et al. [48] elucidated the effect of soil organic matter (OM) on Pb solubility through the formation of complexes during metals' interaction. Kögel-Knabner et al. [49] emphasized that soil OM drives a sizeable amount of Pb concentration by the formation of organo-Pb complexes.

The ion exchange capacity (particularly CEC), pH, ion redox potential, microbial community, texture, mineralogy, and organic matter of soils are the key regulating factors that affect Pb dynamics (e.g., adsorption, solubility, and mobility) in soil and bioavailability to plants.

## 5. Lead Bioavailability/Bioaccessibility in Animals and Humans

Lead toxicity is an important environmental health hazard and its effects on the human body are devastating. Total Pb in a human body is subject to environment, age, and occupation. It is estimated that a person weighing 70 kg will have an average of 120 mg of Pb, with 0.2 mg/L in the blood, 5–50 in their bones (in mg/kg), and 0.2–3 in tissues [55]. The Center for Disease Control and Prevention (USA) has set the standard elevated blood Pb levels for both adults and children (10 μg/dL and5 μg/dL, respectively) [56].

Bioavailability (BA) is an ingested fraction that crosses the gastrointestinal epithelium and is distributed into internal tissues and organs [57]. Bioavailability of Pb was established through in-vivo models such as in mice (*Mus*), monkeys (*Cercopithecidae*), rabbits (*Oryctolagus cuniculus*), rats (*Rattus*), and swine (*Sus scrofa*). However, extrapolation of the in-vivo models into human has not provided a realistic effect due to their physiological differences and species diversity. In-vivo experiments are much simpler than epidemiological studies because they are cheaper, faster, highly reproducible, and do not involve ethical issues. However, critical parameters (e.g., exposure levels, conditions, and absorbed Pb concentration) need to be considered while performing *in-vivo* specimen evaluation. The following key factors are to be considered for decision making in public health issues using in-vivo models: (a) specific features and limitations of the model; (b) targeting the human population in the design of animal studies at developmental stage; (c) the use of acceptable environmental doses, and (d) Pb speciation. In-vitro studies such as Relative Bioavailability Leaching Procedure (RBALP), Unified Bio-accessibility Research Group Europe Method (UBM), Solubility Bio-accessibility Research Consortium assay (SBRC), Physiologically Based Extraction Test (PBET), In Vitro Gastrointestinal (IVG) Method, and the In Vitro Digestion Model (RIVM) can be used to measure Pb bioaccessibility [58]. Pb relative bioavailability (RBA) refers to the comparative bioavailability of different Pb forms that are available in source substance [58]. For estimating the relative bioavailability of Pb, a reference material such as Pb acetate can be used. Lead RBA in soil can be measured by either blood or tissues (kidney, liver, and femur) [58,59]. Deshommes et al. [60] conducted an in-vivo experiment on Pb particles (especially particulate Pb forms including those in paint and dust and those in drinking water supply systems) and stated that the relative bio-accessibility leaching procedure (RBLP) offers the highest degree of validation and simplicity in animal models.

Literature suggests that due to unavailability of data and the existing model (e.g., animal model), we could not predict/estimate human risk assessment and human absorption of Pb particles, particularly for childhood exposure assessment, e.g., neuro-behavioral and neuro-developmental deficiencies, and the effects on growth, hearing, and blood pressure.

## 6. Lead Transportation, Toxicity, and Bioaccumulation Through Food Chain Contamination

Lead is one of the most toxic and frequently encountered heavy metals in the environment [34]. Different quantitative indices are currently being used to estimate Pb toxicity at trophic levels in the food chain (Table 2). Once Pb reaches the soil by any source and penetrates into the plant root system, it may accumulate there or may be translocated to aerial plant parts (APP). Pb mostly accumulates (≥95%) in the roots of plant species and only a small fraction is translocated to APP. Some of the studied plants species with respect to Pb transportation, toxicity, and bioaccumulation are *Allium sativum* [61], *Avicennia marina* [62], *Pisumsativum*, *Phaseolus vulgaris* and *Vicia faba* [34,63,64], *Lathyrus sativus* [65], *Nicotiana tabacum* [66], *Sedum alfredii* [67], *V. unguiculata* [68], and *Zea mays* [69].

Generally, plants uptake metal ions from soils through their roots [17,18,20]. Pb from the soil solution is adsorbed (unevenly) through roots and is bound with the uronic acid/polysaccharide of rhizoderm in many plant species such as *Brassica juncea* [70], *Festuca rubra* [71], *Funaria hygrometrica* [72,73], *Lactuca sativa* [74], and *Vigna unguiculata* [68]. This adsorbed Pb passively enters in roots and is transported through xylem. A concentration gradient was observed near the root apex, except for root cap, where cells are young and have thin cell walls with the lowest rhizodermic pH, which enhances Pb solubility in soil solution.

**Table 2.** Different indices used to quantify Lead toxicity at trophic levels in the food chain [18].

| SN | Factors | Equations | References |
|---|---|---|---|
| 1 | Trophic transfer factor (TTF) | TTF = Pb conc. in organism tissue/Pb conc. in food | [75] |
| 2 | Transfer factor (TF) | TF = Pb conc. in plant tissue/Pb conc. in soil | [76] |
| 3 | Metal transfer factor (MTF) | MTF = Pb conc. in plant/Pb conc. in soil | [77] |
| 4 | Accumulation factor (AF) | AF = Pb conc. in plant edible part/Pb conc. in soil | [78] |
| 5 | Bioaccumulation factor (BAF) | BAF = Pb conc. in organism tissue/Pb conc. in abiotic medium | [79] |
| 6 | Bio-concentration factor (BCF) | BCF = (Pb conc. in experimental organism tissues − Pb conc. in the control organism tissues)/Pb conc. in water | [80] |
| 7 | Biota-sediments AF (BSAF) | BSAF = Pb conc. in the organism/Pb conc. in sediments | [81] |
| 8 | Biomagnification factor (BMF) | BMF = Pb conc. in the organism/Pb conc. in the organism's diet | [82] |
| 9 | Trophic magnification factor (TMF) | TMF is calculated from the slope of logarithmically transformed Pb conc. in organisms plotted against the trophic levels of the organisms in the food web | [83] |

After entering into the roots, Pb moves by apoplast through water stream until it reaches the endodermis region. The endoderm functions as a physical barrier to Pb translocation as water stream is blocked by casparian strip and, thus, Pb enters into the symplastic movement. The low Pb transportation from root to APP has been reported due to immobilization by negatively charged pectins within the root cell wall [2,84]. Insoluble Pb salts precipitate in intercellular spaces of root cells [70,84]. Similarly, Pb accumulation in plasma membranes of root cells [61,84] or sequestration in the vacuoles of rhizodermal and cortical cells of roots is reported [68,84]. The major portion of the absorbed lead is sequestered/excreted from endodermis cells during the plant detoxification process. However, the above reasons are not sufficient to explain the low Pb translocation from root to APP as plant species such as *Brassica pekinensis* and *Pelargonium* potentially translocate Pb to APP, without affecting metabolic functions [85,86]. The lead hyper accumulator plant species can accumulate >1000 ppm [87]. The roots of hyperaccumulator species dissolve metals in soil [86], increase metal uptake and translocation, and make hyperaccumulator species to tolerate higher Pb ions concentrations. Apart from this, various detoxification mechanisms include selective metal uptake, excretion, complexation by specific ligands, and compartmentalization, which are also support for Pb tolerance.

In addition, Pb translocation to APP increases by organic chelators like ethylene diamine tetra acetate (EDTA) and micro-organisms [2,25]. Liu et al. [88] observed higher translocation to APP with increased soil Pb level in *B. Pekinensis* cultivars. This may be due to the potential of high Pb concentrations to destroy the casparian strip based physical barrier.

Xylem helps in the transportation of metals from plant roots to shoots [89], which is probably supported by transpiration [90]. Arias et al. [2] demonstrated X-ray mapping and found high Pb deposition in xylem and phloem cells on mesquite plants. After penetrating into the central cylinder of the stem, Pb can again be transported via the apoplastic pathway and further translocated to leaf areas through vascular flow [73]. In xylem, Pb can form complexes with amino/organic acids [87]. However, inorganic Pb can also be transferred. Translocation factor (i.e., lead in aerial parts/leading roots) can be implemented to know the degree of Pb translocation [86,88]. After implementing this factor, low numeric values will indicate that lead has been sequestered in the roots system [88].

The molecular mechanism of Pb entrance in roots is not clear yet. It is believed that several pathways can be used by Pb for the same purpose, especially ionic channels. However, Pb uptake is a non-selective phenomenon and is independent of the $H^+$/ATPase pump [91]. Lead absorption is inhibited by calcium [92] as Pb competes with Ca for calcium channels. $Ca^{2+}$-permeable channels are important gateways for Pb to penetrate into the root system [91,93]. The transgenic plant studies reveal that Pb can also penetrate into roots through other alternative non-selective pathways, e.g., cyclic nucleotide-gated ion channels and low-affinity cation transporters [94]. Comprehensive details for the average lead content in different food crop plants are summarized in Table 3. It is noted that higher concentrations of Pb are associated with fruit crops (Table 3).

**Table 3.** Details for the average lead contents in different crop plants.

| Plant Species | Scientific Name | Concentration (mg/kg) | References |
|---|---|---|---|
| | **Vegetable crops** | | |
| Coriander | *Coriandrum sativum* | 4.5 | [95] |
| Spinach | *Spinacia oleracea* | 0.98–9.2 | [96–99] |
| Coriander | *Coriandrum sativum* | 0.4–75.5 | [98,100–105] |
| Cabbage | *Brassica oleracea* | 0.07–12 | [97,104,106–108] |
| Radish leaf | *Raphanus sativus* | 0.4 | [100] |
| Amaranthus | *Amaranthus blitum* | 23.26 | [109] |
| Parsley | *Petroselimum crispum* | 2.31 | [97] |
| Slender amaranth | *Amaranthus viridis* | 2.56 | [101] |
| Sugar beet | *Beta vulgaris L* | 149.5 | [102] |
| Slender amaranth | *Amaranthus viridis* | 5.44 | [110] |
| Tomato | *Solanum lycopersicum* | 5.5 | [99] |
| Brinjal | *Solanum melongena* | 2.1 | [95] |
| Cucumber | *Cucumis sativus* | 1.5 | [95] |
| Brinjal | *Solanum lycopersicum* | 2.2 | [98] |
| Raddish | *Raphanus sativus* | 0.75 | [111] |
| Eggplant | *Solanum melongena* | 4.93 | [112] |
| Brinjal | *Solanum tuberosum* | 6.19 | [112] |
| Pumpkin | *Cucurbita maxima* | 0.25 | [113] |
| Chilli | *Capsicum annuum* | 0.17 | [113] |
| Carrot | *Daucus carota* | 0.72–7.8 | [95–97] |
| Sugar beet | *Beta vulgaris* L. | 26.35 | [109] |
| Potato | *Solanum tuberosum* | 0.012–2.58 | [106,107] |
| Cauliflower | *Brassica oleracea* | 0.36–6.1 | [95,97,104] |
| | **Spices Crops** | | |
| Aniseed | *Pimpinella anisum* | 0.26–5.68 | [114,115] |
| Bay leaf | *Cinnamomum tamala* | 0.98–3.58 | [116–118] |
| Cardamom | *Elettaria cardamomum* | 0.583 | [115] |
| Cassia | *Cinnamonum cassia* | 4.159 | [115] |
| Curry | *Murraya koenigii* | 3.617 | [117] |
| Dill | *Anethum graveolens* L. | 0.81 | [119] |
| Fennel | *Foeniculum vulgare* | 0.316 | [115] |
| Fenugreek | *Trigonella foenum-graecum* L. | 9.38 | [114] |
| Rosemary | *Rosmarinus officinalis* | 10.8 | [120] |
| Tulsi | *Ocimum sanctum* | 4.59 | [116] |
| | **Fruit Crops** | | |
| Mango | *Magnifera indica* | 0.642–1.620 | [121,122] |
| Orange | *Citrus sinensis* | 26 | |
| Pomegranate | *Punica granatum* | 28 | |
| Grapes | *Vitis vinifera* | 24 | |
| Lemon | *Citrus limon* | 29 | [123] |
| Strawberry | *Fragaria ananassa* | 10 | |
| Buckthorn | *Hippophae rhamnoides* | 20 | |
| Peaches | *Prunus persica* | 11 | |

**Table 3.** *Cont.*

| Plant Species | Scientific Name | Concentration (mg/kg) | References |
|---|---|---|---|
| Banana | *Musa* sp. | 0.003–0.05 | [122,124] |
| Jackfruit | *Artocarpus heterophyllus* | 0.017 | |
| Orange | *Citrus sinensis* | 0.106 | [125] |
| Trengerine | *Citrus tangernia* | 0.097 | |
| Banana | *Musa* | 0.118 | |
| Papaw | *Carica papaya* | 0.072 | |
| **Cereals and Legumes Crops** | | | |
| Pearl millet | *Pennisetum glaucum* | 0.12 | [126] |
| Sorghum | *Sorghum bicolor* | 0.18 | [126] |
| Wheat | *Triticum aestivum* | 0.40 | [127] |
| | | 0.47 | [128] |
| Barley | *Hordeum vulgare* | 0.22 | [129] |
| Quinoa | *Chenopodium quinoa* | 0.37 | [130] |
| Maize | *Zea mays* | 0.50 | [131] |
| | | 0.34 | [132] |
| | | 0.31 | [133] |
| Rice | *Oryza sativa* | 0.52 | [134] |
| | | 0.89 | [135] |
| Black gram | *Vigna mungo* | 0.60 | [133] |
| Lentil | *Lens culinaris* | 0.55 | [133] |
| Common bean | *Phaseolus vulgaris* | 0.12 | [136] |
| Soybean | *Glycine max* | 0.08 | [137] |
| Safflower | *Carthamus Tinctorius* | 0.80 | [138] |
| Rapeseed | *Brassica napus* | 0.51 | [138] |
| Sunflower | *Helianthus annus* | 0.57 | [131] |

Accidental soil ingestion is a major Pb exposure pathway for humans inhabited in a Pb polluted area [9,139]. However, the intake of Pb contaminated plants has been an important exposure to humans and animals [9,139,140]. Edible/wild plants cultivated/grown in the vicinity of phosphate industries can be Pb bio-indicators of toxic metals [9]. Inhabitants and workers of these industries/provinces may be exposed to Pb contamination. The Pb exposures and blood concentration to these closely inhabited/living populations is subject to the season as well as industrial activity. The children's blood lead levels (BLLs) were observed to be higher during the summer and early fall [141]. The BLLs are highly significant, are evident in multiple locations, periods, and ages, and are population-specific [142,143]. Higher levels were observed (10–60%) in warm-weather and levels increased in 2-year-old children, more so than 1 or over 4-year-olds [142,143]. Zahran et al. [143] emphasized that lead seasonality must be considered for Pb risk analysis. One health concept was proposed to take care of animal, human, and environmental health all together [144,145].

## 7. Mechanistic Understanding of Pb Toxicity and Tolerance in Plants and Humans

Lead causes a broad range (physiological, morphological, and biochemical) of toxic effects on living organism. In plants, Pb toxicity is characterized with impaired chlorophyll (Chl a) production, cell division, elongation of root, lamellar organization in the chloroplast, plant growth, seed germination, seedling development, and transpiration [67,87]. However, the magnitude of the effects varies and/or depends on Pb levels, exposure time, plant stress intensity, and plant developmental stage. Plants have internal detoxification mechanisms to deal with Pb toxicity, i.e., complexation by specific ligands, selective metal uptake, excretion, and compartmentalization [18,21,61,87].

Lead induced oxidative stress is reported to produce reactive oxygen species (ROS) in plants [146,147]. These ROS synthesized as a result of oxidative stress in plants can cause deleterious effects such as lipid peroxidation, disrupted cell membrane, DNA and protein damage, inhibition of photosynthesis, and inhibition of ATP production [148]. To overcome the adverse effects of ROS, plants produce a variety of antioxidative enzymes. Lead imposed changes in antioxidative enzyme production of various food crops have been well established (Table 4). The activity of antioxidative enzymes, such as superoxide dismutase, peroxidase, and ascorbate peroxidase, were positively correlated with Pb content, while Catalase, Glutathione reductase, and Glutathione peroxidise were decreased in both leaf and root tissues (Table 4).

Lead poisoning cases in humans are mostly the result of oral ingestion and absorption via the gut [149]. Pb absorption from the gastrointestinal tract is subject to physical characteristics (such as age, pregnancy, fasting, and Fe and Ca status) and the physico-chemical nature of the material ingested (e.g., size of particles, solubility, mineralogy, and Pb species) [150]. The Pb absorbed in the intestine is further carried to soft tissue, e.g., in the liver, kidneys, and bone tissue, where it accumulates over time [149]. The main transport processfor Pb to different body tissues from the intestine is via red blood cells, where binding takes place between Pb and haemoglobin (HB). Nearly 99% of the Pb in blood is observed in erythrocytes, with approximately 1% in both serum and plasma. Distribution of Pb in the organs (the lungs, spleen, brain, aorta, renal cortex, bones, and teeth) relies greatly on Pb concentration in plasma rather than on the whole blood. The half-life of Pb in blood is estimated to be 35 days, whereas the half-life of Pb in soft tissue is estimated to be 40 days. Pb can be resident in bone for up to 30 years and concentrations of Pb in teeth and bone grow in proportion to age [149]. The Pb biological half-time is believed to be significantly greater in children than in adults. Lead creates chemical bonds with thiol groups of proteins and Pb toxicity is believed to inhibit enzymes and subsequently interfere with homeostasis of Mg, Ca, and Zn. Lead-induced oxidative stress is caused due to Pb poisoning as it disrupts the pro-oxidant/antioxidant cell defence system. Antioxidant nutrients, such as vitamins E, C, $B_6$, and B-carotene, and also Zn and Se, are believed to combat Pb-induced oxidative stress [151].

High levels of Pb absorption are found in children rather than in adults. It is approximated that adults may absorb 3–10% of an oral dose of water-soluble Pb, whereas for children, it may be as high as 40–50%. Higher Pb concentrations are found in the blood of children who are Fe- or Ca-deficient than those with replete Fe or Ca. Pb absorption may raise during the pregnancy period and over 95% of Pb deposits in skeletal bones as insoluble phosphate [149]. According to autopsy studies, cortical bone and teeth together account for 90–95% of the body's Pb burden. The total Pb body burden in the skeleton is 80–95% in adults and about 73% in children [149]. Mothers may transfer Pb to the foetus and also to infants during the period of breastfeeding [152]. Pb toxicity principally targets the human central nervous system and children's ingestion of large amounts of Pb from the environment, particularly when anaemic, is linked to lower intelligence and impaired motor function [149].

The Joint FAO/WHO Expert Committee on Food Additives (JECFA) made an estimation of tolerable weekly intake based on dose-response analyses and concluded that the provisional tolerable weekly intake (PTWI) is linked to a reduction in children's IQ of at least 3 points and systolic blood pressure of approximately 3 mmHg (0.4 kPA) higher in adults [149]. When observed in terms of a shift in IQ distribution or blood pressure in a population, these changes assume greater importance. The JECFA's conclusion, therefore, was that the PTWI is no longer adequately protective of health and they withdrew it. The lack of an indication of a threshold level for key effects of Pb based on the dose-response analysis led the JECFA to conclude that a new PTWI considered as health-protective could not be established. The JECFA reiterated that foetuses, infants, and children are the subgroups that have the highest sensitivity to Pb [150,153] due to the neuro-developmental effects. The European Commission [154] has set guidelines for maximum permissible levels of Pb in some foodstuffs (Table 5). Interventions such as eliminating leaded petrol, banning the use of Pb in wine bottles, and the discontinuation of soldered cans are seen as an important factor in successful reduction of Pb in food. In children, Pb toxicity symptoms are loss of appetite, anemia, behavioral changes, delayed mental growth and learning, fatigue, headaches, hyperactivity, insomnia, metallic taste, reduced nerve conduction, weight loss, and possibly neuron disorders [155]. The behavior changes are irreversible and untreatable as the cerebrum of *Homo sapiens* has little capability for reparation. A daily Pb intake of up to 7 μg/kg body weight or 490 μg of Pb for an adult was accepted by WHO, FAO. However, no such guideline is given for infants and children, who are relatively more sensitive to low Pb levels [156].

In broilers that have high Pb acetate (200 mg/kg) exposure in their diet, these show anorexia, greenish diarrhea, leg paresis, weight loss, wing droop, and lethargy symptoms including gross change in kidney, spleen, and liver function, gizzard lining, hemorrhages on muscles [157], etc. Gao et al. [157] concluded that Pb could alter the expression of selenoprotein related genes in the cartilage tissue of broilers. Rahman and Joshi [158] revealed that Pb acetate (i.e., 250–400) in drinking water could lead to reduced feed intake and growth indices in broilers due to higher oxidative stress. Pb-induced oxidative stress can also reduce antioxidant activities such as catalase, glutathione superoxide dismutase [159], etc. and erythrocytes burst due to lipid peroxidation in erythrocytes membranes and may cause hemolytic anemia [160]. Pb could also respond to change in the activities and expression of antioxidant enzyme–related genes [161,162]. Most animal experiments confirm that Pb transportation in the body occurs through blood circulation and accumulates in soft tissues, bones, and other pivotal organs [163]. Bones are a major sink of Pb (~90%) and mostly replace calcium, thus decreasing in bone mineral density (BMD) due to Pb exposure [164].

**Table 4.** Effects of Pb toxicity on activities of different antioxidant enzymes in different plants [16].

| | Enzymes | | Pb Exposure Level | Duration | References |
|---|---|---|---|---|---|
| | **Enhanced** | **Reduced** | | | |
| *Sedum alfredii* | SOD | APX | 0–200 μM | 14 | [63] |
| *Triticum aestivum* | SOD, POX, APX | CAT | 0, 0.15, 0.3, 1.5, 3.0 mM | 6 | [165] |
| | SOD, POX, CAT, APX | - | 0, 1, 2, 4 mM | 3 | [166] |
| | SOD, CAT | APX, GPX, GR | 0, 8, 40 mg $L^{-1}$ | 5 | [167] |
| | SOD | GPX | 0, 500, 1000, 2500 μM | 7 | [168] |
| *Oryza sativa* | SOD | CAT, POD | 0, 50, 100, 200 M | 16 | [169] |
| *Triticum aestivum* | SOD, CAT | APX, GPX, GR | 0, 50, 100, 250, 500 μM | 4 | [170] |
| *Zea mays* | SOD, APX, GPX, GR | CAT | 0, 16, 40, 80 mg $L^{-1}$ $Pb^{2+}$ | 8 | [146] |
| | APX, DHAR, MDHAR | - | 0, 16, 40, 80 mg $L^{-1}$ $Pb^{2+}$ | 1 | [171] |
| *Oryza sativa* | SOD, APX, GR | CAT | 0, 10, 50 μM | 4 | [172] |

SOD: Superoxide dismutase; POX: Peroxidase; APX: Ascorbate peroxidase; CAT: Catalase; GPX: Glutathione peroxidise; GR: Glutathione reductase; MDHAR: monodehydroascorbate reductase; DHAR: dehydroascorbate reductase.

**Table 5.** Maximum permissible level of Pb in foodstuffs (mg/kg Fresh Weight).

| Lead in Food Stuffs (mg/kg Fresh Weight) | Maximum Permissible Level |
|---|---|
| Food of Plant Origin | |
| Rye, grain | 0.20 |
| Wheat, grain | 0.20 |
| Bread | - |
| Miscellaneous cereals | - |
| Cabbage | 0.30 |
| Carrot and potatoes | 0.10 |
| Apple | 0.10 |
| Milk chocolate | - |
| Food of animal origin | |
| Carcass meat | 0.10 |
| Offal | - |
| Fish | 0.30 |
| Fresh water fish, | 0.30 |
| Eggs | - |
| Milk | 0.02 |
| Dairy products | - |

## 8. Human Health Effects Due to the Consumption of Pb Contaminated Foodstuffs

Lead enters into the body through pathways like inhalation of wind-blown Pb-laden dust, ingestion of Pb contaminated soils, oral intake of Pb contaminated water, and food grown in Pb-contaminated areas. Pb accumulation in livestock tissues may also pose a major risk to human health through livestock meat consumption [173,174]. After absorption, Pb is distributed in the body through red blood cells (RBC). Pb is mostly bound to hemoglobin rather than RBC membrane after entering the cell [175]. The hematopoietic is a sensitive system for critical Pb toxicity and may lead to anemia [160]. Histopathological observations confirmed that Pb ions are transported to the liver, where they can induce chronic damage to the liver. Pb toxicity also increases blood enzyme levels and reduces protein synthesis [176–178]. Pb imposes toxic effects on kidneys through structural damage and changes in the excretory function [176,177,179]. The other organ and tissue systems affected due to lead toxicity are the nervous, cardiovascular, and reproductive systems [160,175,180]. Pb toxicity imposes mineralizing of bones and teeth, which is a major body burden [3]. The International Agency for Research on Cancer (IARC) stated that inorganic Pb is probably carcinogenic to humans (Group 2A) based on limited evidence in humans and sufficient evidence in animals [181]. Generalized clinical symptoms of Pb poisoning in humans are comprehensively summarized in Table 6.

**Table 6.** Generalized clinical symptoms of Pb poisoning in humans.

| SL No. | Body Organ/System | Clinical Symptoms of Pb Poisoning |
|---|---|---|
| 1 | Eyes | Blindness of parts of visual field<br>Hallucinations |
| 2 | Ears | Hearing loss |
| 3 | Mouth | Unusual taste<br>Slurred speech<br>Blue line along the gum |
| 4 | Kidney | Structural damage and failure<br>Changes in the excretory function |
| 5 | Liver | Jaundice<br>Lead-induced oxidative stress<br>Decreased liver function<br>Microvesicular and macrovesicular steatosis<br>Hemosiderosis and cholestasis |
| 6 | Skin | Pallor and/or lividity |
| 7 | Central nervous system (CNS) | Insomnia<br>Loss of appetite<br>Decreased libido<br>Depression<br>Irritability<br>Cognitive deficits<br>Memory loss<br>Headache<br>Personality changes<br>Delirium<br>Coma<br>Encephalopathy |
| 8 | Reproductive organs | Sperm dysfunctions<br>Pregnancy complications<br>Preterm birth |
| 9 | Abdomen/Stomach | Pain<br>Nausea<br>Diarrhoea<br>Constipation |

**Table 6.** *Cont.*

| SL No. | Body Organ/System | Clinical Symptoms of Pb Poisoning |
|---|---|---|
| 10 | Blood | Anaemia |
| 11 | General | Malaise<br>Fatigue<br>Weight loss |
| 12 | Neuro- muscular | Tremor<br>Pain<br>Delayed reaction times<br>Loss of coordination<br>Convulsions<br>Foot or ankle drop<br>Seizers<br>Weakness |
| 13 | Bones | Mineralizing bones and teeth<br>Decreased bone density |

## 9. Pb Remediation approaches

Innovative and site-specific Pb remediation technologies for efficient clean-up of contaminated sites are prerequisites for a healthy life and safe food production. There are different (physical, chemical, and biological) processes developed to reduce total Pb concentration and Pb bioavailability to mitigate Pb accumulation in the food chain [182,183].

### *9.1. Physical Approaches*

#### 9.1.1. Replacement of the Medium (Soil/Water)

In this method, the complete or partial replacement of the contaminated resources (soil/water) is done based on the magnitude of the contamination [183]. This method of remediation is very useful at a small scale at the local level. The biggest challenge for this method is the safe disposal of the contaminated soil/water in a cost-effective manner.

#### 9.1.2. Vitrification

This method can be applied through both in-situ and ex-situ remediation mechanisms. In vitrification methods, soil is melted with the help of a high-temperature process and Pb sequestration achieved in solidified vitreous mass [183,184]. Vitrification can be used long-term and effective low volume can be obtained for reuse [183]. This is a costly method and may not be suitable for applications in large areas. Dellisanti et al. [185] carried out the vitrification of Pb-rich ceramic waste. Wang et al. [186] treated fly ash from a municipal solid waste incinerator to radiated heavy metals including Pb. Navarro et al. [187] applied vitrification for remediating the hazardous mine wastes from old mercury and Ag-Pb mines in Spain.

#### 9.1.3. Electrokinetic Remediation

Electrokinetic remediation is achieved by applying current in the field. This process involves techniques such as electrophoresis, electric seepage/electro-migration, electro-osmosis, and electrolysis [188]. Kim et al. [189] has shown that contaminated rice soil could be cleaned using an electrokinetic technique, which reduces Pb contamination by 19.4% in 4 weeks. Jeon et al. [190] remediated a soil contaminated with Pb in a paddy rice crop using EDTA as an electrolyte. The electrokinetic remediation technique generates almost nil waste. Electrokinetic remediation is applicable for saturated soils with low groundwater flow, requires short repair time and low energy, and provides a complete repair [183]. The heterogeneity of soil and treatment depths are the two important limitations of this method.

### *9.2. Chemical Approaches*

Various chemical amendments are widely used for immobilization of lead in soil and ground water at the field scale (Table 7).

#### 9.2.1. Chemical stabilization

This method is used to decrease the mobility, bioavailability, and bio-accessibility of heavy metals in soil. The immobilizing agents, i.e., biochar (Wheat, Rice, Miscanthus straw biochar, Sugarcane bagasse biochar, Holm oak chips biochar), clay minerals (Sepiolite with limestone, Palygorskite, and Bentonite), liming materials (Oyster shells and eggshells), metal oxides (Mn oxides and Ferric oxyhydroxide powder with limestone), organic composts (Biosolid), and phosphate compounds (Phosphate rock, Calcium magnesium Phosphate, and Single superphosphate) were previously used in the chemical stabilization process [183], details of which are given in Table 7. Chemical stabilization is a simple, quick, relatively cost-effective chemical approach by which Pb can be immobilized by adsorption, chemical precipitation, ion exchange, and surface complexation mechanisms to limit Pb transport and bioavailability. However, in this process, Pb remains in the soil and hence, long-term immobility is recommended.

#### 9.2.2. Solidification/Stabilization

Solidification/stabilization (waste fixation) is relatively low cost, low risk, easily implemented, and highly resistant to biodegradation with abroad engineering applicability [183]. Soil solidification refers to the encapsulation of waste materials in a monolithic solid with high structural integrity [183,191]. Soil stabilization is achieved by stabilization of soil contaminants through chemical interaction between Pb and binding reagents [192]. Wang et al. [193] and Antemir et al. [194] demonstrated the potential cement-based binders in remediating heavy metals including Pb in England. Navarro-Blasco et al. [195] assessed the Pb adsorption capacities of calcium aluminate cement. Voglar and Lestan [196] used calcium aluminate cement and sulfate resistant Portland cement as binders for Pb immobilization in Slovenian soil. Wang et al. [197] assessed Portland cement, ground granulated blast furnace slag, pulverized fuel ash, MgO, and zeolite for on-site soil solidification/stabilization of Pb in UK soil.

#### 9.2.3. Soil Washing

The soil washing process is achieved by Pb leaching from soil matrix using reagents/extractants such as chelating agents, inorganic acids, organic acids, surfactants, and water. The soil mixing with respective reagents/extractants is done where extractants transfer Pb from soil to the liquid phase through chelation or desorption, chemical dissolution, and ion exchange mechanisms [198]. Soil washing is a rapid, permanent, effective chemical method for Pb remediation with long term liability [199]. Hu et al. [200] achieved removal of 73% Pb using EDTA as a chelating agent. Wang et al. [201] used iminodisuccinic acid, glutamate-N, N-diacetic acid, glucomonocarbonic acid, and polyaspartic acid to extract 53% and 55% Pb from Pb-Zn contaminated soil. However, these technologies have certain hurdles for their practical utility due to change in soil properties, loss of nutrients, adverse effect of washing chemicals, generation of wastewater, and cost of chemicals and their negative impact on the environment.

### *9.3. Biological Approaches*

Biological methods for Pb remediation are the most eco-friendly alternatives to remediate Pb from the contaminated resources. Biological remediation can be referred as direct utilization of any natural/genetically engineered living organism and their product for Pb detoxification to restore soil function and quality.

**Table 7.** Chemical amendments for immobilization of lead in soil and groundwater at the field scale [183].

| SN | Amendments | Immobilization Mechanisms | Observations | Reference |
|---|---|---|---|---|
| 1. | | Clay minerals | | |
| | Sepiolite + limestone | Chemical precipitation and surface complexation | The treatment decreased exchangeable Pb (99.8%) and reduced Pb in brown rice (81.2%). | [202] |
| | | | The treatment significantly increased soil pH and CEC, decreased Pb exchangeable fractions, and inhibited Pb accumulation in rice. | [203] |
| | Palygorskite | | Significantly reduced water leachable Pb fractions (50%). | [204] |
| | Bentonite | | Reduced Pb exchangeable fractions (20.3–49.3%). Increased residual portions (6.73–10.0%). Pb concentrations in the rice roots (5.13–26.7%) and shoot (3.73–7.8%) were reduced. | [205] |
| 2. | | Phosphate compounds | | |
| | Phosphate rock $(Ca_{10}(PO_4)6Cl_2$ Calcium magnesium phosphate $(Ca_3(PO_4)_2)$ Single superphosphate $(Ca(H_2PO_4)_2)$ | Pb: Pb phosphate precipitation, especially pyromorphite-like mineral; | P fertilizers decreased water soluble and exchangeable Pb fractions (22.03–81.4%) and reduced Pb uptakes (16.03–58.0%) by a Chinese green vegetable. | [206] |
| 3. | | Liming materials | | |
| | Oyster shells and egg shells | Chemical precipitation | TCLP-leachable Pb was effectively reduced. | [207] |
| 4. | | Organic composts | | |
| | Biosolid | Surface complexation and chemical precipitation | The treatment enhanced soil pH, cation exchange capacity, and humic acids, with improved soil sorption capacity. The readily soluble Pb forms were reduced. | [208] |
| 5. | | Metal oxides | | |
| | Ferric oxyhydroxide powder + limestone | Specific sorption, co-precipitation, and inner-sphere complex | Pb decreased by 97% in pore water. Pb was transformed into residual mineral. | [209] |
| | Mn oxides | | Pb immobilization. | [210] |
| 6. | | Biochar | | |
| | Wheat Straw Biochar | Increase in soil pH, total organic carbon, abundant functional groups, and complex structures of biochar leads to reduction in heavy metals extractable fractions | The soil extractable Pb was decreased. As a result, Pb in root tissues was significantly reduced. | [211] |
| | | | Biochar significantly transformed the exchangeable Pb fractions into relatively stable fractions. | [212] |
| | Sugarcane bagasse biochar | | The exchangeable Pb was reduced and the organically-bound fraction increased with increased biochar input. Pb bioavailability to plant shoots and roots decreased with increasing biochar input. | [213] |

**Table 7.** *Cont.*

| SN | Amendments | Immobilization Mechanisms | Observations | Reference |
|---|---|---|---|---|
| | Holm oak chips biochar | | Biochar stabilized Pb and reduced its accumulation in barley grain. | [214] |
| | Rice straw biochar | | Rice straw biochar decreased. Pb bioavailability and reduced Pb contents in vegetables. | [215] |
| | Miscanthus (Miscanthus giganteus) straw biochar | | $CaCl_2$-extractability of Pb significantly decreased with increased biochar input. | [216] |

#### 9.3.1. Phytoremediation

Phytoremediation is an environmentally-friendly, attractive, aesthetically pleasing, noninvasive, energy-efficient, and cost-effective technology that can remediate Pb in low to moderate contaminated soil. It includes phytostabilization and phytoextraction. Phytostabilization decreases the mobility/bioavailability of Pb through adsorption by roots, chemical precipitation, and complexation in the root zone. Phytostabilization is only effective up to the root depth of plants. Cheng et al. [217] observed the Pb phytoremediation potential of *Miscanthus floridulus.* Yang et al. [218] ascertained the phyto-extraction potential of a co-planting system of *Pteris vittata* L. and the *Ricinus communis* L. in Pb contaminated soil and observed an increased yield of *P. vittata* after Pb uptake. Metal hyper accumulater plant species such as *Eichhornia crassipes, Lemna* sp., *and Pistia stratiotes* have been widely used to remediate Pb from diversified environments (Table 8).

#### 9.3.2. Microbial Remediation

Microbial remediation refers to decreasing the availability of Pb in the environment using indigenous/exotic microbes. Bacterial species such as *Alcaligenes* sp., *Bacillus firmus, Bacillus licheniformis, Enterobacter cloacae, Escherichia coli, Micrococcus luteus, Pseudomonas fluorescens*, and *Salmonella typhi* show adsorption potential of Pb from the contaminated resources [219–223]. Wang et al. [224] concluded that bacterial strain B38 (mutant of *Bacillus subtilis*) has immense potential to remediate heavy metals including Pb in China. Zeng et al. [225] observed that *Aspergillus niger* strain SY1 effectively removed Pb (99.5%) from contaminated sediment through bioleaching. The fungal biomass of *Lepiotahystrix, Aspergillus niger, Aspergillus terreus*, and *Trichoderma longibrachiatum* are reported as potential bio-sorbents [223,226,227]. The algal species i.e., *Palmaria palmate, Spirulina maxima, Spirogyra hyaline, Cystoseira barbata, Cladophora* sp., *Chara aculeolata, Nitella opaca*, and *Ulva lactuca* are were identified to be efficient bio-sorbents [223,228,229]. Microbial remediation is considered to be a natural, safe, and effective eco-friendly technology with low energy and low operation cost inputs [183]. Most importantly, microbial remediation does not impose any environmental and health hazards. The process depends on the environmental condition and inputs such as nutrients, oxygen, and other amendments to stimulate microbial activity for Pb remediation [183].

#### 9.3.3. Microbial Assisted Phytoremediation

Many approaches including molecular fingerprinting techniques *viz.* length heterogeneity analysis by PCR (LH-PCR), terminal restriction fragment length polymorphism (T-RFLP), denaturing gradient gel electrophoresis (DGGE), single strand conformation polymorphism (SSCP), ribosomal intergenic spacer analysis (RISA), cloning, and In Situ Hybridization (ISH/FISH) were used to identify the potent microbial community involved in phytoremediation [230–234]. This approach is based on the rhizosphere associated microbes such as *Bacillus, Beijerinckia, Burkholderia, Enterobacter, Erwinia, Flavobacterium, Gluconacetobacter, Klebsiella, Pseudomonas, and Serratia* [235–237]. Babu et al. [235] inoculated soil with rhizospheric bacteria *Pinus sylvestris* and found significant increases in biomass, chlorophyll content, nodule number, and Pb accumulation in *Alnus firma* seedlings.

**Table 8.** Phytoremediation potential of different plant species for Pb contaminated water and soil [21].

| Species | Treatments | Observation | Findings | References |
|---|---|---|---|---|
| *Ceratophyllum demersum* | Artificial wastewater | Removal rate 92.0–95.0% | Maximum BCF of 1284.35 in 4 mg/L of Pb 12th Day. | [238] |
| *Leptodictyum riparium* | Artificial wastewater | Removal rate 96.7% | Having high resistance and effectiveness for Pb accumulation. | [239] |
| *Scirpus grossus* | 600 L spiked water in Pb (10, 30, and 50 mg/L), duration 98 days | Pb concentration in water decreased up to 99% after 28 days and highest Pb uptake: 1343, 4909, 3236 mg/kg for the treatment of 10, 30, and 50 mg/L Pb, respectively | Highest BCF and TF were 485, 261, and 2.52 on day 42 of Pb treatment at 30 mg/L concentration in 70 days retention time. | [240] |
| *Pistia stratiotes* | Greenhouse condition using glass pots with a defined amount of added HMs | Pb removal was >90% in the first week | No enhancement of Pb removal efficiency with increased Pb concentrations. | [241] |
| *Eichhornia crassipes* | Operation in up-flow anaerobic packed bed reactors system | Pb Removal rate: 98% | In the coupled pond system, water hyacinth was observed to have enhanced Pb removal efficiency by accumulating Pb into root | [242,243] |
| *Eichhornia crassipes* | Stock solutions with initial concentration of 20 g/L | Pb Removal rate: 98.33% | Powdered root of water hyacinth absorbed higher Pb. | [244,245] |
| *Brassica oleracea var. Acephala* | Treatments of different Concentration Pb = 0, 1, 5, and 10 mg/kg | Phytoremediation of saline soils with 10 and 16 mg/kg Pb | Negatively correlated with plant fresh and dry weights. | [246] |
| *Posidonia oceanica* | Sediments | Pb Levels (mg/kg) in root: 4.52 ± 0.55, | Ability of *Posidonia oceanic* to accumulate and detoxify Pb rather than being attributed to differences in ecological and morpho-anatomical characteristics. | [247] |
| *Datura inoxia* | Concentrations of 0.5, 1.0, 3.0, 5.0, 10, 15, 20, 25, 30, 35, 40, 45, 50 mg/L metal | Survival rate = 50% | *Datura* exhibits phytoremediation potential. | [248] |
| *Magnolia grandiflora* | Soil | Pb Accumulation rate: 63.4%, | Relationship between heavy metal concentrations in soils and washed new and old leaves. | [249] |
| *Pistia stratiotes* | HMs from steel effluents: 120 g of plant in 10 L effluent | Removal rates: Pb = 70.7%, | *E. crassipes* more efficient than *P. stratiotes*. | [250] |
| *Lemna* sp. | Artificial by concentration of 2, 5, and 10 mg/L | Pb removal rates by *Lemna gibba*: 60.1% at 2 mg/L at pH 9, 98.1% at 10 mg/L at pH 7, | BCF and metal uptake yield per unit of dry biomass for Pb is 403–738. | [251] |

**Table 8.** *Cont.*

| Species | Treatments | Observation | Findings | References |
|---|---|---|---|---|
| *Pistia stratiotes* | Stock solution (2000 mg/L) | 96% removal of Pb(II) from 25 mL of solution in 60 min by only 0.125 g of biomass | Results consistent with the Langmuir model by maximum biosorption capacity of 122.70 mg Pb (II)/g of biomass. | [252] |
| *Mixture of Typha angustifolia and Limnocharis flava* | Wastewater oxidation pond | Removal rate: Pb = 62.07% | Positive relation between retention time and heavy metal removal. | [253] |
| *Lemna* sp. | 200 g fresh plant in mixed sewage of industrial and municipal effluents | Pb Removal efficiency >80% | BCFs for Pb = 523, indicating that this plant is a moderate accumulator of Pb. | [254] |
| *Lemna* sp. | Artificial: Pb = 0.25 mg/L | Removal rates: Pb = 36% | Removal efficiency up to 80% at higher metal loading rate where 24 h light and pre-treatment steps required. | [242,243] |
| *Eichhornia crassipes* | Mining wastewater | Accumulation in leaves (mg/kg): Pb = 3.40–5.06 | BCF: Pb = 242–506 | [255] |
| *Mixture of P. australis and T. latifolia* | Urban sewage mixed with industrial effluents | Removal rate: Pb = 61.0 ± 1.2% | - | [256] |

*9.4. Biotechnological and Genetic Approaches*

Genomics, metagenomics, metabolomics, proteomics, transcriptomics, nanoparticles, and isotope probing are modern technologies to understand Pb phytoremediation [234,235,257]. The biotechnology and genetic approaches to remediate Pb from the contaminated resources have great potential and have been proved in some plants [19]. Hattab et al. [258] observed a significant increase in ROS and cellular oxidative stress in *Medicago sativa* through influencing the expression of CuZn-SOD, GSH synthase (GS), and GPX against Pb stress. Fan et al. [259] observed an unknown protein, product of PSE1 (Pb-sensitive1) gene with NC domain, which is localized in cytoplasm and has potential for Pb tolerance in *A. thaliana*. Jiang et al. [260] studied the role of PDR12 knockout *Arabidopsis* and under Pb stress conditions and concluded that PDR12 is responsible for the activation of a Pb exclusion mechanism. ABC transporter of the mitochondria 3 (ATM3) [260–262], acyl-CoA-binding protein [263], and leucine-rich repeat2 (LRR2) and ethylene-insensitive 2 (EIN2) [264] are also important to regulate Pb transportation to the exterior of the cell [19,265]. A cytosol-localized malate dehydrogenase (CMDH4) protein functions as regulation of Pb tolerance mechanisms [19,266]. Pb is easily affected by GSH reductase in the plant cell [267]. Pb-mediated increased expression phytochelatins were also observed in *Salvinia minima* [268]. *M. sativa* plants showed 23-fold increased expression of PCS gene in the presence of Pb [258]. Furthermore, GMO plants develop efficient metabolic processes and over express genes/enzymes that are capable of bioremediation specific pollutant. Different omic-approaches help to explore different potential solutions targeting precise pollutants. For utilizing the omic-approaches below, certain research should be covered:

(a) Identification of candidate genes for effective and efficient removal of Pb contaminants.
(b) Diversity and phylogenetic studies of gene and protein sequences which control Pb bioremediation.
(c) Development of Genetically modified organism (GMO) plants through transgenesis.

GMO plants are capable of remediating various waste effluents and polluted lands and could be advantageous for bioremediation practical applications. Moreover, information on the fundamental omic-approaches concerned in bioremediation can also contribute towards the development of efficient bioremediation systems. Besides that, analysis of comparative genomic and proteomic study, their functional variations, as well as evolutionary relationships existing between them can contribute towards designing new efficient bioremediation systems. Systems biology information like molecular pathways, gene ontogeny analysis, co-expression, and protein-protein interactions can influence the Pb bioremediation processes. Therefore, with the help of bioinformatic analyses and modern biotechnological techniques, one can evaluate and justify the need for genetically modified organisms for the development of efficient remediation systems in the near future (Figure 1).

*9.5. Nano-Technological Approaches*

Recent scientific development in nanoscience research opens the way to cost-effective, eco-friendly, and sustainable remediation approaches. A nano technological approach has been successfully used in soil, sediments, solid waste, and a wastewater remediation [18,269] process. Nano-materials are dynamic, efficient, and broadly applicable with economic expediency [18,270]. The characteristic features of nano-materials such as Nanocatalysts, CNTs, graphenes, nano-scale metal oxides, nanomembranes, carbon nanotubes, nanobiological processes and zero-valent iron (FeO), $Fe_2O_3$, $Fe_3O_4$, $TiO_2$, $SiO_2$, and $Al_2O_3$ are summarized in Table 9. Nanoparticles (1–100 nm) provide very high adaptability for both in-situ and ex-situ remediation approaches [18]. Nanomaterials, nanoadsorbents, and nanosized compounds (quantum dots, nanofilms, nanoparticles, nanotubes, nanowires, and other various colloids) used for Pb remediation are listed in Table 9. Nanoparticles (less than 50 nm) have high potential as Pb adsorbents. Nano-adsorbents, i.e., activated carbon, alginate biopolymer, clay materials, silica, magnetic iron oxide nanoparticles (MNPs), metal oxides, nano-titanates, etc. have been utilized to remove Pb [18,271,272]. The researchers showed that nano-material can enhance the accumulation of metals by improving the cell wall permeability, co-transportation of nanomaterials with heavy metals, and transporter gene regulation [18,273].

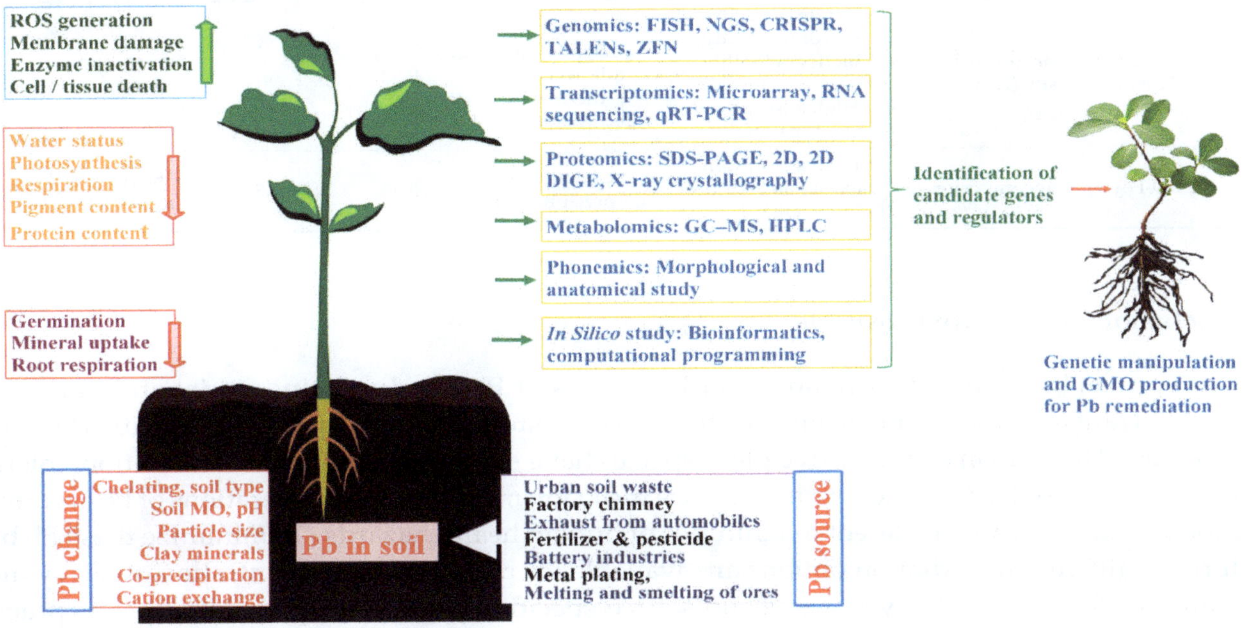


**Figure 1.** Biotechnological and Genetic Approaches for the development of efficient remediation systems.

**Table 9.** Characteristics of nano-particles in Pb removal [18].

| SN | Nano-Particles | Characters | NP Synthesis | Absorbent Dose | Optimum pH | Removal Efficiency | References |
|---|---|---|---|---|---|---|---|
| 1 | Iron oxides NPs | Magnetite nanoparticles | Co-precipitation from a mixture of Fe(II) and (III) salts with aqueous $NH_3$ and KOH | 50 mg/20cm$^3$ | 5.31–9.37 | Pb(II)—76–92% | [274] |
| 2 | Ferrite nano particles | - | Modified co-precipitation synthesis | 0.008 | - | Pb(II) up to 38.1% | [275] |
| 3 | Activated carbon NPs | High surface area and greater adsorption capacity | - | 0.02 | 2–10 | Pb up to 87% | [276] |
| 4 | Nano scale zero valentiron (nZVI) | High surface area and cation exchange capacity | Reduction of Fe(II) using borohydride | - | - | - | [277] |
| 5 | Starch stabilized zero valent Iron nanoparticles(nZVI-Starch) | Larger surface area for sorption reactions | Chemical reduction method | 1 g/kg soil | 4.2 | 100% | [278] |
| 6 | Zeolite materials obtained from fly ash | Greater specific area | Hydrothermal process | 6.0 | 5.6–6.6 | >80% | [279] |
| 7 | Pyromellitic acid dianhydride/N-(3-(trimethoxysilyl) propylethylene diamine(PMDA/TMSPEDA) | Bound heavy metal ions via co-ordinate and electrostatic interactions | Ring opening polymerization and sol-gel reaction | 0.01 | 7 | Pd(II)—79.60% | [280] |
| 8 | Ag and Zn nanoparticles functionalized cellulose | High catalytic activity, great biocompatibility, high adsorption capacity, high surface-area, reusability, and greater dispersion degree | Co-precipitation method | 0.5 and 1.0 | 5.5 | - | [281] |

**Table 9.** *Cont.*

| SN | Nano-Particles | Characters | NP Synthesis | Absorbent Dose | Optimum pH | Removal Efficiency | References |
|---|---|---|---|---|---|---|---|
| 9 | ZnO@Chitosancoreshell Nanocomposite (ZOCS) | Hydrophilicity, biocompatibility, biodegradability, non-toxicity, and High adsorption capacity | Direct precipitation followed by thermal decarbonation | 0.02 | 6 | Pb(II) up to 99% | [282] |
| 10 | ZnO-$Fe_3O_4$ nanocomposites | High adsorption capacity and surface area | Chemical co-precipitation | 0.50 | 5.5 | Pb(II) up to 39.2% | [283] |

## 10. Conclusion and Future Prospects

The source, bioaccumulation, and health hazards of Pb are due to industrial and agricultural activities. Translocation of Pb from soil to a crop system is a complex and species dependent phenomenon. The human consumptive plant species have shown different bioaccumulation, tolerance, and toxicity levels for lead. Based on the tolerance mechanism, different concentrations of Pb accumulate in the food chain and cause different magnitudes of human health hazards. To minimize these Pb based health risks, different remediation options are available for reducing the concentration of heavy metals in soil and the food chain. However, site and source-specific integrated approaches must be practiced to formulate suitable remediation strategies. Biological remediation, such as phytoremediation and PGPR, can be an environmentally friendly and cost-effective strategy for alleviating Pb toxicity in moderately contaminated soils. Eco-feasible technological innovations such as nano-tools and awareness among farmers' fraternity could possibly boost local economies and livelihoods with certain financial guarantees. Similarly, suggestive measures should be taken to ensure the sustained efficacy of Pb remediation such as the development of promising plants/mechanisms suitable for Pb phytoremediation. Exploitation of molecular approaches is required to manipulate Pb transporters and their cellular targeting to specific cell types. Development of transgenic plants with enhanced plant-microbe interaction is also a viable option to enhance phyto-remediation of Pb.

**Author Contributions:** The authors have made the following declaration about their contributions. Conceptualization: A.K. (Amit Kumar) [1], A.K. (Amit Kumar) [2], A.A.S., and A.K.C.; Methodology and structure of review: A.K. (Amit Kumar) [1], A.K. (Amit Kumar) [2], A.K.C., and C.-P.M.M.S.; Formal analysis of literature: A.K. (Amit Kumar) [1], G.S., and D.K.G.; Data curation and table preparation: D.K.G., S.K.M., C.-P.M.M.S., S.S.K., S.A.K., G.S., and K.K.Y.; Contribution in their respective specialization in different sections: A.K. (Amit Kumar) [1], A.K. (Amit Kumar) [2], C.-P.M.M.S., A.K.C., A.A.S., G.S., R.M., D.K.G., S.K.M., S.S.K., S.A.K., and K.K.Y.; The original draft preparation: A.K. (Amit Kumar) [1], A.K. (Amit Kumar) [2], A.K.C., G.S., R.M., and D.K.G.; Edited and finalized the manuscript: A.K. (Amit Kumar) [2], A.K.C., G.S., C.-P.M.M.S., D.K.G., and A.A.S. All authors have read and agreed to the published version of the manuscript.

## References

1. World Health Organisation (WHO). Action Is Needed on Chemicals of Major Public Health Concern. *Public Health Environ.* **2010**, 1–4. Available online: https://www.who.int/ipcs/assessment/public_health/chemicals_phc/en/ (accessed on 27 November 2019).
2. Arias, J.A.; Peralta-Videa, J.R.; Ellzey, J.T.; Ren, M.; Viveros, M.N.; Gardea-Torresdey, J.L. Effects of *Glomus deserticola* inoculation on Prosopis: Enhancing chromium and lead uptake and translocation as confirmed by X-ray mapping, ICP-OES and TEM techniques. *Environ. Exp. Bot.* **2010**, *68*, 139–148. [CrossRef]
3. ATSDR. Case Estudies in EnvironmentalL Medicine (CSEM) Lead Toxicity. Available online: https://www.atsdr.cdc.gov/csem/lead/docs/CSEM-Lead_toxicity_508.pdf (accessed on 27 November 2017).
4. Pais, I.; Jones, J.B. *The Handbook of Trace Elements*; Saint Lucie Press: Boca Raton, FL, USA, 1997.

5. Assi, M.A.; Hezmee, M.N.M.; Haron, A.W.; Sabri, M.Y.; Rajion, M.A. The detrimental effects of lead on human and animal health. *Vet. World* **2016**, *9*, 660–671. [CrossRef] [PubMed]
6. IARC. Agents Classified by the IARC Monographs. Volume 1–123. Available online: https://monographs.iarc.fr/wp-content/uploads/2019/02/List_of_Classifications.pdf (accessed on 24 November 2019).
7. EPA-IRIS. Lead and Compounds (Inorganic); CASRN 7439-92-1. Integrated Risk Information System (IRIS), Chemical Assessment Summary. Available online: https://cfpub.epa.gov/ncea/iris/iris_documents/documents/subst/0277_summary.pdf (accessed on 30 December 2019).
8. WHO. Lead Poisoning and Health. Available online: https://www.who.int/newsroom/fact-sheets/detail/lead-poisoning-and-health (accessed on 1 September 2019).
9. Saba, D.; Manouchehri, N.; Besançon, S.; El Samad, O.; Bou Khozam, R.; Nafeh Kassir, L.; Kassoufe, A.; Chebibe, H.; Ouainib, N.; Cambier, P. Bioaccessibility of lead in Dittrichia viscosa plants and risk assessment of human exposure around a fertilizer industry in Lebanon. *J. Environ. Manag.* **2019**, *250*, 109537. [CrossRef] [PubMed]
10. Sandin, G.; Peters, G.M. Environmental impact of textile reuse and recycling—A review. *J. Clean. Prod.* **2018**. [CrossRef]
11. Cabral-Pinto, M.M.S.; Inácio, M.; Neves, O.; Almeida, A.A.; Pinto, E.; Oliveiros, B.; Ferreira da Silva, E.A.F. Human Health Risk Assessment Due to Agricultural Activities and Crop Consumption in the Surroundingsof an Industrial Area. *Expo. Health* **2019**, 1–12. Available online: https://cerena.ist.utl.pt/news/paper-published-human-health-risk-assessment-due-agricultural-activities-and-crop-consumption (accessed on 23 November 2019).
12. Cabral-Pinto, M.M.S.; Ordens, C.M.; de Melo, M.T.C.; Inácio, M.; Almeida, A.; Pinto, E.; da Silva, E.A.F. An Inter-disciplinary Approach to Evaluate Human Health Risks Due to Long-Term Exposure to Contaminated Groundwater Near a Chemical Complex. *Expo. Health* **2019**, 1–16. Available online: https://espace.library.uq.edu.au/view/UQ:996aa4e (accessed on 15 December 2019). [CrossRef]
13. Cabral Pinto, M.; Ferreira da Silva, E. Heavy Metals of Santiago Island (Cape Verde) alluvial deposits: Baseline value maps and human health risk assessment. *Int. J. Environ. Res. Public Health* **2019**, *16*, 2. [CrossRef]
14. Lee, J.W.; Choi, H.; Hwang, U.K.; Kang, J.C.; Kang, Y.J.; Kim, K.; Kim, J.H. Toxic effects of lead exposure on bioaccumulation, oxidative stress, neurotoxicity, and immune responses in fish: A review. *Environ. Toxi.Pharma.* **2019**, *68*, 101–108. [CrossRef]
15. Hindarwati, Y.; Soeprobowati, T.R.; Sudarno. Heavy metal content in terraced rice fieldsat Sruwen Tengaran Semarang—Indonesia. *E3S Web Conf.* **2018**, *31*, 03009. [CrossRef]
16. Zulfiqar, U.; Farooq, M.; Hussain, S.; Maqsood, M.; Hussain, M.; Ishfaq, M.; Ahmad, M.; Anjum, M.J. Lead toxicity in plants: Impacts and remediation. *J. Environ. Manag.* **2019**, *250*, 109557. [CrossRef] [PubMed]
17. Kumar, A.; Chaturvedi, A.K.; Yadav, K.; kumar, K.P.; Malyan, S.K.; Raja, P.; Kumar, R.; Khan, S.A.; Yadav, K.K.; Rana, K.L.; et al. Fungal Phytoremediation of Heavy Metal-Contaminated Resources: Current Scenario and Future Prospects. Recent Advancement in White Biotechnology through Fungi. In *Fungal Biology*; Yadav, A., Singh, S., Mishra, S., Gupta, A., Eds.; Springer: Berlin, Germany, 2019.
18. Kumar, S.; Prasad, S.; Yadav, K.K.; Shrivastava, M.; Gupta, N.; Nagar, S.; Bach, Q.V.; Kamyab, H.; Khan, S.A.; Yadav, S.; et al. Hazardous heavy metals contamination of vegetables and food chain: Role of sustainable remediation approaches—A review. *Environ. Res.* **2019**, *179*, 108792. [CrossRef] [PubMed]
19. Kumar, A.; Prasad, M.N.V. Plant genetic engineering approach for the Pb and Zn remediation: Defense reactions and detoxification mechanisms. In *Transgenic Plant Technology for Remediation of Toxic Metals and Metalloids*; Elsevier: San Diego, CA, USA, 2019; Volume 17, pp. 359–380.
20. Gupta, N.; Yadav, K.K.; Kumar, V.; Kumar, S.; Chadd, R.P.; Kumar, A. Trace elements in soil-vegetables interface: Translocation, bioaccumulation, toxicity and amelioration: A review. *Sci. Total Environ.* **2019**, *651*, 2927–2942. [CrossRef] [PubMed]
21. Yadav, K.K.; Gupta, N.; Kumar, A.; Reece, L.M.; Singh, N.; Rezania, S.; Khan, S.A. Mechanistic understanding and holistic approach of phytoremediation: A review on application and future prospects. *Ecol. Eng.* **2018**, *120*, 274–298. [CrossRef]
22. Sammut, M.; Noack, Y.; Rose, J.; Hazemann, J.; Proux, O.; Depoux Ziebel, M.; Fiani, E. Speciation of Cd and Pb in dust emitted from sinter plant. *Chemosphere* **2010**, *78*, 445–450. [CrossRef] [PubMed]

23. Vega, F.; Andrade, M.; Covelo, E. Influence of soil properties on the sorption and retention of cadmium, copper and lead, separately and together, by 20 soil horizons: Comparison of linear regression and tree regression analyses. *J. Hazard. Mater.* **2010**, *174*, 522–533. [CrossRef] [PubMed]
24. Cecchi, M.; Dumat, C.; Alric, A.; Felix-Faure, B.; Pradere, P.; Guiresse, M. Multi-metal contamination of a calcic cambisol by fallout from a lead-recycling plant. *Geoderma* **2008**, *144*, 287–298. [CrossRef]
25. Punamiya, P.; Datta, R.; Sarkar, D.; Barber, S.; Patel, M.; Das, P. Symbiotic role of glomus mosseae in phytoextraction of lead in vetiver grass. *J. Hazard. Mater.* **2010**, *177*, 465–474. [CrossRef]
26. Ratul, A.K.; Hassan, M.; Uddin, M.K.; Sultana, M.S.; Akbor, M.A.; Ahsan, M.A. Potential health risk of heavy metals accumulation in vegetables irrigated with polluted riverwater. *Int. Food Res. J.* **2018**, *25*, 329–338.
27. Chaoua, S.; Boussaa, S.; El Gharmali, A.; Boumezzough, A. Impact of irrigation withwastewater on accumulation of heavy metals in soil and crops in the region of Marrakech inMorocco. *J. Saudi Soc. Agric. Sci.* **2018**, *18*, 429–436.
28. Yang, J.; Ma, S.; Zhou, J.; Song, Y.; Li, F. Heavy metal contamination in soils andvegetables and health risk assessment of inhabitants in Daye, China. *J. Int. Med. Res.* **2018**, *46*, 3374–3387. [CrossRef] [PubMed]
29. Rahimi, G.; Kolahchi, Z.; Charkhabi, A. Uptake and translocation of some heavy metalsby rice crop (*oryza sativa*) in paddy soils. *Agriculture* **2017**, *63*, 163–175.
30. Showqi, I.; Lone, F.A.; Naikoo, M. Preliminary assessment of heavy metals in water, sediment and macrophyte (*Lemna minor*) collected from Anchar Lake, Kashmir, India. *Appl. Water Sci.* **2018**, *8*, 3–11. [CrossRef]
31. Sawut, R.; Kasima, N.; Maihemuti, B.; Hue, L.; Abliz, A.; Abdujappar, A.; Kurbana, M. Pollution characteristics and health risk assessment of heavy metals in the vegetable bases ofnorthwest China. *Sci. Total Environ.* **2018**, *642*, 864–878. [CrossRef]
32. Latif, A.; Bilal, M.; Asghar, W.; Azeem, M.; Ahmad, M.I. Heavy metal accumulation invegetables and assessment of their potential health risk. *J. Environ. Anal. Chem.* **2018**, *5*, 234. [CrossRef]
33. Kumar, V.; Chopra, A.K. Heavy metals accumulation in soil and agricultural crops grown in the province of Asahi India Glass Ltd., Haridwar (Uttarakhand), India. *Adv. Crop Sci. Technol.* **2015**, *4*, 1. [CrossRef]
34. Shahid, M.; Pinelli, E.; Pourrut, B.; Silvestre, J.; Dumat, C. Lead-induced genotoxicity to *Vicia faba* L. roots in relation with metal cell uptake and initial speciation. *Ecotoxicol. Environ. Saf.* **2011**, *74*, 78–84. [CrossRef]
35. Tabelin, C.; Igarashi, T. Mechanisms of arsenic and lead release from hydrothermally altered rock. *J. Hazard. Mater.* **2009**, *169*, 980–990. [CrossRef]
36. Lawal, O.; Sanni, A.; Ajayi, I.; Rabiu, O. Equilibrium, thermodynamic and kinetic studies for the biosorption of aqueous lead(II) ions onto the seed husk of *Calophyllum inophyllum*. *J. Hazard. Mater.* **2010**, *177*, 829–835. [CrossRef] [PubMed]
37. Bi, X.; Ren, L.; Gong, M.; He, Y.; Wang, L.; Ma, Z. Transfer of cadmium and lead from soil to mangoes in an uncontaminated area, Hainan Island, China. *Geoderma* **2010**, *155*, 115–120. [CrossRef]
38. Cenkci, S.; Cigerci, I.H.; Yildiz, M.; Ozay, C.; Bozdag, A.; Terzi, H. Lead contamination reduces chlorophyll biosynthesis and genomic template stability in *Brassica rapa* L. *Environ. Exp. Bot.* **2010**, *67*, 467–473. [CrossRef]
39. Padmavathiamma, P.K.; Li, L.Y. Phytoavailability and fractionation of lead and manganese in a contaminated soil after application of three amendments. *Bioresour. Technol.* **2010**, *101*, 5667–5676. [CrossRef]
40. Kopittke, P.M.; Asher, C.J.; Kopittke, R.A.; Menzies, N.W. Prediction of Pb speciation in concentrated and dilute nutrient solutions. *Environ. Pollut.* **2008**, *153*, 548–554. [CrossRef] [PubMed]
41. Komjarova, I.; Blust, R. Effect of Na, Ca and pH on simultaneous uptake of Cd, Cu, Ni, Pb, and Zn in the water flea *Daphnia magna* measured using stable isotopes. *Aquat. Toxicol.* **2009**, *94*, 81–86. [CrossRef] [PubMed]
42. Mager, E.M.; Esbaugh, A.J.; Brix, K.V.; Ryan, A.C.; Grosell, M. Influences of water chemistry on the acute toxicity of lead to Pimephales promelas and Ceriodaphnia dubia. *Comp. Biochem. Physiol. C Toxicol. Pharmacol.* **2011**, *153*, 82–90. [CrossRef] [PubMed]
43. Levin, R.; Vieira, C.L.Z.; Mordarski, D.C.; Rosenbaum, M.H. Lead seasonality in humans, animals, and the natural environment. *Environ. Res.* **2019**, *180*. [CrossRef]
44. Esbaugh, A.J.; Brix, K.V.; Mager, E.M.; De Schamphelaere, K.; Grosell, M. Multilinear regression analysis preliminary biotic ligand modeling, and cross species comparison of the effects of water chemistry on chronic lead toxicity in invertebrates. *Comp. Biochem. Physiol. C Toxicol. Pharmacol.* **2012**, *155*, 423–431. [CrossRef]
45. Kamel, M.M.; Ibrahm, M.A.; Ismael, A.M.; Motaleeb, M.A.E. Adsorption of some heavy metal ions from aqueous solutions by using kaolinite clay. Ass. Univ. *Bull. Environ. Res.* **2004**, *7*, 101–109.

46. Suzuki, T.; Niinae, M.; Koga, T.; Akita, T.; Ohta, M.; Choso, T. EDDS-enhanced electrokinetic remediation of heavy metal-contaminated clay soils under neutral pH conditions. *Colloids Surf. Physicochem. Eng. Asp.* **2014**, *440*, 145–150. [CrossRef]
47. Mao, L.C.; Bailey, E.H.; Chester, J.; Dean, J.; Ander, E.L.; Chenery, S.R.; Young, S.D. Lability of Pb in soil: Effects of soil properties and contaminant source. *Environ. Chem.* **2014**, *11*, 690–701. [CrossRef]
48. Li, X.; Meng, D.; Li, J.; Yin, H.; Liu, H.; Liu, X. Response of soil microbial communities and microbial interactions to long-term heavy metal contamination. *Environ. Pollut.* **2017**, *231*, 908–917. [CrossRef] [PubMed]
49. Kögel-Knabner, W.; Amelung, Z.; Cao, S.; Fiedler, P.; Frenzel, R. Biogeochemistry of paddy soils. *Geoderma* **2010**, *157*, 1–14. [CrossRef]
50. O'Reilly, S.E.; Hochella, M.F. Lead sorption efficiencies of natural and synthetic Mn and Fe-oxides. *Geochem. Cosmochim. Acta* **2003**, *67*, 4471–4487. [CrossRef]
51. Tao, J.; Liu, X.; Liang, Y.; Niu, J.; Xiao, Y.; Gu, Y. Maize growth responses to soil microbes and soil properties after fertilization with different green manures. *Appl. Microbiol. Biotechnol.* **2016**, *13*, 1–11. [CrossRef]
52. Xu, X.; Hui, D.; King, A.W.; Song, X.; Thornton, P.E.; Zhang, L. Convergence of microbial assimilations of soil carbon, nitrogen, phosphorus, and sulfur in terrestrial ecosystems. *Sci. Rep.* **2015**, *5*, 17445. [CrossRef]
53. Frank, J.J.; Poulakos, A.G.; Tornero-Velez, R.; Xue, J. Systematic review and meta-analyses of lead (Pb) concentrations in environmental media (soil, dust, water, food, and air) reported in the United States from 1996 to 2016. *Sci. Tot. Envir.* **2019**, *694*, 133489. [CrossRef]
54. Zhao, F.J.; Ma, Y.B.; Zhu, Y.G.; Tang, Z.; McGrath, S.P. Soil contamination in China: Current status and mitigation strategies. *Environ. Sci. Technol.* **2015**, *49*, 750–759. [CrossRef]
55. Emsley, J. *Nature's Building Blocks: An AZ Guide to the Elements*; Oxford University Press: Oxford, UK, 2011.
56. CDC. *Advisory Committee on Childhood Lead Poisoning Prevention (ACCLPP)*; Jessica Kingsley Publishers: London, UK, 2012.
57. United States Environmental Protection Agency (USEPA). *Estimation of Relative Bioavailability of Lead in Soil and Soil-Like Materials Using In Vivo and In Vitro Methods*; Report No.: OSWER 9285.7-77; Office of Solid Waste and Emergency Response: Washington, DC, USA, 2007.
58. Yan, K.; Dong, Z.; Wijayawardena, M.A.A.; Liu, Y.; Naidu, R.; Semple, K. Measurement of soil lead bioavailability and influence of soil types and properties: A review. *Chemosphere* **2017**, *184*, 27–42. [CrossRef]
59. Denys, S.; Caboche, J.; Tack, K.; Rychen, G.; Wragg, J.; Cave, M.; Jondreville, C.; Feidt, C. In Vivo Validation of the Unified BARGE Method to Assess the Bioaccessibility of Arsenic, Antimony, Cadmium, and Lead in Soils. *Environ. Sci. Technol.* **2012**, *46*, 6252–6260. [CrossRef]
60. Deshommes, E.; Tardif, R.; Edwards, M.; Sauvé, S.; Prevost, M. Experimental determination of the oral bioavailability and bioaccessibility of lead particles. *Chem. Cent. J.* **2012**, *6*, 138–154. [CrossRef]
61. Jiang, W.; Liu, D. Pb-induced cellular defense system in the root meristematic cells of *Allium sativum* L. *BMC Plant Biol.* **2010**, *10*, 1–40. [CrossRef] [PubMed]
62. Yan, Z.Z.; Ke, L.; Tam, N.F.Y. Lead stress in seedlings of *Avicennia marina*, a common mangrove species in South China, with and without cotyledons. *Aquat. Bot.* **2010**, *92*, 112–118. [CrossRef]
63. Piechalak, A.; Tomaszewska, B.; Baralkiewicz, D.; Malecka, A. Accumulation and detoxification of lead ions in legumes. *Phytochemistry* **2002**, *60*, 153–162. [CrossRef]
64. Małecka, A.; Piechalak, A.; Morkunas, I.; Tomaszewska, B. Accumulation of lead in root cells of Pisum sativum. *Acta Physiol. Plant.* **2008**, *30*, 629–637. [CrossRef]
65. Brunet, J.; Varrault, G.; Zuily-Fodil, Y.; Repellin, A. Accumulation of lead in the roots of grass pea (*Lathyrus sativus* L.) plants triggers systemic variation in gene expression in the shoots. *Chemosphere* **2009**, *77*, 1113–1120. [CrossRef]
66. Gichner, T.; Znidar, I.; Száková, J. Evaluation of DNA damage and mutagenicity induced by lead in tobacco plants. *Mutat. Res. Genet. Toxicol. Environ. Mutagen.* **2008**, *652*, 186–190. [CrossRef]
67. Gupta, D.; Huang, H.; Yang, X.; Razafindrabe, B.; Inouhe, M. The detoxification of lead in Sedum alfredii H. is not related to phytochelatins but the glutathione. *J. Hazard. Mater.* **2010**, *177*, 437–444. [CrossRef]
68. Kopittke, P.M.; Asher, C.J.; Kopittke, R.A.; Menzies, N.W. Toxic effects of $Pb^{2+}$ on growth of cowpea (Vigna unguiculata). *Environ. Pollut.* **2007**, *150*, 280–287. [CrossRef]
69. Metanat, K.; Ghasemi-Fasaei, R.; Ronaghi, A.; Yasrebi, J. Lead Phytostabilization and Cationic Micronutrient Uptake by Maize as Influenced by Pb Levels and Application of Low Molecular Weight Organic Acids. *Commun. Soil Sci. Plant Anal.* **2019**, *50*, 1–10. [CrossRef]

70. Meyers, D.E.R.; Auchterlonie, G.J.; Webb, R.I.; Wood, B. Uptake and localisation of lead in the root system of Brassica juncea. *Environ. Pollut.* **2008**, *153*, 323–332. [CrossRef]
71. Ginn, B.R.; Szymanowski, J.S.; Fein, J.B. Metal and proton binding onto the roots of Fescue rubra. *Chem Geol.* **2008**, *253*, 130–135. [CrossRef]
72. Krzeslowska, M.; Lenartowska, M.; Mellerowicz, E.J.; Samardakiewicz, S.; Wozny, A. Pectinous cell wall thickenings formation–a response of moss protonemata cells to lead. *Environ. Exp. Bot.* **2009**, *65*, 119–131. [CrossRef]
73. Krzesłowska, M.; Lenartowska, M.; Samardakiewicz, S.; Bilski, H.; Wozny, A. Lead deposited in the cell wall of Funaria hygrometrica protonemata is not stable–a remobilization can occur. *Environ. Pollut.* **2010**, *158*, 325–338. [CrossRef] [PubMed]
74. Uzu, G.; Sobanska, S.; Sarret, G.; Munoz, M.; Dumat, C. Foliar lead uptake by lettuce exposed to atmospheric fallouts. *Environ. Sci. Technol.* **2010**, *44*, 1036–1042. [CrossRef] [PubMed]
75. De Forest, D.K.; Brix, K.V.; Adams, W.J. Assessing metal bioaccumulation in aquatic environments: The inverse relationship between bioaccumulation factors, trophic transfer factors and exposure concentration. *Aquat. Toxicol.* **2007**, *84*, 236–246. [CrossRef] [PubMed]
76. Bhatia, A.; Singh, S.; Kumar, A. Heavy metal contamination of soil, irrigation water and vegetables in peri-urban agricultural areas and markets of Delhi. *Water Environ. Res.* **2015**, *87*, 2027–2034. [CrossRef] [PubMed]
77. Jan, F.A.; Ishaq, M.; Khan, S. A comparative study of human health risks via consumptionof food crops grown on wastewater irrigated soil (Peshawar) and relatively clean waterirrigated soil (lower Dir). *J. Hazard. Mater.* **2010**, *179*, 612–621. [CrossRef]
78. Balkhair, K.S.; Ashraf, M.A. Field accumulation risks of heavy metals in soil andvegetable crop irrigated with sewage water in western region of Saudi Arabia. *Saudi J. Biol. Sci.* **2016**, *23*, 32–44. [CrossRef]
79. Maurya, P.K.; Malik, D.S.; Yadav, K.K.; Kumar, A.; Kumar, S.; Kamyab, H. Bioaccumulation and potential sources of heavy metal contamination in fish species in River Ganga basin: Possible human health risks evaluation. *Toxicol. Rep.* **2019**, *6*, 472–481. [CrossRef]
80. Chalkiadaki, O.; Dassenakis, M.; Lydakis-Simantiris, N. Bioconcentration of Cd and Ni invarious tissues of two marine bivalves living in different habitats and exposed to heavilypolluted seawater. *Chem. Ecol.* **2014**, *30*, 726–742. [CrossRef]
81. Ziyaadini, M.; Yousefiyanpour, Z.; Ghasemzadeh, J. Biota-sediment accumulation factorand concentration of heavy metals (Hg, Cd, As, Ni, Pb and Cu) in sediments and tissues ofChiton lamyi (Mollusca: Polyplacophora: Chitonidae) in Chabahar Bay, Iran. *Iran. J. Fish. Sci.* **2017**, *16*, 1123–1134.
82. Yarsan, E.; Yipel, M. The important terms of marine pollution biomarkers andbiomonitoring, bioaccumulation, bioconcentration, biomagnification. *J. Mol. Biomark. Diagn.* **2013**, *1*, 1–4. [CrossRef]
83. Conder, J.M.; Gobas, F.A.P.C.; Borga, K. Use of trophic magnification factors and relatedmeasures to characterize bioaccumulation potential of chemicals. *Integr. Environ. Assess. Manag.* **2012**, *8*, 85–97. [CrossRef]
84. Zhang, C.; Wang, X.; Ashraf, U.; Qiu, B.; Ali, S. Transfer of lead (Pb) in the soil-plant-mealybug-ladybird beetle food chain, a comparison between two host plants. *Ecotoxicol. Environ. Saf.* **2017**, *143*, 289–295. [CrossRef]
85. Xiong, Z.; Zhao, F.; Li, M. Lead toxicity in Brassica pekinensis Rupr: Effect on nitrate assimilation and growth. *Environ. Toxicol.* **2006**, *21*, 147–153. [CrossRef] [PubMed]
86. Arshad, M.; Silvestre, J.; Pinelli, E.; Kallerhoff, J.; Kaemmerer, M.; Tarigo, A.; Shahid, M.; Guiresse, M.; Pradere, P.; Dumat, C. A field study of lead phytoextraction by various scented Pelargonium cultivars. *Chemosphere* **2008**, *71*, 2187–2192. [CrossRef] [PubMed]
87. Maestri, E.; Marmiroli, M.; Visioli, G.; Marmiroli, N. Metal tolerance and hyperaccumulation: Costs and trade-offs between traits and environment. *Environ. Exp. Bot.* **2010**, *68*, 1–13. [CrossRef]
88. Liu, W.; Zhou, Q.; Zhang, Y.; Wei, S. Lead accumulation in different Chinese cabbage cultivars and screening for pollution-safe cultivars. *J. Environ. Manag.* **2010**, *91*, 781–788. [CrossRef]
89. Verbruggen, N.; Hermans, C.; Schat, H. Molecular mechanisms of metal hyperaccumulation in plants. *New Phytol.* **2009**, *181*, 759–776. [CrossRef]
90. Liao, Y.; Chien, S.C.; Wang, M.; Shen, Y.; Hung, P.; Das, B. Effect of transpiration on Pb uptake by lettuce and on water soluble low molecular weight organic acids in rhizosphere. *Chemosphere* **2006**, *65*, 343–351. [CrossRef]

91. Wang, H.; Shan, X.; Wen, B.; Owens, G.; Fang, J.; Zhang, S. Effect of indole-3-acetic acid on lead accumulation in maize (*Zea mays* L.) seedlings and the relevant antioxidant response. *Environ. Exp. Bot.* **2007**, *61*, 246–253. [CrossRef]
92. Kim, Y.Y.; Yang, Y.Y.; Lee, Y. Pb and Cd uptake in rice roots. *Physiol. Plant.* **2002**, *116*, 368–372. [CrossRef]
93. Pourrut, B.; Perchet, G.; Silvestre, J.; Cecchi, M.; Guiresse, M.; Pinelli, E. Potential role of NADPH-oxidase in early steps of lead-induced oxidative burst in Vicia faba roots. *J. Plant Physiol.* **2008**, *165*, 571–579. [CrossRef] [PubMed]
94. Wojas, S.; Ruszczynska, A.; Bulska, E.; Wojciechowski, M.; Antosiewicz, D.M. $Ca^{2+}$-dependent plant response to $Pb^{2+}$ is regulated by LCT1. *Environ. Pollut.* **2007**, *147*, 584–592. [CrossRef]
95. Sonawane, V.Y. Analysis of Heavy metals in vegetables collected from selected area around Dhulia, North Maharashtra, Maharashtra, India. *Analysis* **2015**, *8*, 1935–1939.
96. Jolly, Y.N.; Islam, A.; Akbar, S. Transfer of metals from soil to vegetables and possible health risk assessment. *SpringerPlus* **2013**, *2*, 385. [CrossRef]
97. Ali, M.H.; Al-Qahtani, K.M. Assessment of some heavy metals in vegetables, cereals and fruits in Saudi Arabian markets. *Egypt. J. Aquat. Res.* **2012**, *38*, 31–37. [CrossRef]
98. Labhade, K.R. Assessment of heavy metal contamination in vegetables grown in and around Nashik City, Maharashtra State, India. *IOSR J. Appl. Chem.* **2013**, *5*, 9–14.
99. Mohod, C.V. A review on the concentration of the heavy metals in vegetable samples like spinach and tomato grown near the area of AmbaNalla of Amravati City. *Int. J. Innov. Res. Sci. Eng.* **2015**, *4*, 2788–2792. [CrossRef]
100. Maleki, A.; Amini, H.; Nazmara, S.; Zandi, S.; Mahvi, A.H. Spatial distribution of heavy metals in soil, water, and vegetables of farms in Sanandaj, Kurdistan, Iran. *J. Environ. Health Sci. Eng.* **2014**, *12*, 136. [CrossRef]
101. Gupta, S.; Jena, V.; Jena, S.; Davić, N.; Matić, N.; Radojević, D.; Solanki, J.S. Assessment of heavy metal contents of green leafy vegetables. *Croat. J. Food Sci. Technol.* **2013**, *5*, 53–60.
102. Ramesh, H.L.; YoganandaMoorthy, V.N. Assessment of heavy metal contamination in green leafy vegetables grown in Bangalore urban district of Karnataka. *Adv. Life Sci. Technol.* **2012**, *6*, 40–51.
103. Ramteke, S.; Sahu, B.L.; Dahariya, N.S.; Patel, K.S.; Blazhev, B.; Matini, L. Heavy metal contamination of vegetables. *J. Environ. Prot.* **2016**, *7*, 996–1004. [CrossRef]
104. Guerra, F.; Trevizam, A.R.; Muraoka, T.; Marcante, N.C.; Canniatti-Brazaca, S.G. Heavy metals in vegetables and potential risk for human health. *Sci. Agric.* **2012**, *69*, 54–60. [CrossRef]
105. Anwar, S.; Nawaz, M.F.; Gul, S.; Rizwan, M.; Ali, S.; Kareem, A. Uptake and distribution of minerals and heavy metals in commonly grown leafy vegetable species irrigated with sewage water. *Environ. Monit. Assess.* **2016**, *188*, 541. [CrossRef] [PubMed]
106. Gebrekidan, A.; Weldegebriel, Y.; Hadera, A.; Van der Bruggen, B. Toxicological assessment of heavy metals accumulated in vegetables and fruits grown in Ginfelriver near Sheba Tannery, Tigray, Northern Ethiopia. *Ecotoxicol. Environ. Saf.* **2013**, *95*, 171–178. [CrossRef]
107. Deribachew, B.; Amde, M.; Nigussie-Dechassa, R.; Taddese, A.M. Selected heavy metals in some vegetables produced through wastewater irrigation and their toxicological implications in Eastern Ethiopia. *Afr. J. Food Agric. Nutr. Dev.* **2015**, *15*, 10013–10032.
108. Mohammed, N.K.; Khamis, F.O. Assessment of heavy metal contamination in vegetables consumed in Zanzibars. *Nat. Sci.* **2012**, *4*, 588. [CrossRef]
109. Priya, E.S.; Sunil, G.; Shivaiah, K.; Gaddameedi, A.; Kumar, A. Extent of heavy metal contamination in leafy vegetables, soil and water from surrounding of Musi river, Hyderabad, India. *J. Ind. Pollut. Control.* **2014**, *30*, 289–293.
110. Adedokun, A.H.; Njoku, K.L.; Akinola, M.O.; Adesuyi, A.A.; Jolaoso, A.O. Potential human health risk assessment of heavy metals intake via consumption of some leafy vegetables obtained from four market in Lagos metropolis, Nigeria. *J. Appl. Sci. Environ. Manag.* **2016**, *20*, 530–539. [CrossRef]
111. Taghipour, H.; Mosaferi, M. Heavy metals in the vegetables collected from production sites. *Health Promot. Perspect.* **2013**, *3*, 185–193. [PubMed]
112. Soloman, P.E.; Jain, S.; Chauhan, S.S. Bio accretion of heavy metals by Okra and Eggplant grown in polluted areas of Jaipur City and associated health risks. *Int. J. Innov. Res. Sci. Eng. Technol.* **2017**, *6*, 428–434.
113. Islam, M.S.; Hoque, M.F. Concentrations of heavy metals in vegetables around the industrial area of Dhaka city, Bangladesh and health risk assessment. *Int. Food Res. J.* **2014**, *21*, 12–18.

114. Jawad, I. Determination of Heavy Met Herbs Available on the Iraq. *Adv. Environ. Biol.* **2016**, *10*, 66–69.
115. Matloob, M.H. Using Stripping Voltammetry to Determine Heavy Metals in Cooking Spices Used in Iraq. *Pol. J. Environ. Stud.* **2016**, *25*, 2057–2070. [CrossRef]
116. Inam, F.; Deo, S.; Narkhede, N. Analysis of minerals and heavy metals in some spices collected from local market. *J. Pharaceut. Biol. Sci.* **2013**, *8*, 40–43. [CrossRef]
117. Gaya, U.I.; Ikechukwu, S.A. Heavy metal contamination of selected spices obtained from Nigeria. *J. Appl. Sci. Environ. Manag.* **2016**, *20*, 681–688. [CrossRef]
118. Nkansah, M.A.; Amoako, C.O. Heavy metal content of some common spices available in markets in the Kumasi metropolis of Ghana. *Am. J. Sci. Ind. Res.* **2010**, *1*, 158–163. [CrossRef]
119. Esetlili, B.Ç.; Pekcan, T.; Çobanoğlu, Ö.; Aydoğdu, E.; Turan, S.; Anac, D. Essential plant nutrients and heavy metals concentrations of some medicinal and aromatic plants. *J. Agric. Sci.* **2014**, *20*, 239–247.
120. Mosleh, Y.Y.; Mofeed, J.; Almaghrabi, O.A.; Kadasa, N.M.; El-Alzahrani, H.S.; Fuller, M.P. Residues of heavy metals, PCDDs, PCDFs, and DL-PCBs some medicinal plants collected randomly from the Jeddah, central market. *Life Sci. J.* **2014**, *11*, 1–8.
121. Ogunkunle, A.T.J.; Bello, O.S.; Ojofeitimi, O.S. Determination of heavy metal contamination of street-vended fruits and vegetables in Lagos state, Nigeria. *Int. Food Res. J.* **2014**, *21*, 18–27.
122. Shaheen, N.; Irfan, N.M.; Khan, I.N.; Islam, S.; Islam, M.S.; Ahmed, M.K. Presence of heavy metals in fruits and vegetables: Health risk implications in Bangladesh. *Chemosphere* **2016**, *152*, 431–438. [CrossRef] [PubMed]
123. Ibraheen, L.H.; Abed, S.A. Accumulation detection of some heavy metals in some types of fruits in the local market of Al-Diwaniyah City, Iraq. *Rasayan J. Chem.* **2017**, *10*, 339–343.
124. Radwan, M.A.; Salama, A.K. Market basket survey for some heavy metals in Egyptian fruits and vegetables. *Food Chem. Toxicol.* **2006**, *44*, 1273–1278. [CrossRef]
125. Sobukola, O.P.; Adeniran, O.M.; Odedairo, A.A.; Kajihausa, O.E. Heavy metal levels of some fruits and leafy vegetables from selected markets in Lagos, Nigeria. *Afr. J. Food Sci.* **2010**, *4*, 389–393.
126. Dahiru, M.F.; Umar, A.B.; Sani, M.D. Cadmium, copper, lead and zinc levels in sorghum and millet grown in the city of Kano and its environs. *Glob. Adv. Res. J. Environ. Sci. Toxicol.* **2013**, *2*, 82–85.
127. Guo, G.; Lei, M.; Wang, Y.; Song, B.; Yang, J. Accumulation of As, Cd, and Pb in sixteen wheat cultivars grown in contaminated soils and associated health risk assessment. *Int. J. Environ. Res. Public Health* **2018**, *15*, 2601. [CrossRef]
128. Rezapour, S.; Atashpaz, B.; Moghaddam, S.S.; Damalas, C.A. Heavy metal bioavailability and accumulation in winter wheat (*Triticum aestivum* L.) irrigated with treated wastewater in calcareous soils. *Sci. Total Environ.* **2018**, *656*, 261–269. [CrossRef]
129. Sakizadeh, M.; Ghorbani, H. Concentration of heavy metals in soil and staple crops and the associated health risk. *Arch. Hyg. Sci.* **2017**, *6*, 303–313. [CrossRef]
130. Haseeb, M.; Basra, S.M.A.; Afzal, I.; Wahid, A. Quinoa response to lead: Growth and lead partitioning. *Int. J. Agric. Biol.* **2018**, *20*, 338–344. [CrossRef]
131. Kacalkova, L.; Tlustos, P.; Szakova, J. Chromium, nickel, cadmium, and lead accumulation in maize, sunflower, willow, and poplar. *Pol. J. Environ. Stud.* **2014**, *23*, 753–761.
132. Akenga, T.; Sudoi, V.; Machuka, W.; Kerich, E.; Ronoh, E. Heavy metals uptake in maize grains and leaves in different agro ecological zones in Uasin Gishu County. *J. Environ. Prot.* **2017**, *8*, 1435–1444. [CrossRef]
133. Islam, M.S.; Ahmed, M.K.; Habibullah-Al-Mamun, M. Heavy metals in cereals and pulses: Health implications in Bangladesh. *J. Agric. Food Chem.* **2014**, *62*, 10828–10835. [CrossRef] [PubMed]
134. Lee, K.J.; Feng, Y.Y.; Choi, D.H.; Lee, B.W. Lead accumulation and distribution in different rice cultivars. *J. Crop Sci. Biotechnol.* **2016**, *19*, 323–328. [CrossRef]
135. Ihedioha, J.N.; Ujam, O.T.; Nwuche, C.O.; Ekere, N.R.; Chime, C.C. Assessment of heavy metal contamination of rice grains (Oryza sativa) and soil from Ada field, Enugu, Nigeria: Estimating the human health risk. *Hum. Ecol. Risk Assess. Int. J.* **2016**, *22*, 1665–1677. [CrossRef]
136. Kahraman, A.; Onder, M. Accumulation of heavy metals in dry beans sown on different dates. *J. Elem.* **2018**, *23*, 201–216. [CrossRef]
137. Sadrabad, E.K.; Boroujeni, H.M.; Heydari, A. Heavy metal accumulation in soybeans cultivated in Iran, 2015–2016. *J. Nutr. Food Secur.* **2018**, *3*, 27–32.

138. Palizban, A.; Badii, A.; Asghari, G.; Nafchi, M. Lead and cadmium contamination in seeds and oils of Brassica napus L. and Carthamus tinctorius grown in Isfahan province/Iran. *Iran. J. Toxicol.* **2015**, *8*, 1196–1202.
139. Barbillon, A.; Aubry, C.; Nold, F.; Besancon, S.; Manouchehri, N. Health Risks Assessment in Three Urban Farms of Paris Region for Different Scenarios of Urban Agricultural Users: A Case of Soil Trace Metals Contamination. *J. Agric. Sci.* **2019**, *10*, 352–370. [CrossRef]
140. Liu, X.; Song, Q.; Tang, Y.; Li, W.; Xu, J.; Wu, J.; Fan, W.; Brookes, P.C. Human health risk assessment of heavy metals in soil–vegetable system: A multi-medium analysis. *Sci. Total Environ.* **2013**, *463–464*, 530–540. [CrossRef]
141. Laidlaw, M.A.S.; Filippelli, G.M.; Sadler, R.C.; Gonzales, C.R.; Ball, A.S.; Mielke, H.W. Children's Blood Lead Seasonality in Flint, Michigan (USA), and Soil-Sourced Lead Hazard Risks. *Int. J. Environ. Res. Public Health* **2016**, *13*, 358. [CrossRef] [PubMed]
142. Ngueta, G.; Abdous, B.; Tardif, R.; St-Laurent, J.; Levallois, P. Use of a cumulative exposure index to estimate the impact of tap water lead concentration on blood lead levels in 1- to 5-year-old children (Montreal, Canada). *Environ. Health Perspect.* **2016**, *124*, 388–395. [CrossRef] [PubMed]
143. Zahran, S.; Laidlaw, M.A.S.; McElmurry, S.P.; Filippelli, G.M.; Taylor, M. Linking Source and Effect: Resuspended Soil Lead, Air Lead, and Children's Blood Lead Levels in Detroit, Michigan. *Environ. Sci. Technol.* **2013**, *47*, 2839–2845. [CrossRef] [PubMed]
144. Lebov, J.; Grieger, K.; Womack, D.; Zaccaro, D.; Whitehead, N.; Kowalcyk, B.; MacDonald, P.D.M. A framework for One Health research. *One Health* **2017**, *3*, 44–50. [CrossRef] [PubMed]
145. Davisa, M.F.; Rankin, S.C.; Schurer, J.M.; Cole, S.; Conti, L.; Rabinowitz, P. COHERE Expert Review Group. Checklist for One Health Epidemiological Reporting of Evidence (COHERE). *One Health* **2017**, *4*, 14–21. [CrossRef] [PubMed]
146. Kaur, G.; Singh, H.P.; Batish, D.R.; Kohli, R.K. Adaptations to oxidative stress in Zea mays roots under short term $Pb^{2+}$ exposure. *Biologia* **2015**, *70*, 190–197. [CrossRef]
147. Rai, P.K.; Lee, S.S.; Zhang, M.; Tsang, Y.F.; Kim, K.H. Heavy metals in food crops: Health risks, fate, mechanisms, and Management. *Environ. Inter.* **2019**, *125*, 365–385. [CrossRef]
148. Ekmekci, Y.; Tanyolac, D.; Ayhan, B. A crop tolerating oxidative stress induced by excess lead: Maize. *Acta Physiol. Plant.* **2009**, *31*, 319–330. [CrossRef]
149. Kabata-Pendias, A.; Szteke, B. *Trace Elements in Abiotic and Biotic Environments*; CRC Press: Boca Raton, FL, USA, 2015.
150. World Health Organisation (WHO). Safety evaluation of certain food additives and contaminants. Available online: https://apps.who.int/iris/bitstream/handle/10665/44515/WHO_TRS_960_eng.pdf;jsessionid=AFB1D9565016562EA09FFBFF1734B190?sequence=1 (accessed on 25 December 2019).
151. Hsu, P.C.; Guo, Y.L. Antioxidant nutrients and lead toxicity. *Toxicology* **2002**, *180*, 33–44. [CrossRef]
152. Concha, G.; Eneroth, H.; Hallstrom, H.; Sand, S. *Contaminants and Minerals in Foods for Infants and Young Children*; National Food Agency: Uppsala, Sweden, 2013.
153. World Health Organisation (WHO). *Childhood Lead Poisoning*; WHO: Geneva, Switzerland, 2010.
154. EC. Commission of the European Communities. Commission Regulation (EC) No. 1881/2006 Regulation ofsetting maximum levels for certain contaminants infoodstuffs L364-5/L364-24. *Off. J. Eur. Union* **2006**, *364*, 5–24.
155. Bellinger, D.C.; Malin, A.; Wright, R.O. The Neurodevelopmental Toxicity of Lead: History, Epidemiology, and Public Health Implications. *Adv. Neurotoxicol.* **2018**, *2*, 1–26.
156. Wani, A.L.; Ara, A.; Usmani, J.A. Lead toxicity: A review. *Interdiscip. Toxicol.* **2015**, *8*, 55–64. [CrossRef] [PubMed]
157. Gao, H.; Liu, C.P.; Song, S.Q.; Fu, J. Effects of Dietary Selenium Against Lead Toxicity on mRNA Levels of 25 Selenoprotein Genes in the Cartilage Tissue of Broiler Chicken. *Biol. Trace Elem. Res.* **2016**, *172*, 234–241. [CrossRef] [PubMed]
158. Rahman, S.; Joshi, M.V. Effect of lead toxicity on growth and performance of broilers. *Tamilnadu J. Vet. Anim. Sci.* **2009**, *5*, 59–62.
159. Flora, G.; Gupta, D.; Tiwari, A. Toxicity of lead: A review with recent updates. *Interdiscip. Toxicol.* **2012**, *5*, 47–58. [CrossRef]
160. Flora, S.J.S. Nutritional components modify metal absorption, toxic response and chelation therapy. *J. Nutr. Environ. Med.* **2002**, *12*, 53–67. [CrossRef]

161. Kasperczyk, A.; Machnik, G.; Dobrakowski, M.; Sypniewski, D.; Birkner, E.; Kasperczyk, S. Gene expression and activity of antioxidant enzymes in the blood cells of workers who were occupationally exposed to lead. *Toxicology* **2012**, *301*, 79–84. [CrossRef]
162. Kasperczyk, S.; Dobrakowski, M.; Kasperczyk, A.; Ostałowska, A.; Birkner, E. The administration of N-acetylcysteine reduces oxidative stress and regulates glutathione metabolism in the blood cells of workers exposed to lead. *Clin. Toxicol.* **2013**, *51*, 480–486. [CrossRef]
163. Gangoso, L.; Lloret, P.A.; Rodrıguez-Navarro, A.A.B.; Mateo, R.; Hiraldo, F.; Donazar, J.A. Long-term effects of lead poisoning on bone mineralization in vultures exposed to ammunition sources. *Environ. Pollut.* **2009**, *157*, 569–574. [CrossRef]
164. Theppeang, K.; Glass, T.A.; Bandeen-Roche, K.; Todd, A.C.; Rohde, C.A.; Links, J.M.; Schwartz, B.S. Associations of Bone Mineral Density and Lead Levels in Blood, Tibia, and Patella in Urban-Dwelling Women. *Environ. Health Perspect.* **2008**, *116*, 784–790. [CrossRef]
165. Lamhamdi, M.; Bakrim, A.; Aarab, A.; Lafont, R.; Sayah, F. Lead phytotoxicity on wheat (*Triticum aestivum* L.) seed germination and seedlings growth. *Comptes Rendus Biol.* **2011**, *334*, 118–126. [CrossRef] [PubMed]
166. Yang, Y.; Zhang, Y.; Wei, X.; You, J.; Wang, W.; Lu, J.; Shi, R. Comparative antioxidative responses and proline metabolism in two wheat cultivars under short term lead stress. *Ecotoxicol. Environ. Saf.* **2011**, *74*, 733–740. [CrossRef]
167. Kaur, G.; Singh, H.P.; Batish, D.R.; Kohli, R.K. A time course assessment of changes in reactive oxygen species generation and antioxidant defense in hydroponically grown wheat in response to lead ions ($Pb^{2+}$). *Protoplasma* **2012**, *249*, 1091–1100. [CrossRef] [PubMed]
168. Kaur, G.; Singh, H.P.; Batish, D.R.; Kumar, R.K. Growth, photosynthetic activity and oxidative stress in wheat (Triticum aestivum) after exposure of lead to soil. *J. Environ. Biol.* **2012**, *33*, 265–269. [PubMed]
169. Li, X.; Bu, N.; Yueying, L.; Maa, L.; Xin, S.; Zhang, L. Growth, photosynthesis and antioxidant responses of endophyte infected and noninfected rice under lead stress conditions. *J. Hazard. Mater.* **2012**, *213*, 55–61. [CrossRef] [PubMed]
170. Kaur, G.; Singh, H.P.; Batish, D.R.; Kohli, R.K. Lead (Pb)-induced biochemical and ultrastructural changes in wheat (Triticum aestivum) roots. *Protoplasma* **2013**, *250*, 53–62. [CrossRef]
171. Kaur, G.; Kaur, S.; Singh, H.P.; Batish, D.R.; Kohli, R.K.; Rishi, V. Biochemical adaptations in Zea maysroots to short-term $Pb^{2+}$ exposure: ROS generation and metabolism. *Bull. Environ. Contam. Toxicol.* **2015**, *95*, 246–253. [CrossRef]
172. Thakur, S.; Singh, L.; Zularisam, A.W.; Sakinah, M.; Din, M.F.M. Lead induced oxidative stress and alteration in the activities of antioxidative enzymes in rice shoots. *Biol. Plant.* **2017**, *61*, 595–598. [CrossRef]
173. Rumbeiha, W.K.; Braselton, W.E.; Donch, D. A retrospective study of the disappearance of blood lead in cattle with accidental lead toxicosis. *J. Vet. Diagn. Investig.* **2001**, *13*, 373–378. [CrossRef]
174. Sharpe, R.T.; Livesey, C.T. Lead poisoning in cattle and its implications for food safety. *Vet. Rec.* **2006**, *159*, 71–74. [CrossRef]
175. Abadin, H.; Ashizawa, A.; Stevens, Y.W.; Llados, F.; Diamond, G.; Sage, G.; Quinones, A.; Bosch, S.J.; Swarts, S.G. *Toxicological Profile for Lead, Atlanta (GA): Agency for Toxic Substances and Disease Registry (US)*; Lewis Publishers: Boca Raton, FL, USA, 2007.
176. Yuan, G.; Dai, S.; Yin, Z.; Lu, H.; Jia, R.; Xu, J.; Song, X.; Li, L.; Shu, Y.; Zhao, X. Toxicological assessment of combined lead and cadmium: Acute and sub-chronic toxicity study in rats. *Food Chem. Toxicol.* **2014**, *65*, 260–268. [CrossRef] [PubMed]
177. Cobbina, S.J.; Chen, Y.; Zhou, Z.; Wu, X.; Zhao, T.; Zhang, Z.; Feng, W.; Wang, W.; Li, Q.; Wu, X.; et al. Toxicity assessment due to sub-chronic exposure to individual and mixtures of four toxic heavy metals. *J. Hazard. Mater.* **2015**, *294*, 109–120. [CrossRef] [PubMed]
178. Shaban El-Neweshy, M.; Said El-Sayed, Y. Influence of vitamin C supplementation on lead-induced histopathological alterations in male rats. *Exp. Toxicol. Pathol.* **2011**, *63*, 221–227. [CrossRef] [PubMed]
179. Abdou, H.M.; Hassan, M.A. Protective role of omega-3 polyunsaturated fatty acid against lead acetate-induced toxicity in liver and kidney of female rats. *BioMed Res. Int.* **2014**, *2014*, 435857. [CrossRef] [PubMed]
180. Carocci, A.; Catalano, A.; Lauria, G.; Sinicropi, M.S.; Genchi, G. Lead toxicity, antioxidant defense and environment. *Rev. Environ. Contam. Toxicol.* **2016**, *238*, 45–67. [PubMed]
181. International Agency for Research on Cancer (IARC). *Agents Classified by the IARC Monographs, Volumes 1–121*; IARC Monographs: Lyon, France, 2018; pp. 1–25.

182. Bhargava, A.; Carmona, F.F.; Bhargava, M.; Srivastava, S. Approaches for enhanced phytoextraction of heavy metals. *J. Environ. Manag.* **2012**, *105*, 103–120. [CrossRef] [PubMed]
183. Gong, Y.; Zhao, D.; Wang, Q. An overview of field-scale studies on remediation of soil contaminated with heavy metals and metalloids: Technical progress over the last decade. *Water Res.* **2018**, *147*, 440–460. [CrossRef]
184. Mallampati, S.R.; Mitoma, Y.; Okuda, T.; Simion, C.; Lee, B.K. Dynamic immobilization of simulated radionuclide 133Cs in soil by thermal treatment/vitrification with nanometallic Ca/CaO composites. *J. Environ. Radioact.* **2015**, *139*, 118–124. [CrossRef]
185. Dellisanti, F.; Rossi, P.L.; Valdre, G. In-field remediation of tons of heavy metal-rich waste by Joule heating vitrification. *Int. J. Miner. Process.* **2009**, *93*, 239–245. [CrossRef]
186. Wang, Q.; Tian, S.; Wang, Q.; Huang, Q.; Yang, J. Melting characteristics during the vitrification of MSWI fly ash with a pilot-scale diesel oil furnace. *J. Hazard. Mater.* **2008**, *160*, 376–381. [CrossRef]
187. Navarro, A.; Cardellach, E.; Canadas, I.; Rodríguez, J. Solar thermal vitrification of mining contaminated soils. *Int. J. Miner. Process.* **2013**, *119*, 65–74. [CrossRef]
188. Yao, Z.; Li, J.; Xie, H.; Yu, C. Review on remediation technologies of soil contaminated by heavy metals. *Proc. Environ. Sci.* **2012**, *16*, 722–729. [CrossRef]
189. Kim, W.S.; Park, G.Y.; Kim, D.H.; Jung, H.B.; Ko, S.H.; Baek, K. In situ field scale electrokinetic remediation of multi-metals contaminated paddy soil: Influence of electrode configuration. *Electrochim. Acta* **2012**, *86*, 89–95. [CrossRef]
190. Jeon, E.K.; Jung, J.M.; Kim, W.S.; Ko, S.H.; Baek, K. In situ electrokinetic remediation of As-, Cu-, and Pb-contaminated paddy soil using hexagonal electrode configuration: A full scale study. Environ. *Sci. Pollut. Control Ser.* **2015**, *22*, 711–720. [CrossRef] [PubMed]
191. Khan, F.I.; Husain, T.; Hejazi, R. An overview and analysis of site remediation technologies. *J. Environ. Manag.* **2004**, *71*, 95–122. [CrossRef] [PubMed]
192. Chen, Q.Y.; Tyrer, M.; Hills, C.D.; Yang, X.M.; Carey, P. Immobilisation of heavy metal in cement-based solidification/stabilisation: A review. *Waste Manag.* **2009**, *29*, 390–403. [CrossRef]
193. Wang, F.; Wang, H.; Al-Tabbaa, A. Leachability and heavy metal speciation of 17-year old stabilised/solidified contaminated site soils. *J. Hazard. Mater.* **2014**, *278*, 144–151. [CrossRef]
194. Antemir, A.; Hills, C.D.; Carey, P.J.; Magnie, M.C.; Polettini, A. Investigation of 4-year-old stabilised/solidified and accelerated carbonated contaminated soil. *J. Hazard. Mater.* **2010**, *181*, 543–555. [CrossRef]
195. Navarro-Blasco, I.; Duran, A.; Sirera, R.; Fernández, J.M.; Alvarez, J.I. Solidification/stabilization of toxic metals in calcium aluminate cement matrices. *J. Hazard. Mater.* **2013**, *260*, 89–103. [CrossRef]
196. Voglar, G.E.; Leštan, D. Equilibrium leaching of toxic elements from cement stabilized soil. *J. Hazard. Mater.* **2013**, *246–247*, 18–25. [CrossRef]
197. Wang, F.; Wang, H.; Jin, F.; Al-Tabbaa, A. The performance of blended conventional and novel binders in the in-situ stabilisation/solidification of a contaminated site soil. *J. Hazard. Mater.* **2015**, *285*, 46–52. [CrossRef] [PubMed]
198. Ferraro, A.; Fabbricino, M.; van Hullebusch, E.D.; Esposito, G.; Pirozzi, F. Effect of soil/contamination characteristics and process operational conditions on aminopolycarboxylates enhanced soil washing for heavy metals removal: A review. *Rev. Environ. Sci. Biotechnol.* **2015**, *15*, 111–145. [CrossRef]
199. Park, B.; Son, Y. Ultrasonic and mechanical soil washing processes for the removal of heavy metals from soils. *Ultrason. Sonochem.* **2017**, *35*, 640–645. [CrossRef] [PubMed]
200. Hu, P.; Yang, B.; Dong, C.; Chen, L.; Cao, X.; Zhao, J.; Wu, L.; Luo, Y.; Christie, P. Assessment of EDTA heap leaching of an agricultural soil highly contaminated with heavy metals. *Chemosphere* **2014**, *117*, 532–537. [CrossRef] [PubMed]
201. Wang, G.; Zhang, S.; Zhong, Q.; Xu, X.; Li, T.; Jia, Y.; Zhang, Y.; Peijnenburg, W.J.G.M.; Vijver, M.G. Effect of soil washing with biodegradable chelators on the toxicity of residual metals and soil biological properties. *Sci. Total Environ.* **2018**, *625*, 1021–1029. [CrossRef] [PubMed]
202. Wu, Y.J.; Zhou, H.; Zou, Z.J.; Zhu, W.; Yang, W.T.; Peng, P.Q.; Zeng, M.; Liao, B.H. A three-year in-situ study on the persistence of a combined amendment (limestone+sepiolite) for remedying paddy soil polluted with heavy metals. *Ecotoxicol. Environ. Saf.* **2016**, *130*, 163–170. [CrossRef] [PubMed]

203. Zhou, H.; Zhou, X.; Zeng, M.; Liao, B.H.; Liu, L.; Yang, W.T.; Wu, Y.M.; Qiu, Q.Y.; Wang, Y.J. Effects of combined amendments on heavy metal accumulation in rice (*Oryza sativa* L.) planted on contaminated paddy soil. *Ecotoxicol. Environ. Saf.* **2014**, *101*, 226–232. [CrossRef]
204. Zotiadis, V.; Argyraki, A.; Theologou, E. Pilot-scale application of attapulgitic clay for stabilization of toxic elements in contaminated soil. *J. Geotech. Geoenviron. Eng.* **2012**, *138*, 633–637. [CrossRef]
205. Sun, Y.; Li, Y.; Xu, Y.; Liang, X.; Wang, L. In situ stabilization remediation of cadmium (Cd) and lead (Pb) co-contaminated paddy soil using bentonite. *Appl. Clay Sci.* **2015**, *105–106*, 200–206. [CrossRef]
206. Wang, B.; Xie, Z.; Chen, J.; Jiang, J.; Su, Q. Effects of field application of phosphate fertilizers on the availability and uptake of lead, zinc and cadmium by cabbage (*Brassica chinensis* L.) in a mining tailing contaminated soil. *J. Environ. Sci.* **2008**, *20*, 1109–1117. [CrossRef]
207. Lim, J.E.; Ahmad, M.; Lee, S.S.; Shope, C.L.; Hashimoto, Y.; Kim, K.R.; Usman, A.R.A.; Yang, J.E.; Ok, Y.S. Effects of lime-based waste materials on immobilization and phytoavailability of cadmium and lead in contaminated soil. *Clean-Soil Air Water* **2013**, *41*, 1235–1241. [CrossRef]
208. Placek, A.; Grobelak, A.; Kacprzak, M. Improving the phytoremediation of heavy metals contaminated soil by use of sewage sludge. *Int. J. Phytoremediat.* **2016**, *18*, 605–618. [CrossRef] [PubMed]
209. Okkenhaug, G.; Grasshorn Gebhardt, K.A.; Amstaetter, K.; Lassen Bue, H.; Herzel, H.; Mariussen, E.; Rossebø Almås, Å.; Cornelissen, G.; Breedveld, G.D.; Rasmussen, G.; et al. Antimony (Sb) and lead (Pb) in contaminated shooting range soils: Sb and Pb mobility and immobilization by iron based sorbents, a field study. *J. Hazard. Mater.* **2016**, *307*, 336–343. [CrossRef] [PubMed]
210. McCann, C.M.; Gray, N.D.; Tourney, J.; Davenport, R.J.; Wade, M.; Finlay, N.; Hudson-Edwards, K.A.; Johnson, K.L. Remediation of a historically Pb contaminated soil using a model natural Mn oxide waste. *Chemosphere* **2015**, *138*, 211–217. [CrossRef] [PubMed]
211. Bian, R.; Joseph, S.; Cui, L.; Pan, G.; Li, L.; Liu, X.; Zhang, A.; Rutlidge, H.; Wong, S.; Chia, C.; et al. A three-year experiment confirms continuous immobilization of cadmium and lead in contaminated paddy field with biochar amendment. *J. Hazard. Mater.* **2014**, *272*, 121–128. [CrossRef]
212. Cui, L.; Pan, G.; Li, L.; Bian, R.; Liu, X.; Yan, J.; Quan, G.; Ding, C.; Chen, T.; Liu, Y.; et al. Continuous immobilization of cadmium and lead in biochar amended contaminated paddy soil: A five-year field experiment. *Ecol. Eng.* **2016**, *93*, 1–8. [CrossRef]
213. Nie, C.; Yang, X.; Niazi, N.K.; Xu, X.; Wen, Y.; Rinklebe, J.; Ok, Y.S.; Xu, S.; Wang, H. Impact of sugarcane bagasse-derived biochar on heavy metal availability and microbial activity: A field study. *Chemosphere* **2018**, *200*, 274–282. [CrossRef]
214. Moreno-Jimenez, E.; Fernandez, J.M.; Puschenreiter, M.; Williams, P.N.; Plaza, C. Availability and transfer to grain of As, Cd, Cu, Ni, Pb and Zn in a barley agri-system: Impact of biochar, organic and mineral fertilizers. *Agric. Ecosyst. Environ.* **2016**, *219*, 171–178. [CrossRef]
215. Niu, L.; Jia, P.; Li, S.; Kuang, J.; He, X.; Zhou, W.; Liao, B.; Shu, W.; Li, J. Slash-and char: An ancient agricultural technique holds new promise for management of soils contaminated by Cd, Pb and Zn. *Environ. Pollut.* **2015**, *205*, 333–339. [CrossRef]
216. Houben, D.; Evrard, L.; Sonnet, P. Mobility bioavailability and pH-dependent leaching of cadmium, zinc and lead in a contaminated soil amended with biochar. *Chemosphere* **2013**, *92*, 1450–1457. [CrossRef]
217. Cheng, S.F.; Huang, C.Y.; Chen, K.L.; Lin, S.C.; Lin, Y.C. Phytoattenuation of lead-contaminated agricultural land using *Miscanthus floridulus*—An in situ case study. *Desalin. Water Treat.* **2016**, *57*, 7773–7779. [CrossRef]
218. Yang, J.; Yang, J.; Huang, J. Role of co-planting and chitosan in phytoextraction of as and heavy metals by Pteris vittata and castor bean—A field case. *Ecol. Eng.* **2017**, *109*, 35–40. [CrossRef]
219. Basha, S.A.; Rajaganesh, K. Microbial Bioremediation of Heavy Metals from Textile Industry Dye Effluents using Isolated Bacterial Strains. *Int. J. Curr. Microbiol. Appl. Sci.* **2014**, *3*, 785–794.
220. Kang, C.H.; So, J.S. Heavy metal and antibiotic resistance of ureolytic bacteria and their immobilization of heavy metals. *Ecol. Eng.* **2016**, *97*, 304–312. [CrossRef]
221. Puyen, Z.M.; Villagrasa, E.; Maldonado, J.; Diestra, E.; Esteve, I.; Solé, A. Biosorption of lead and copper by heavy-metal tolerant Micrococcus luteus DE2008. *Bioresour. Technol.* **2012**, *126*, 233–237. [CrossRef]
222. Jin, Y.; Luan, Y.; Ning, Y.; Wang, L. Effects and mechanisms of microbial remediation of heavy metals in soil: A critical review. *Appl. Sci.* **2018**, *8*, 1336. [CrossRef]

223. Jacob, J.M.; Karthik, C.; Saratale, R.G.; Kumar, S.S.; Prabakar, D.; Kadirvelu, K.; Pugazhendhi, A. Biological approaches to tackle heavy metal pollution: A survey of literature. *J. Environ. Manag.* **2018**, *217*, 56–70. [CrossRef]
224. Wang, T.; Sun, H.; Mao, H.; Zhang, Y.; Wang, C.; Zhang, Z.; Wang, B.; Sun, L. The immobilization of heavy metals in soil by bioaugmentation of a UV-mutant Bacillus subtilis 38 assisted by NovoGro biostimulation and changes of soil microbial community. *J. Hazard. Mater.* **2014**, *278*, 483–490. [CrossRef]
225. Zeng, X.; Wei, S.; Sun, L.; Jacques, D.A.; Tang, J.; Lian, M.; Xu, Z. Bioleaching of heavy metals from contaminated sediments by the Aspergillus niger strain SY1. *J. Soils Sediments* **2015**, *15*, 1029–1038. [CrossRef]
226. Dursun, A.Y.; Uslu, G.; Cuci, Y.; Aksu, Z. Bioaccumulation of copper (II), lead (II) and chromium (VI) by growing Aspergillus niger. *Process. Biochem.* **2003**, *38*, 1647–1651. [CrossRef]
227. Kariuki, Z.; Kiptoo, J.; Onyancha, D. Biosorption studies of lead and copper using rogers mushroom biomass Lepiota hystrix. *S. Afr. J. Chem. Eng.* **2017**, *23*, 62–70. [CrossRef]
228. Sooksawat, N.; Meetam, M.; Kruatrachue, M.; Pokethitiyook, P.; Nathalang, K. Phytoremediation potential of charophytes: Bioaccumulation and toxicity studies of cadmium, lead and zinc. *J. Environ. Sci.* **2013**, *25*, 596–604. [CrossRef]
229. Ibrahim, W.M.; Abdel Aziz, Y.S.; Hamdy, S.M.; Gad, N.S. Comparative Study for Biosorption of Heavy Metals from Synthetic Wastewater by Different Types of Marine Algae. *J. Bioremed. Biodegrad.* **2018**, *9*, 1–425. [CrossRef]
230. Jafari, M.; Younes, R.D.; Yubert, G. Molecular Techniques in Fungal Bioremediation. In *Fungi as Bioremediators*; Springer: Berlin/Heidelberg, Germany, 2013; pp. 453–465.
231. Subrahmanyam, G.; Hu, H.W.; Zheng, Y.M.; Gattupalli, A.; He, J.Z.; Liu, Y.R. Response of ammonia oxidizing microbes to the stresses of arsenic and copper in two acidic alfisols. *Appl. Soil Ecol.* **2014**, *77*, 59–67. [CrossRef]
232. Subrahmanyam, G.; Shen, J.P.; Liu, Y.R.; Archana, G.; Zhang, L.M. Effect of long-term industrial waste effluent pollution on soil enzyme activities and bacterial community composition. *Environ. Monit. Assess.* **2016**, *188*, 112. [CrossRef]
233. Subrahmanyam, G.; Sharma, R.K.; Kumar, G.N.; Archana, G. *Vigna radiata* var. GM4 plant growth enhancement and root colonization by a multi-metal-resistant plant growth-promoting bacterium *Enterobacter* sp. C1D in Cr (VI)-amended soils. *Pedosphere* **2018**, *28*, 144–156. [CrossRef]
234. Jaiswal, S.; Singh, D.K.; Shukla, P. Gene editing and systems biology tools for pesticide bioremediation: A review. *Front. Microbiol.* **2019**, *10*, 87. [CrossRef]
235. Babu, A.G.; Kim, J.D.; Oh, B.T. Enhancement of heavy metal phytoremediation by Alnus firma with endophytic Bacillus thuringiensis GDB1. *J. Hazard. Mater.* **2013**, *25*, 477–483. [CrossRef]
236. Sheng, X.F.; Xia, J.J.; Jiang, C.Y.; He, L.Y.; Qian, M. Characterization of heavy metal-resistant endophytic bacteria from rape (Brassica napus) roots and their potential in promoting the growth and lead accumulation of rape. *Environ. Pollut.* **2008**, *156*, 1164–1170. [CrossRef]
237. Tak, H.I.; Ahmad, F.; Babalola, O.O. Advances in the Application of Plant Growth-Promoting Rhizobacteria in Phytoremediation of Heavy Metals. In *Reviews of Environmental Contamination and Toxicology*; Whitacre, D.M., Ed.; Springer: New York, NY, USA, 2013; p. 223.
238. Lima, L.K.S.; Pelosi, B.T.; Silva, M.G.C.; Vieira, M.G.A. Lead and chromium biosorption by Pistia stratiotes biomass. *Chem. Eng. Trans.* **2013**, *32*, 1045–1050.
239. Basile, A.; Sorbo, S.; Conte, B.; Cobianchi, R.C.; Trinchella, F.; Capasso, C.; Carginale, V. Toxicity, accumulation, and removal of heavy metals by three aquatic macrophytes. *Int. J. Phytoremediat.* **2012**, *14*, 374–387. [CrossRef] [PubMed]
240. Tangahu, B.V.; Abdullah, S.R.S.; Basri, H.; Idris, M.; Anuar, N.; Mukhlisin, M. Phytoremediation of wastewater containing lead (Pb) in pilot reed bed using *Scirpus grossus*. *Int. J. Phytoremediat.* **2013**, *15*, 663–676. [CrossRef] [PubMed]
241. Veselý, T.; Tlustos, P.; Száková, J. The use of water lettuce (*Pistia stratiotes* L.) for rhizofiltration of a highly polluted solution by cadmium and lead. *Int. J. Phytoremediat.* **2011**, *13*, 859–872. [CrossRef] [PubMed]
242. Sekomo, C.B.; Rousseau, D.P.L.; Saleh, S.A.; Lens, P.N.L. Heavy metal removal in duckweed and algae ponds as a polishing step for textile waste water treatment. *Ecol. Eng.* **2012**, *44*, 102–110. [CrossRef]
243. Sekomo, C.B.; Kagisha, V.; Rousseau, D.; Lens, P. Heavy metal removal by combining anaerobic upflow packed bed reactors with water hyacinth ponds. *Environ. Technol.* **2012**, *33*, 1455–1464. [CrossRef] [PubMed]

244. Li, Q.; Chen, B.; Lin, P.; Zhou, J.; Zhan, J.; Shen, Q.; Pan, X. Adsorption of heavy metal from aqueous solution by dehydrated root powder of long-root Eichhornia crassipes. *Int. J. Phytoremediat.* **2016**, *18*, 103–109. [CrossRef]
245. Li, R.; Zhou, Z.; Xie, X.; Li, Y.; Zhang, Y.; Xu, X. Effects of dissolved organic matter on uptake and translocation of lead in Brassica chinensis and potential health risk of Pb. *Int. J. Environ. Res. Public Health* **2016**, *13*, 687. [CrossRef]
246. Haghighi, M.; Kafi, M.; Pessarakli, M.; Sheibanirad, A.; Sharifinia, M.R. Using kale (Brassica oleracea var. acephala) as a phytoremediation plant species for lead (Pb) and cadmium (Cd) removal in saline soils. *J. Plant Nutr.* **2016**, *39*, 1460–1471. [CrossRef]
247. Bonanno, G.; Borg, J.A.; Martino, V.D. Levels of heavy metals in wetland and marine vascular plants and their biomonitoring potential: A comparative assessment. *Sci. Total Environ.* **2017**, *576*, 796–806. [CrossRef]
248. Waoo, A.A. Comparative effect of four heavy metals, Cd, Pb, Ni, and Cr, on Datura inoxia in Tissue Culture. *Int. J. Adv. Eng. Technol. Sci.* **2016**, *2*, 7–11.
249. Liang, J.; Fang, H.L.; Zhang, T.L.; Wang, X.X.; Liu, Y.D. Heavy metal in leaves of twelve plant species from seven different areas in Shanghai, China. *Urban For. Urban Green.* **2017**, *27*, 390–398. [CrossRef]
250. Aurangzeb, N.; Nisa, S.; Bibi, Y.; Javed, F.; Hussain, F. Phytoremediation potential of aquatic herbs from steel foundry effluent. *Braz. J. Chem. Eng.* **2014**, *31*, 881–886. [CrossRef]
251. Verma, R.; Suthar, S. Lead and cadmium removal from water using duckweed-Lemna gibba L.: Impact of pH and initial metal load. *Alex. Eng. J.* **2015**, *54*, 1297–1304. [CrossRef]
252. Volf, I.; Rakoto, N.G.; Bulgariu, L. Valorization of Pistia stratiotes biomass as biosorbent for lead (II) ions removal from aqueous media. *Sep. Sci. Technol.* **2015**, *50*, 1577–1586. [CrossRef]
253. Syukor, A.R.A.; Sulaiman, S.; Siddique, M.N.I.; Zularisam, A.W.; Said, M.I.M. Integration of phytogreen for heavy metal removal from wastewater. *J. Clean. Prod.* **2016**, *112*, 3124–3131. [CrossRef]
254. Bokhari, S.H.; Ahmad, I.; Mahmood-Ul-Hassan, M.; Mohammad, A. Phytoremediation potential of Lemna minor L. for heavy metals. *Int. J. Phytoremediat.* **2016**, *18*, 25–32. [CrossRef]
255. Prasad, B.; Maiti, D. Comparative study of metal uptake by Eichhornia crassipes growing in ponds from mining and nonmining areas—A field study. *Bioremediat. J.* **2016**, *20*, 144–152. [CrossRef]
256. Kumari, M.; Tripathi, B.D. Efficiency of Phragmites australis and Typha latifolia for heavy metal removal from wastewater. *Ecotoxicol. Environ. Saf.* **2015**, *112*, 80–86. [CrossRef]
257. Lorenzo-Gutiérrez, D.; Gómez-Gil, L.; Guarro, J.; Roncero, M.I.G.; Fernández-Bravo, A.; Capilla, J.; López-Fernández, L. Role of the Fusarium oxysporum metallothionein Mt1 in resistance to metal toxicity and virulence. *Metallomics* **2019**, *11*, 1230–1240. [CrossRef]
258. Hattab, S.; Flores-Casseres, M.L.; Boussetta, H.; Doumas, P.; Hernandez, L.E.; Banni, M. Characterisation of lead-induced stress molecular biomarkers in Medicago sativa plants. *Environ. Exp. Bot.* **2016**, *123*, 1–12. [CrossRef]
259. Fan, T.; Yang, L.; Wu, X.; Ni, J.; Jiang, H.; Zhang, Q.A.; Fang, L.; Sheng, Y.; Ren, Y.; Cao, S. The PSE1 gene modulates lead tolerance in Arabidopsis. *J. Exp. Bot.* **2016**, *67*, 4685–4695. [CrossRef]
260. Jiang, L.; Wang, W.; Chen, Z.; Gao, Q.; Xu, Q.; Cao, H. A role for APX1 gene in lead tolerance in Arabidopsis thaliana. *Plant Sci.* **2017**, *256*, 94–102. [CrossRef]
261. Kim, D.Y.; Bovet, L.; Kushnir, S.; Noh, E.W.; Martinoia, E.; Lee, Y. AtATM3 is involved in heavy metal resistance in Arabidopsis. *Plant Physiol.* **2006**, *140*, 922–932. [CrossRef] [PubMed]
262. Jiang, Y.; Wang, W.; Xie, Q.; Liu, N.; Liu, L.; Wang, D.; Zhang, X.; Yang, C.; Chen, X.; Tang, D.; et al. Plants transfer lipids to sustain colonization by mutualistic mycorrhizal and parasitic fungi. *Science* **2017**, *356*, 1172–1175. [CrossRef] [PubMed]
263. Xiao, S.; Gao, W.; Chen, Q.F.; Ramalingam, S.; Chye, M.L. Overexpression of membrane-associated acyl-CoA-binding protein ACBP1 enhances lead tolerance in Arabidopsis. *Plant J.* **2008**, *54*, 141–151. [CrossRef] [PubMed]
264. Cao, S.; Chen, Z.; Liu, G.; Jiang, L.; Yuan, H.; Ren, G.; Bian, X.H.; Jian, H.; Ma, X. The Arabidopsis Ethylene-Insensitive 2 gene is required for lead resistance. *Plant Physiol. Biochem.* **2009**, *47*, 308–312. [CrossRef] [PubMed]
265. Zhu, F.Y.; Li, L.; Lam, P.Y.; Chen, M.X.; Chye, M.L.; Lo, C. Sorghum extracellular leucine-rich repeat protein SbLRR2 mediates lead tolerance in transgenic Arabidopsis. *Plant Cell Physiol.* **2013**, *54*, 1549–1559. [CrossRef]

266. Yang, L.; Fan, T.; Guan, L.; Ren, Y.; Han, Y.; Liu, Q.; Liu, Y.; Cao, S. CMDH4 encodes a protein that is required for lead tolerance in Arabidopsis. *Plant Soil* **2016**, *412*, 317–330. [CrossRef]
267. Anjum, M.; Miandad, R.; Waqas, M.; Gehany, F.; Barakat, M.A. Remediation ofwastewater using various nano-materials. *Arab. J. Chem.* **2016**, *11*, 1–23.
268. Estrella-Gomez, N.; Mendoza-Cozatl, D.; Moreno-Sanchez, R.; Gonzalez-Mendoza, D.; Zapata-Perez, O.; Martınez-Hernandez, A.; Santamaria, J.M. The Pb-hyperaccumulator aquatic fern Salvinia minima Baker, responds to Pb21 by increasing phytochelatins via changes in SmPCS expression and in phytochelatin synthase activity. *Aquat. Toxicol.* **2009**, *91*, 320–328. [CrossRef]
269. Adeleye, A.S.; Conway, J.R.; Garner, K.; Huang, Y.; Su, Y.; Keller, A.A. Engineerednanomaterials for water treatment and remediation: Costs, benefits, and applicability. *Chem. Eng. J.* **2016**, *286*, 640–662. [CrossRef]
270. Wernisch, S.; Trapp, O.; Lindner, W. Application of cinchona-sulfonate-based chiralzwitterionic ion exchangers for the separation of proline-containing dipeptide rotamers anddetermination of on-column isomerization parameters from dynamic elution profiles. *Anal. Chim. Acta* **2013**, *795*, 88–98. [CrossRef] [PubMed]
271. Yong-Mei, H.; Man, C.; Zhong-Bo, H. Effective removal of Cu (II) ions from aqueoussolution by amino-functionalized magnetic nanoparticles. *J. Hazard. Mater.* **2010**, *184*, 392–399.
272. Yadav, K.K.; Singh, J.K.; Gupta, N.; Kumar, V. A review of nano-bioremediationtechnologies for environmental cleanup: A novel biological approach. *J. Mater. Environ. Sci.* **2017**, *8*, 740–757.
273. Srivastav, A.; Yadav, K.K.; Yadav, S.; Gupta, N.; Singh, J.K.; Katiyar, R.; Kumar, V. Nano-phytoremediation of pollutants from contaminated soil environment: Current scenario and future prospects. In *Phytoremediation*; Springer: Cham, Switzerland, 2018; pp. 383–401.
274. Bobik, M.; Korus, I.; Dudek, L. The effect of magnetite nanoparticles synthesis conditionson their ability to separate heavy metal ions. *Arch. Environ. Prot.* **2017**, *43*, 3–9. [CrossRef]
275. Klekotka, U.; Wińska, E.; Zambrzycka-Szelewa, E.; Satuła, D.; Kalska-Szostko, B. Heavy-metal detectors based on modified ferrite nanoparticles. *Beilstein J. Nanotechnol.* **2018**, *9*, 762–770. [CrossRef]
276. Hegazi, H.A. Removal of heavy metals from wastewater using agricultural and industrialwastes as adsorbents. *HBRC J.* **2013**, *9*, 276–282. [CrossRef]
277. Mohammed, A.A.; Brouers, F.; Sadi, S.I.A.; Al-Musawi, T.J. Role of $Fe_3O_4$ magnetitenanoparticles used to coat bentonite in zinc (II) ions sequestration. *Environ. Nanotechnol. Monit. Manag.* **2018**, *10*, 17–27.
278. Okuo, J.; Emina, A.; Stanley, O.; Anegbe, B. Synthesis, characterization and applicationof starch stabilized zerovalent iron nanoparticles in the remediation of Pb-acid battery soil. *Environ. Nanotechnol. Monit. Manag.* **2018**, *9*, 12–17.
279. Visa, M. Synthesis and characterization of new zeolitematerials obtained from fly ash forheavy metals removal inadvanced wastewater treatment. *Powder Technol.* **2016**, *294*, 338–347. [CrossRef]
280. Alsohaimi, I.H.; Wabaidur, S.M.; Kumar, M.; Khan, M.A.; Alothman, Z.A.; Abdalla, M.A. Synthesis, characterization of PMDA/TMSPEDA hybrid nano-composite and itsapplications as an adsorbent for the removal of bivalent heavy metals ions. *Chem. Eng. J.* **2015**, *270*, 9–21. [CrossRef]
281. Ali, A.; Mannan, A.; Hussain, I.; Hussain, I.; Zi, M. Effective removal of metal ions fromaquous solution by silver and zinc nanoparticles functionalized cellulose: Isotherm, kineticsand statistical supposition of process. *Environ. Nanotechnol. Monit. Manag.* **2018**, *9*, 1–11.
282. Saad, A.H.A.; Azzam, A.M.; El-Wakeel, S.T.; Mostafa, B.B.; El-latif, M.B.A. Removal oftoxic metal ions from wastewater using ZnO@Chitosan core shell Nanocomposite. *Environ. Nanotechnol. Monit. Manag.* **2018**, *9*, 67–75.
283. Goyal, P.; Chakraborty, S.; Misra, S.K. Multifunctional $Fe_3O_4$-ZnO nanocomposites forenvironmental remediation applications. *Environ. Nanotechnol. Monit. Manag.* **2018**, *10*, 28–35.

# 10

# Associations between Trace Elements and Cognitive Decline: An Exploratory 5-Year Follow-Up Study of an Elderly Cohort


**Bianca Gerardo [1,2,*], Marina Cabral Pinto [3], Joana Nogueira [1,2], Paula Pinto [2], Agostinho Almeida [4], Edgar Pinto [4,5], Paula Marinho-Reis [3,6], Luísa Diniz [4], Paula I. Moreira [7,8], Mário R. Simões [1,2] and Sandra Freitas [1,2]**

1 Center for Research in Neuropsychology and Cognitive and Behavioral Intervention (CINEICC), Faculty of Psychology and Educational Sciences (FPCEUC), Univ Coimbra, 3000-115 Coimbra, Portugal; joananogueira.f@gmail.com (J.N.); simoesmr@fpce.uc.pt (M.R.S.); sandrafreitas0209@gmail.com (S.F.)
2 Psychological Assessment and Psychometrics Laboratory (PsyAssessmentLab), Faculty of Psychology and Educational Sciences (FPCEUC), Univ Coimbra, 3000-115 Coimbra, Portugal; paulapinto663@hotmail.com
3 Geobiotec Research Centre, Department of Geosciences, University of Aveiro, 3810-193 Aveiro, Portugal; marinacp@ua.pt (M.C.P.); pmarinho@dct.uminho.pt (P.M.-R.)
4 LAQV/REQUIMTE, Laboratory of Applied Chemistry, Faculty of Pharmacy, University of Porto, 4050-313 Porto, Portugal; aalmeida@ff.up.pt (A.A.); ecp@ess.ipp.pt (E.P.); luisa2diniz@gmail.com (L.D.)
5 Department of Environmental Health, School of Health, P.Porto, CISA/Research Center in Environment and Health, 4200-072 Porto, Portugal
6 Departamento de Ciências da Terra, Instituto de Ciências da Terra, Polo da Universidade do Minho, Campus de Gualtar, 4710-057 Braga, Portugal
7 Center for Neuroscience and Cell Biology (CNC), University of Coimbra, 3004-504 Coimbra, Portugal; pimoreira@fmed.uc.pt
8 Institute of Physiology, Faculty of Medicine, University of Coimbra, 3000-548 Coimbra, Portugal
* Correspondence: bianca.s.gerardo94@gmail.com.



**Abstract:** Trace elements (TE) homeostasis is crucial in normal brain functioning. Although imbalances have the potential to exacerbate events leading neurodegenerative diseases, few studies have directly addressed the eventual relationships between TE levels in the human body and future cognitive status. The present study aimed to assess how different TE body-levels relate to cognitive decline. This exploratory research included a study-group (RES) of 20 elderly individuals living in two Portuguese geographical areas of interest (Estarreja; Mértola), as well as a 20 subjects neuropsychological control-group (CTR). Participants were neuropsychologically assessed through the Mini Mental State Examination (MMSE) and the Montreal Cognitive Assessment (MoCA) and the RES group was biomonitored for TE through fingernail analysis. After 5 years, the cognitive assessments were repeated. Analyses of the RES neuropsychological data showed an average decrease of 6.5 and 5.27 points in MMSE and MoCA, respectively, but TE contents in fingernails were generally within the referenced values for non-exposed individuals. Higher levels of Nickel and Selenium significantly predicted lesser cognitive decline within 5 years. Such preliminary results evidence an association between higher contents of these TE and higher cognitive scores at follow-up, suggesting their contribution to the maintenance of cognitive abilities. Future expansion of the present study is needed in order to comprehensively assess the potential benefits of these TE.



**Keywords:** cognitive decline; longitudinal study; risk for dementia; trace elements; nickel; selenium; human tissues; industrial area; mining area

## 1. Introduction

The rapid aging of the world's population is a current phenomenon that imposes great social and economic challenges. It is estimated that by 2050, there will be 1.5 billion people with ages above 65 years worldwide, corresponding to 16% of the population [1]. It is further foreseen that this proportion will be much higher in specific countries, such as Italy, Portugal, Greece, Japan, and Korea, where it will exceed the one-third mark [2]. Specifically, in Portugal, the estimations predict a ratio of 464 elderly (≥65 years old) per 100 young individuals by 2060, the quadruple of the ratio registered in 2001 [3]. Such aging rebounds in a growing number of people living with dementia—a condition characterized by the gradual loss of cognitive and functional capacities that interferes with the performance and execution of daily-living activities—as its prevalence/incidence increases sharply with age (from 2.3% among 65–69 years old individuals to 42% among elderly aged 90 or older) [2]. With a forecast of 41 million cases across OECD countries by 2050 [2], dementia is currently a major public health priority [4].

Strict genetic etiology is rare in dementia, being the vast majority of cases sporadic in origin [5–7]. Among elderly people, Alzheimer's disease (AD) is the most frequent form of the condition and is mainly expressed by an accentuated loss of memory, as well as by the progressive decline of other cognitive and functional capacities. At the cellular level, AD pathology is characterized by neurofibrillary tangles (abnormal accumulations of hyperphosphorylated tau protein) and senile plaques (aggregations of amyloid-β protein) [8]. Several factors have been reported as contributors to the development of sporadic AD—besides aging, genetics is another well-established risk factor, with the presence of the APOE ε4 allele being the strongest genetic elicitor of increased risk for the disease [9]. Meanwhile, environmental exposure to toxic metals, to certain chemical compounds (such as pesticides and industrial chemicals) and to air pollutants have also been proposed as a risk factor [5,7,10].

Although not all environmental contaminants/toxins have been tested in regards to their effects on the central nervous system, it is hypothesized that there may be a risk for older adults to develop AD (and other neurodegenerative diseases such as Parkinson's disease) that is associated with the neurologic impairments resulting from such environmental exposures [10–13]. Trace elements homeostasis is crucial in normal brain functioning, and disturbances may exacerbate AD-associated events [14,15]. Amongst trace metals, the relationship between AD and aluminum (Al) seems to be the best understood, with larger environmental studies, epidemiological studies and case-control evidence supporting an association between high levels of exposure to Al and the risk for AD [11,16]. For instance, this association has been suggested by studies reporting high rates of AD among older adults exposed to high concentrations of the metal [17], as well as by evidence of higher blood Al levels among AD patients (e.g., [18–20]). Copper (Cu) is another widely investigated element. Post-mortem case-control studies report a significant decrease of Cu in brain tissues of AD patients [21,22], particularly in region such as the hippocampus, amygdala [23], frontal cortex [24], cerebellum, motor and sensory cortex, cingulate gyrus, temporal gyrus, and entorhinal cortex [25]. In turn, research on iron (Fe) evidences higher contents of this metal in the same biological matrix [26], namely, in the Brodmann area [21], frontal cortex [27], and amygdala [23]. Furthermore, epidemiological studies report an association between higher levels of Fe in soil and the diagnosis of AD [28] as well as between these concentrations and AD-related mortality [29]. Indeed, this metal seems to be responsible for oxidative stress and increases amyloid deposition [30]. Abnormalities of brain copper (Cu), iron (Fe) and zinc (Zn) homeostasis in AD were reported [31].

Systematic exposure to trace elements (TE) is a common occurrence worldwide. Human contamination may result from various pathways, namely, through inhalation of contaminated air/ambient particulate matter, ingestion of contaminated soil/dust, water and foodstuffs (such as agriculture crops, meat, and seafood), and dermal absorption of TE present in soil/dust [10,32,33]. The toxicity of each TE is dependent on the route and absorbed dose, which is in turn dependent on the concentration and duration of exposure [34]. Furthermore, each TE has a specific half-life, and the elimination of different elements from the body widely varies [35]. Previous studies report the significance of occupational exposure to specific TE and their effects on the development of cognitive disorders and dementia [5,6,35–37].

Despite the increasing evidence reporting the role of environmental exposures in the development of neurodegenerative diseases, the real effects of long-term exposure to TE in cognition remain unclear—literature on the possible pathogenic role of different metals do not provide direct evidence to define a causal relationship between environmental exposure to certain metals and AD [15,16]. Thus, the study of the influence of TE in the cognition of elderly people is necessary. The crossing of biological and neuropsychological assessment data to address this matter is an interesting approach. On the one hand, a comprehensive neuropsychological assessment is an essential component to identify subtle cognitive and behavioral deficits associated with neurotoxic exposure [38–40]. On the other hand, the use of fingernails for biomonitoring purposes provides a valid measurement of relatively long-term TE intake/exposure in a single specimen analysis [41]. Human nails are mainly constituted by keratin-rich proteins, and as they remain isolated from further metabolic activities after growth, they retain a high content of most TE. Moreover, the incorporation of TE in the nails' matrix occurs in a time-integrated fashion and in proportion to the dietary intakes and other exposures to which an individual may be subjected [41,42]. Bioaccumulation ratios show that nails have higher mass/mass ratios (mass of TE per mass of sample) compared to blood and urine [32,41,43]. As nail samples are easy to collect (nail clipping is a non-invasive procedure), transport, and store [44], fingernails are useful and cost-effective specimens for element profiling.

Considering this translational approach, the present exploratory study comprises the cognitive assessment of a group of elderly individuals residing in potential environmentally-toxic geographic regions 5 years after the TE biomonitoring, in order to assess whether nail TE levels are associated with performance scores. The main aim is to investigate the effects of TE on future cognitive status.

## 2. Materials and Methods

### 2.1. Participants and Study Design

The present research comprised a study-group (RES) and a neuropsychological control-group (CTR). Criteria for participants' inclusion accounted: (i) being a native Portuguese speaker, (ii) being over 50 years of age (given the likelihood of this cohort to report cognitive complaints and/or being involved in pathological aging processes of the dementia spectrum), (iii) having resided in the study areas for at least the past 5 years (prior to this study, in order to ensure minimum exposure time [45]; applicable for the RES group), (iv) having maintained the same occupation/professional activity.

The exclusion criteria were the following: (a) having a history of neurological disease (other than of the dementia spectrum); (b) having a history of psychiatric illness, including depression with the exception of stable mild depressive symptoms (depression symptomatology was assessed through the application of the Geriatric Depression Scale—30 [46–48]; individuals with total scores $\geq 21$ were excluded); (c) having a significant visual, auditory, or language impairment that would negatively affect their ability to satisfactorily complete cognitive tests or understand test instructions; (d) current or prior alcohol, drugs, and other substances abuse; (e) current or prior use of antipsychotic medication; (f) being a smoker; (g) current or prior intake of supplements containing any of the analyzed TE; (h) significant changes in diet.

All subjects enrolled in the RES group were permanent residents from the municipality of Estarreja or from the municipality of Mértola and were voluntarily recruited through convenience sampling. Participants from the CTR group were randomly selected from an already existing longitudinal cohort composed of 250 community-residents recruited from several social institutions, aging associations and primary health care centers, who have been neuropsychologically assessed for the past 10 years. This cohort has been stratified according to several sociodemographic variables with a distribution similar to that observed in the Portuguese population, resulting in a cohort representative of the population of the country.

The RES group was initially composed of 76 participants who were cognitively assessed by an experienced neuropsychologist and whose fingernail clippings were collected and analyzed for biomonitoring of twenty selected TE: Al, arsenic (As), barium (Ba), cadmium (Cd), cobalt (Co), chromium (Cr), Cu, Fe, mercury (Hg), lithium (Li), manganese (Mn), nickel (Ni), lead (Pb), antimony (Sb), selenium (Se), tin (Sn), strontium (Sr), titanium (Ti), vanadium (V), and zinc (Zn). Out of the initial group, a total of 20 individuals were available for the second neuropsychological assessment, performed 5 years after the biomonitoring study. The neuropsychological data was paired with the biological data resulting from the nails TE profiling. In turn, the CTR group was composed of 20 subjects randomly selected from the mentioned pre-existing cohort of 250 subjects, previously matched by educational level with the RES group. For comparability reasons, since this pre-existing cohort has been followed for 10 years, the data considered for the CTR group referred to the 5-year follow-up benchmark.

All cognitive performances were assessed through the application of two cognitive screening tools: the Mini-Mental State Examination (MMSE [49,50]) and the Montreal Cognitive Assessment (MoCA [51,52]). The referred educational pairing between the two groups was performed based on previous evidences that report educational level as the variable that most significantly contributes to the prediction of scores on cognitive screening tests, such as MMSE and MoCA [53–57].

Ethical approval for the present research was obtained from the National Committee for Data Protection (No. 11726/2017). This study complies with the ethical guidelines for human experimentation stated in the Declaration of Helsinki committee and all participants gave written informed consent prior to participation after the aims and research procedures were fully explained by a member of the study group.

### 2.2. *Geographical Areas of Interest*

The study included residents from two distinct areas in Portugal—a northern industrial region (Estarreja) and a southern region associated with the mining industry (Mértola). Both areas display interesting characteristics to study the potential effect of long-term environmental exposure to chemicals.

Estarreja, a municipality located near one of the largest Portuguese cities, Aveiro, hosts one of the largest chemical complexes of the country. In this municipality, located near the small town with the same name, the Estarreja Chemical Complex (ECC) has been intensively operating since the 1950s in the production of aniline and derivatives, chlorine-alkalis, sodium and chlorate compounds through electrolysis using Hg cathodes, polyvinyl chloride resins, and polymeric methyl diphenyl isocyanate. In the past, the ECC also produced ammonium sulphate, ammonium nitrate, and sulphuric acid [58]. Until the 21st century such activity had serious deleterious effects on the environment with regards to soils/agricultural fields, surface waters, groundwater, and atmosphere [59–63]. Toxic solid wastes and liquid effluents were discharged directly into manmade, permeable water channels, without any previous treatment [58]. Environmental remediation works only began in 1998 [63].

Mértola is a municipality located in Baixo Alentejo, a southern area with several abandoned manganese mines spatially related with the Iberian Pyrite Belt. Mining releases large amounts of potentially toxic elements into the environment, which contaminate soils, surface waters, and groundwater. Right next to Mértola are the São Domingos Mines, the biggest Portuguese mining exploration until their shutdown in 1996. Close to the mines, the abandoned tailings deposits remain exposed to the weathering conditions that promote mineral leaching, with the release of elements such as As, Pb, Cd, and Cr. In addition, depending on the weather conditions, fine and ultrafine mineral particles are re-suspended and added to the ambient particulate matter, being transported to the neighboring areas [64–66].

*2.3. Neuropsychological Assessment*

All participants were inquired in regards of their sociodemographic features (i.e., age, educational level, nationality, number of years residing in the regions of interest, marital status) through a sociodemographic questionnaire; their current and past clinical history (i.e., weight, height, medical records) through a clinical structured interview; and their professional activity and consumption habits (i.e., main profession, working-time in agriculture/factories/mines, use of pesticides, consumption of home-grown foodstuff, and source of water for consumption/irrigation), through specifically developed questionnaires. Furthermore, all participants underwent a neuropsychological assessment performed by an experienced neuropsychologist. The screening tools of the assessment battery were administrated in a fixed order and were the following:

1. The Mini Mental State Examination (MMSE [49,50]), which is a brief cognitive screening tool widely used in the assessment of cognitive impairment [53]. It is composed of 30 dichotomous items (0—incorrect; 1—correct) and assesses 5 cognitive domains: Orientation, Memory, Attention and Calculus, Language, and Visuoconstruction. Higher scores indicate better cognitive performance.
2. The Montreal Cognitive Assessment (MoCA [51,52]), which is a screening tool developed to detect milder forms of cognitive decline, such as Mild Cognitive Impairment (MCI [67]). This instrument assesses 6 cognitive domains—Executive Functions; Visuospatial Abilities; Memory; Language; Attention Concentration and Working Memory; Temporal and Spatial Orientation—on a scale of 30 points, whose higher global scores translate better cognitive functioning.
3. The Geriatric Depression Scale—30 (GDS-30 [46–48]), which is a brief scale specifically developed for the screening of depressive symptoms in advanced adulthood. It is composed of 30 yes/no questions regarding the affective and cognitive domains of depression. Greater scores indicate more severe symptomatology.

*2.4. Fingernail Samples and Analysis*

Fingernail clippings from the RES participants were collected individually with a synthetic quartz knife and put into decontaminated plastic bags. Plastic forceps were used to remove visible exogenous material when needed.

The procedure used is described elsewhere [68]. Fingernail clippings were properly washed in order to remove exogenous contamination while preserving endogenous TE content. Samples were then dried at 95 °C in a laboratory drying oven (Raypa, Spain) until a constant weight was reached (ca. 40 h). The dried samples (approximately 0.1 mg) were mineralized in a Milestone (Italy) MLS 1200 Mega high-performance microwave digestion unit equipped with an HPR 1000/10 rotor, through a microwave-assisted acid digestion procedure using 1 mL of concentrated $HNO_3$ (>69.0% m/m; TraceSELECT®, Fluka, France) and 0.5 mL of $H_2O_2$ (≥30% *v/v*; TraceSELECT®, Fluka, Germany). The used microwave oven program (W/min) was the following: 250/1, 0/2, 250/5, 400/5 and 600/5. After cooling, ultrapure water was added to the samples digest, and the volume was brought to 10 mL The solutions were then stored at 4 °C in closed propylene tubes until analysis.

All labware materials were decontaminated by immersion for at least 24 h in a 10% $HNO_3$ bath followed by thorough rinsing with ultrapure water. The ultrapure water (resistivity > 18.2 MΩ·cm at 25 °C) was produced in an aarium® pro (Sartorius, Germany) water purification system. For each digestion run (10 samples), a sample blank was prepared, and the average blank level was subtracted from the samples' values. The certified reference material ERM-DB001—Trace Elements in Human Hair was used for analytical quality control, digested and analyzed through the same procedures as for the study samples. TE concentrations were determined through Inductively Coupled Plasma-Mass Spectrometry (ICP-MS) using a Thermo Fisher Scientific (Waltham, MA, USA) iCAP™ Q instrument equipped with a MicroMist™ nebulizer (Glass Expansion, Port Melbourne, Australia), a Peltier-cooled baffled cyclonic spray chamber, a standard quartz torch, and a two-cone interface design (nickel sample and skimmer cones) operated under the instrumental conditions presented in Table S1. High-purity argon (99.9997%; Gasin, Portugal) was used as nebulizer and plasma gas. The instrument was tuned before each analytical series for maximum sensitivity and signal stability and minimum formation of oxides and double-charged ions. Calibration standards and the internal standard solution were prepared from commercially available multi-element stock solutions (Plasma CAL, SCP Science, Baie-D'Urfe, QC, Canada, and ICP-MS Internal Std, Isostandards, Material, Madrid, Spain, respectively).

The following elemental isotopes were monitored for analytical determinations: $^{7}Li$, $^{27}Al$, $^{48}Ti$, $^{51}V$, $^{52}Cr$, $^{55}Mn$, $^{57}Fe$, $^{59}Co$, $^{60}Ni$, $^{63}Cu$, $^{66}Zn$, $^{75}As$, $^{82}Se$, $^{8}8Sr$, $^{111}Cd$, $^{118}Sn$, $^{121}Sb$, $^{137}Ba$, $^{202}Hg$, and $^{208}Pb$. The elemental isotopes $^{45}Sc$, $^{89}Y$, $^{115}In$, and $^{159}Tb$ were used as internal standards. Limits of detection were calculated as the concentration corresponding to three times the standard deviation of 10 replicate measurements of the blank solution (2% *v*/*v* $HNO_3$) and are presented in Table S2. Results for the certified reference material are presented in Table S3. The reference material used provides data for a limited set of TE, but the analytically most problematic elements are included. Since the analytical procedure has been validated for this set, accuracy problems are not expected for the other determined elements.

### 2.5. Statistical Analysis

Descriptive statistics were used to characterize both groups, RES and CTR, as well as the analytical results. Differences between the cognitive assessment scores of RES subjects at baseline and past five years were studied through a Wilcoxon signed-rank test analysis. Group differences between the cognitive decline registered for the RES group and for the CTR group were assessed through a Mann–Whitney U test. This option for non-parametric tests was based on the small size of the follow-up study group (the RES group). A probability of ≤0.05 was assumed as significant in testing the null hypotheses of no differences between the two moments of assessment and of no differences between the two groups. Potential relationships between TE contents in fingernails and neuropsychological data from the follow-up were firstly assessed through correlation analyses. Linear regression models (simple and multiple) were then computed to further investigate these relationships and assess whether TE significantly predicted future cognitive performance. A significance level of 95% was also used as criteria in both correlation and regression analyses. Comparisons between regression models were based on changes in the variance explained by each model, indexed by the adjusted coefficient of determination ($R^2_{adjusted}$). All statistical analyses were conducted using the Statistical Package for the Social Sciences (SPSS) for Windows, version 22 [69].

## 3. Results

### 3.1. Sample Characterization

The baseline and the follow-up assessments were separated by a 5-year interval (M = 5.30; SD = 0.47). Out of the 76 RES participants initially enrolled in the study, a total of 20 individuals (26.3%) could be included in the follow-up. Drop-outs were mainly due to death (42 out of 56 cases). Of the 42 deceases registered, 10 were due to cardiovascular disorders, 8 to respiratory infections, 8 to cancer, 5 to stroke, 4 to dementia complications (1 case of AD, 1 case of Parkinson's disease, and 2 cases of non-specified dementia, diagnosed by general practitioners), 1 to gastrointestinal hemorrhage, and 6 due to unknown reasons. Other reasons for experimental death were medical conditions (2 cases), unreachability (9 cases), and unavailability (3 cases). Most RES participants enrolled in the follow-up were female (80%), with an average age of 83.60 ± 6.98 years, 2.45 ± 1.73 years of formal education (6 were illiterate), and a mean time of residence in the study areas of 61.93 ± 25.08 years. As for the education-matched control group (CTR), this was composed of 20 subjects, 55% of which were females, with a mean age of 74.70 ± 4.51 years and an average educational level of 3.30 ± 0.80 years ($p > 0.05$). Descriptive statistics of the two groups regarding marital status, occupation/professional activity, and medical history are presented in Table 1.

**Table 1.** Descriptive statistics of the total sample in terms of marital status, occupation/professional activity, and medical history, discriminated by group.

| | RES $n = 20$ | CTR $n = 20$ |
|---|---|---|
| **Marital Status, $n$ (%)** | | |
| Single | 2 (10) | 1 (5) |
| Married | 2 (10) | 12 (60) |
| Divorced | 3 (15) | 1 (5) |
| Widowed | 13 (65) | 6 (30) |
| **Occupation/professional activity, $n$ (%)** | | |
| Agriculture/fishery | 6 (30) | 1 (5) |
| Industry/construction | 4 (20) | 3 (15) |
| Commerce/Services | 8 (40) | 14 (70) |
| Housewife | 2 (10) | 3 (10) |
| **Medical History, $n$ (%)** | | |
| Diabetes | 4 (20) | 2 (10) |
| Dyslipidemia | 12 (60) | 9 (45) |
| Cardiovascular diseases | 16 (80) | 15 (75) |

### *3.2. Neuropsychological Data*

Overall, RES participants exhibited worse cognitive capacities at follow-up, comparatively to the baseline. Subjects scored on average less 6.5 points in MMSE (Baseline: M = 23.80; SD = 5.06; Follow-up: M = 17.30; SD = 6.94; Z = −3.628, $p < 0.001$, $r = -0.811$) and less 5.27 points in MoCA (Baseline: M = 16.00; SD = 4.472; follow-up: M = 10.73; SD = 5.985; Z = −2.317, $p = 0.020$, $r = 0.699$) after 5 years. When compared to the Portuguese normative data established according to age and educational level, the percentage of individuals classified as "with cognitive deficit" (i.e., with scores at 1 standard deviation or more below the expected mean considering age and education) raised from 65.0% to 85.0% according to the MMSE [70] and from 78.6% to 81.8% according to the MoCA [67]. The percentage of individuals with scores below the cut-off defined for AD, Frontotemporal Dementia (FTD), and Vascular Dementia (VD) raised from 60.0% to 85.0% according to the MMSE (*cut-off* of <26) [71–73] and from 71.4% to 83.3% according to the MoCA (*cut-off* of <17) [71–73].

In order to assess whether the cognitive decline of RES participants significantly differed from the cognitive decline observed in the general Portuguese population, differences between the RES group and the CTR group were explored based on the decline observed in MMSE and MoCA scores (decline = baseline score—follow-up score). While the CTR group exhibited a mean score difference of 1.40 ± 1.59 points in the MMSE and of 2.06 ± 1.35 points in the MoCA, the RES group presented score differences of 6.50 ± 4.27 and 5.27 ± 5.20 points, respectively. The observed decline was significantly different between the two groups for both the MMSE (U = 56.500, $p < 0.001$, $r = -0.616$) and the MoCA (U = 19.000, $p = 0.005$, $r = -0.583$).

### *3.3. TE Content in Fingernails*

Descriptive statistics for TE content in fingernails of the RES group are summarized in Table 2. Observed values were compared to values reported in the literature for non-exposed people [74] and healthy centenarians [75]. For most TE, the obtained results were within the reported ranges, with Cd, Co, Li, and Pb contents falling in the lower part of the intervals reported for non-exposed individuals. The same was observed for Co and Mn contents when compared to the intervals reported for healthy centenarians. The contents of Ba, Sr, and Se fell within the reported range for non-exposed people but were below (Ba and Sr) and above (Se) the reference range for healthy centenarians.

**Table 2.** Descriptive statistics for the observed trace elements (TE) contents in fingernails (research study-group (RES)), expressed as µg/g; values reported in the literature are presented for comparison purposes.

| | RES Group (µg/g) | | | | Literature Data | | | | | | | |
|---|---|---|---|---|---|---|---|---|---|---|---|---|
| | | | | | Non-Exposed Individuals [74] | | | | Healthy Centenarians [75] | | | |
| | Min-Max | *M* | *SD* | *Md* | Min-Max | *M* | *SD* | *Md* | Min-Max | *M* | *SD* | *Md* |
| Al | 3.57–54.61 | 18.93 | 14.22 | 14.42 | 12.00–137.00 | 36.00 | 22.00 | 32.00 | - | - | - | - |
| As | 0.07–0.30 | 0.12 | 0.06 | 0.11 | 0.07–1.09 | 0.27 | 0.19 | 0.22 | - | - | - | - |
| Ba | 0.11–2.21 | 0.64 | 0.73 | 0.31 | 0.28–3.99 | 1.34 | 1.35 | 0.89 | 0.94–22.92 | 5.10 | 3.92 | 3.85 |
| Cd | 0.002–0.06 | 0.01 | 0.01 | 0.01 | 0.01–0.44 | 0.11 | 0.18 | 0.06 | 0.004–0.19 | 0.03 | 0.03 | 0.02 |
| Co | 0.003–0.04 | 0.01 | 0.01 | 0.01 | 0.01–0.12 | 0.03 | 0.03 | 0.02 | 0.01–0.64 | 0.10 | 0.10 | 0.07 |
| Cr | 0.25–0.93 | 0.61 | 0.20 | 0.60 | 0.22–3.20 | 1.16 | 1.05 | 0.76 | 0.08–2.51 | 0.82 | 0.44 | 0.82 |
| Cu | 2.66–7.40 | 4.67 | 1.32 | 4.83 | 4.20–17.00 | 8.40 | 3.50 | 7.60 | 2.02–8.53 | 3.71 | 0.99 | 3.55 |
| Fe | 7.31–49.84 | 24.01 | 13.25 | 20.06 | 12.00–189.00 | 42.00 | 30.00 | 37.00 | 16.52–692.00 | 154.40 | 124.80 | 116.70 |
| Hg | 0.12–0.85 | 0.42 | 0.20 | 0.42 | 0.03–0.31 | 0.12 | 0.098 | 0.098 | - | - | - | - |
| Li | 0.005–0.18 | 0.04 | 0.04 | 0.02 | 0.01–0.25 | 0.07 | 0.07 | 0.05 | 0.02–2.07 | 0.31 | 0.32 | 0.23 |
| Mn | 0.04–2.93 | 0.58 | 0.79 | 0.17 | 0.19–3.30 | 0.90 | 0.75 | 0.65 | 0.21–15.40 | 3.09 | 2.18 | 2.62 |
| Ni | 0.09–2.98 | 0.93 | 1.04 | 0.36 | 0.14–6.95 | 1.65 | 2.20 | 0.84 | 0.02–3.67 | 0.95 | 0.85 | 0.66 |
| Pb | 0.07–1.73 | 0.38 | 0.38 | 0.30 | 0.27–4.75 | 1.38 | 1.14 | 1.06 | 0.13–9.61 | 1.86 | 1.81 | 1.33 |
| Sb | 0.008–0.13 | 0.04 | 0.04 | 0.03 | 0.01–0.13 | 0.05 | 0.05 | 0.04 | - | - | - | - |
| Se | 0.60–1.03 | 0.803 | 0.13 | 0.80 | 0.62–1.53 | 0.94 | 0.21 | 0.93 | 0.24–0.70 | 0.44 | 0.11 | 0.44 |
| Sn | 0.01–1.20 | 0.33 | 0.36 | 0.17 | 0.11–2.56 | 0.63 | 0.51 | 0.48 | - | - | - | - |
| Sr | 0.07–2.24 | 0.48 | 0.53 | 0.27 | 0.17–1.39 | 0.43 | 0.21 | 0.39 | 1.40–18.50 | 6.20 | 2.47 | 5.80 |
| Ti | 3.68–6.75 | 5.14 | 0.77 | 5.15 | 0.94–16.10 | 4.46 | 5.01 | 2.71 | - | - | - | - |
| V | 0.05–0.38 | 0.120 | 0.07 | 0.10 | 0.02–0.48 | 0.08 | 0.05 | 2.71 | - | - | - | - |
| Zn | 88.65–219.31 | 144.10 | 39.15 | 137.34 | 80.00–191.00 | 120.00 | 29.00 | 116.00 | 93.00–326.00 | 148.00 | 36.00 | 138.00 |

Note: Min = minimum; Max = maximum; *M* = mean; *SD* = standard deviation; *Md* = median; Al = aluminum; As = arsenic; Ba = barium; Cd = cadmium; Co = cobalt; Cr = chromium; Cu = copper; Fe = iron; Hg = mercury; Li = lithium; Mn = manganese; Ni = nickel; Pb = lead; Sb = antimony; Se = selenium; Sn = tin; Sr = strontium; Ti = titanium; V = vanadium; Zn = zinc.

### *3.4. Relationship between Fingernail TE Content and Cognitive Performance*

In order to assess the relationship between TE in fingernails and the results of the neuropsychological assessment (total scores obtained with the MMSE and MoCA), a correlation analysis was performed. Correlation coefficients for the relationship between MMSE scores and Ni content ($r = 0.489$), MoCA scores and Ni content ($r = 0.626$), and MoCA scores and Se content ($r = 0.647$) were the only ones that reached statistical significance ($p < 0.05$), evidencing a moderate to strong positive association between these elements and the cognitive screening tests' scores [76,77].

To further investigate whether TE contents could predict the cognitive performance of the RES group 5 years after the biomonitoring, linear regression analyses were run. The computed linear models assumed as predictor variables the TE contents of Ni and Se (the elements that led to significant correlation coefficients with the cognitive scores). Models are reported in Table 3.

**Table 3.** Linear regression models.

| | Linear Regression Models | $R^2_{adjusted}$ | Model Significance |
|---|---|---|---|
| **MMSE-1** | MMSE = 14.258 + 3.259 Ni | 0.197 | $p = 0.029$ |
| **MMSE-2** | MMSE = 10.329 + 1.436 Education + 3.700 Ni | 0.288 | $p = 0.022$ |
| **MoCA-1** | MoCA = 7.503 + 3.369 Ni | 0.331 | $p = 0.029$ |
| **MoCA-2** | MoCA = −9.258 + 26.821 Se | 0.361 | $p = 0.023$ |
| **MoCA-3** | MoCA = −9.460 + 2.851 Ni + 23.017 Se | 0.622 | $p = 0.005$ |
| **MoCA-4** | MoCA = −16.147 + 1.556 Education + 3.525 Ni + 24.022 Se | 0.679 | $p = 0.007$ |

Note: MMSE = Mini Mental State Examination; MoCA = Montreal Cognitive Assessment; MMSE-n = regression models assuming MMSE total scores as the dependent variable; MoCA-n = regression models assuming MoCA total scores as the dependent variable.

Initially, simple linear regressions showed Ni as a significant predictor of MMSE total scores ($\beta = 0.489$, $t = 2.381$, $p = 0.029$), with the regression model (MMSE-1) explaining 19.7% of the variance of the scores. Regarding MoCA total scores, both Ni ($\beta = 0.626$, $t = 2.538$, $p = 0.029$) and Se ($\beta = 0.647$, $t = 2.687$, $p = 0.023$) showed as relevant predictors. After gathering both TE in a single regression model, the statistical significance for both variables persisted, with Se being highlighted as the best predictor variable (explaining 55.6% of MoCA scores, vs. the 52.9% for Ni). The resulting statistically significant model (MoCA-3; $p = 0.005$) could explain 62.2% of the MoCA total scores variance (an increase of 29.1% and 26.1% of the variance explained by the models MoCA-1 and -2, respectively).

Because education greatly influences cognitive performance on screening tests [53¯57,78] and since aging seems to be associated with cognitive decline [63,64], age and educational level were included in the regression equations in order to account for these two confounding variables. Models for MMSE and MoCA with the inclusion of age showed a lack of statistical significance, and age was not a relevant predictor of the tests' total scores ($p > 0.05$). Contrastingly, when the educational level was included into the equations, Ni and Se remained as relevant predictors, with the regression models showing statistical significance. Compared to the MMSE-1, a 9.1% increase in the variance explained by the model was observed, despite education being shown as a non-significant predictor ($\beta = 0.358$, $t = 1.818$, $p > 0.05$). In the resulting MMSE-2 model, both Ni and education explained together 28.8% of MMSE total scores. The same was true when the educational level was added to the MoCA-3 model. Although education was not a significant predictor ($\beta = 0.304$, $t = 1.605$, $p > 0.05$) of MoCA total scores, its inclusion caused an increase of 5.7% of the variance explained by the regression model. The resulting MoCA-4 model explained 67.9% of MoCA total scores. Therefore, MMSE-2 (where Ni: $\beta = 0.556$, $t = 2.821$, $p = 0.012$) and MoCA-4 (where Ni: $\beta = 0.655$, $t = 3.441$, $p = 0.009$; Se: $\beta = 0.580$, $t = 3.329$, $p = 0.010$) were selected as the final linear models. In MoCA-4 model, Ni is highlighted as the best predictor variable. In both MMSE-2 and MoCA-4 models, higher contents of Ni/Se in fingernails predict better cognitive performances in screening tests 5 years after biomonitoring analyses.

## 4. Discussion

Despite the increasing evidence reporting the role of environmental exposures in the development of neurodegenerative diseases, literature on the possible pathogenic role of different metals does not provide direct evidence to define a causal relationship between environmental exposure to certain metals and AD [15,16]. Therefore, research on the influence of TE on the cognition of elderly people is warranted. The present study aimed to explore potential relationships between TE levels in the human body and cognition. Specifically, the goal was to assess how TE content in fingernails relate to future cognitive performance.

Analyses of the RES neuropsychological data revealed an average decrease of 6.5 points and 5.27 points on the global scores of MMSE and MoCA, from baseline to follow-up (5 years later), as well as an increase in the portion of elderly individuals classified as "with cognitive deficit"—from 65.0% to 85.0% and from 78.6% to 81.8%, respectively. When compared to the CTR group, the decrease observed in MMSE and MoCA scores was significantly greater among the RES participants, indicating that the residents of the geographical areas of interest suffered a steeper cognitive decline than the general Portuguese population of the same educational level (within a 5-years period).

In terms of the RES participants with a cognitive performance similar to patients diagnosed with dementia (AD, FTD, and VD), they constituted 85.0% and 83.3% of the study group, according to the *cut-offs* of the screening tools [71–73]. These percentages of dementia-like performances are greater than expected considering the prevalence rate of the condition estimated for the Portuguese population (9.2% based on the 10/66 *Dementia Research Group*' dementia diagnostic algorithm [79]), which indicates a higher frequency of cognitive impairment across the study group in comparison to the general Portuguese population older than 65 years. Literature systematically reports aging as the primary risk factor for neurodegenerative diseases, as this process implies a set of biological changes at the cellular level, from genomic instability to altered intercellular communication [80]. Indeed, cognitive decline seems to be intimately associated with age, and the prevalence of dementia cases increases exponentially after 65 years of age [80,81]. Considering that the average age of the RES elders included in this study was 83.6 years (with the youngest participant being 68 years old) and that the average age reported by Gonçalves-Pereira et al. for their study group was 74.9 years [79], it is plausible to admit that the difference observed in the frequency of cognitive impairment between both studies can be at least partially explained by the approximately 9-years gap between both groups of elders. However, the observed percentages of dementia-like performances seem to remain disproportioned even when compared to the prevalence estimated for the sub-group of individuals with ages above 80 years (18.37–19.61% based on the 10/66 *Dementia Research Group*'s dementia diagnostic algorithm [79]) assessed by Gonçalves-Pereira et al., suggesting that such differences are not attributed to age discrepancies.

The comparison between fingernails TE content in the RES group and values reported in the literature for non-exposed individuals [74] showed an overall good fit, suggestive of non-significant exposure to environmental contaminants. Nevertheless, it is worth noting that the mean age of the study sample (approximately 84 years) is substantially different from the average age of the population assessed by Rodushkin and Axelsson [74], which mean age was 33 years (range: 1–76 years).

With regards to the potential association between TE content in fingernails and cognitive performance, we found positive significant correlations between Ni and Se levels and the global scores of MMSE and MoCA. Regression analysis showed Ni and Se as significant predictors of future cognitive performances and revealed that, together with educational level, Ni explains 33% of the variance observed in MMSE total scores after 5 years, while Ni and Se explained 68% of the variance in MoCA. It is important to note that age proved to be a non-significant, non-relevant, predictor of the tests' total scores, which is in line with the previous hypothesis that the cognitive decline observed among RES participants is not age-derived.

Nickel is a silver-white metal that belongs to the ferromagnetic elements group and that is widely distributed across the environment. It can be absorbed through the respiratory track, digestive system, and skin [82], and besides being essential for some plants and animal species [83], this metal is also a micronutrient essential for proper functioning of the human body (as it increases hormonal activity and is involved in lipid metabolism [82]). Depending on the duration, intensity, and pathway of exposure, Ni can act as an immunotoxin, as well as a carcinogenic agent, and cause several health problems (e.g., [84–87]). Furthermore, Ni accumulates in the brain and can act as a neurotoxicant [88]. Despite the molecular mechanisms not being well understood, its actions seem to be linked with oxidative stress, mitochondrial dysfunctions, and Ni-induced apoptosis (please see [83] and [88] for more details). Defective apoptotic processes may lead to excessive cell death, which underlies several neurodegenerative conditions. Toxicological studies in rats have shown alterations in the neuronal morphology of rat's brain after Ni administration [89]. A significant reduction of intact neurons in the hippocampus and striatum was observed, as well as ultrastructural alterations in neurons of the hippocampus, striatum, and cortex [89,90]. Mitochondria seemed to be the key target in Ni-induced neurodegeneration [90]. Cognitive and motor behavior were also compromised [89].

Literature reports tend to portrait Ni as an element with neurotoxic repercussions that may affect cognition and contribute to the development of neurodegenerative diseases. Our study seems to contradict such evidence, as the results obtained showed that fingernail Ni content significantly predicts a lower cognitive decay within 5 years. However, the observation that Ni content in fingernails was within the interval reported for both non-exposed individuals [74] and healthy centenarians [75] is relevant, as it not only suggests that our study group was not significantly exposed to this TE but also that our results are limited to normal (non-toxic) content levels. Therefore, our findings suggest that, in spite of its potential neurotoxic effects, Ni may also have beneficial effects on long-term cognitive performance maintenance, provided it is maintained within certain concentration levels.

The role of Se in the human organism seems to be much better understood. Se exists in several chemical species [91–93], being a TE of interest for both nutritional and toxicological reasons. While it is an essential element for the biosynthesis of selenoproteins when in the selenocysteine-bound organic form [94], it is also a toxicant when in the selenomethionine and inorganic forms [93]. Furthermore, Se has a very narrow range of safe exposure [95], and both its deficiency and excess are associated with important adverse health effects [96]. In the human body, Se plays an essential antioxidant role [97] and is very important for the maintenance of the homeostasis of the central nervous system [98]. Its beneficial effects are mediated by selenium-containing proteins (the selenoproteins) such as glutathione peroxidases (GPx), the plasma Se-transport protein (SePP), and thioredoxin reductases (TrxR) [99]. One of the roles of these selenoproteins is to regulate oxidative stress, which has been linked with an increased risk of cognitive decline [100]. For GPX specifically, whose main function is to eliminate peroxides (therefore maintaining lower levels of reactive oxygen species), it is also known that is expressed in neurons and glia cells [101]. Additionally, Se also appear to be inversely associated with systemic inflammation, a critical process in age-associated cognitive decline and dementia [102,103]. In vitro studies observed this negative relationship in several cells, including neurons [104] and macrophages [105]. Such features are essential for neuroprotection and therefore brain metabolism of this TE is different from other organs [106]. In case of deficiency, the brain is the last organ to be depleted and is the first one to return to normal store levels when Se is replenished [107].

Given its beneficial effects, Se has been widely studied within the scope of healthy aging and cognitive impairment. Reduced levels of Se are believed to lead to neurons destruction and increased risk of cognitive decline/dementia [108,109]. Studies show that, compared to healthy elder adults, AD patients exhibited reduced levels of Se in plasma, erythrocyte, blood, cerebrospinal fluid (CSF), and nails [97,110–112]. Specifically, the levels of this TE seem to be decreased in the temporal, hippocampal, and cortex regions of AD patients [113]. Higher CSF levels of selenite (a common inorganic Se species) seem to predict progression to AD in non-vascular MCI patients [114]. Studies on Chinese elderly report an association between lower levels of Se in nails and poorer cognitive performances and that apoE4 carriers have significantly lower nail content of Se than non-carriers [111,113]. Concordantly to this evidence, our findings highlight the beneficial effects of Se in cognition, as higher fingernail Se contents predict lesser cognitive decline in 5 years. The determined Se content in fingernails was also within the reported interval for non-exposed individual [74], which suggests non-excessive exposure and lines up with the well-known U-shaped relationship between Se levels in the human body and adverse health effects. Interestingly, Se levels were higher in our participants when compared with healthy centenarians [75]. As age increases, the decrease of organs' physiological functioning can greatly affect the absorption, distribution, metabolism, and function of Se in the human organism [115]. Given the differences in age between our study group (with a mean age of approximately 84 years) and the centenarians, this may explain the difference in Se levels. Indeed, previous studies report a reduction of Se levels in red blood cells not only in AD and MCI but also in normal elderly people [116,117]. Nevertheless, information on this topic is inconsistent, as studies have also shown higher levels of Se in the plasma among centenarians, when compared to septuagenarians [118] and nonagenarians [119]. An important variable contributing for the heterogeneity of results is the choice of different tissues and fluids when assessing TE levels. Literature findings show how the relationship between Se and certain diseases may vary across studies based on different tissues [120].

The main limitation of the present research lies on the number of drop-outs, mainly due to participants passing away (55% of the initial study sample), which led to a reduced sample size that potentially may have prevented results from reaching statistical significance. Notwithstanding, this limitation stems from a longitudinal methodology based on a 5-years follow-up that was implemented in a cohort whose average age is higher than the life expectancy of its population (estimated to be 80.8 years for Portugal [121]). Furthermore, the present study does not include the measurement of TE in the CTR group, and therefore, results should be interpreted in a preliminary/exploratory framework.

## 5. Conclusions

The present exploratory study provides original preliminary evidence of an existing association between the concentration levels of TE in the human body and future cognitive status—our findings show that fingernail contents of Ni and Se predict lesser cognitive decline within 5 years. Future expansion of the present research is warranted. More comprehensive samples and individual-level measurements of TE for control groups must be included in order to assess the real beneficial potential of these TE in the maintenance of cognitive capacities. Further research on humans is needed, and similar translational approaches based on crossing other biomarkers with neuropsychological data should be considered. Such contributions have the potential to open new avenues for a more successful prevention of cognitive impairment and dementia.

**Supplementary Materials**

Table S1: Operating conditions for the iCAP$^{TM}$Q ICP-MS instrument. Table S2: Limits of detection (LOD) for the analyzed trace elements. Table S3: Results obtained in the analysis of the certified reference material ERM®—DB001 Trace Element in Human Hair.

**Author Contributions:** Conceptualization, B.G.; formal analysis, A.A. and E.P.; funding acquisition, B.G., M.C.P., and S.F.; investigation, B.G., M.C.P., J.N., P.P., and S.F.; methodology, B.G., M.C.P., J.N., P.P., P.M.-R., L.D., M.R.S., and S.F.; project administration, B.G., M.C.P., M.R.S., and S.F.; supervision, M.C.P., M.R.S., and S.F.; writing—original draft, B.G.; writing—review and editing, B.G., M.C.P., A.A., P.I.M., M.R.S., and S.F. All authors have read and agreed to the published version of the manuscript.

**Acknowledgments:** A special thank you to all participants involved in this research, as well to the local institutions of social solidarity for the partnership (Santa Casa da Misericórdia de Estarreja, Associação Humanitária de Salreu, Centro Paroquial Social São Tomé de Canelas, Centro Paroquial Social de Avanca, Fundação Cónego Filipe Figueiredo, Centro Paroquial de Pardilhó, and Santa Casa da Misericórdia de Mértola).

## References


1. WHO. Global Health and Aging. Available online: https://www.who.int/ageing/publications/global_health.pdf (accessed on 9 March 2020).
2. OECD. Health at a Glance 2019. Available online: https://www.oecd-ilibrary.org/social-issues-migration-health/health-at-a-glance-2019_4dd50c09-en (accessed on 9 March 2020).
3. Statistics Portugal. *Censos 2011, XV Recenseamento Geral da População, V Recenseamento Geral da Habitação, Resultados Definitivos, Portugal [2011 Census, XV General Population Census, V General Housing Census, Definitive Results, Portugal]*; INE: Lisbon, Portugal, 2015; ISBN 978-989-25-0181-9.
4. Wimo, A.; Guerchet, M.; Ali, G.C.; Wu, Y.T.; Prina, A.M.; Winblad, B.; Jönsson, L.; Liu, Z.; Prince, M. The worldwide costs of dementia 2015 and comparisons with 2010. *Alzheimers Dement.* **2017**, *13*, 1–7. [CrossRef] [PubMed]
5. Tartaglione, A.M.; Venerosi, A.; Calamandrei, G. Early-life toxic insults and onset of sporadic neurodegenerative diseases—An overview of experimental studies. In *Neurotoxin Modeling of Brain Disorders-Life-Long Outcomes in Behavioral Teratology*; Kostrzewa, R.M., Archer, T., Eds.; Springer: Cham, Switzerland; Boston, MA, USA, 2015; Volume 29, pp. 231–264. ISBN 978-3-319-34136-1.
6. Yan, D.; Zhang, Y.; Liu, L.; Yan, H. Pesticide exposure and risk of Alzheimer's disease: A systematic review and meta-analysis. *Sci. Rep.* **2016**, *6*, 32222. [CrossRef] [PubMed]
7. Paglia, G.; Miedico, O.; Cristofano, A.; Vitale, M.; Angiolillo, A.; Chiaravalle, A.E.; Corso, G.; Di Costanzo, A. Distinctive pattern of serum elements during the progression of Alzheimer's disease. *Sci. Rep.* **2016**, *6*, 22769. [CrossRef] [PubMed]
8. Perl, D.P. Neuropathology of Alzheimer's disease. *Mt. Sinai J. Med.* **2010**, *77*, 32–42. [CrossRef] [PubMed]
9. Dourlen, P.; Kilinc, D.; Malmanche, N.; Chapuis, J.; Lambert, J.C. The new genetic landscape of Alzheimer's disease: From amyloid cascade to genetically driven synaptic failure hypothesis? *Acta Neuropathol.* **2019**, *138*, 221–226. [CrossRef]
10. Yegambaram, M.; Manivannan, B.; Beach, T.G.; Halden, R.U. Role of environmental contaminants in the etiology of Alzheimer's disease: A review. *Curr. Alzheimer Res.* **2015**, *12*, 116–146. [CrossRef]
11. Killin, L.O.; Starr, J.M.; Shiue, I.J.; Russ, T.C. Environmental risk factors for dementia: A systematic review. *BMC Geriatr.* **2016**, *16*, 1–28. [CrossRef]
12. McAllum, E.J.; Finkelstein, D.I. Metals in Alzheimer's and Parkinson's disease: Relevance to dementia with lewy bodies. *J. Mol. Neurosci.* **2016**, *60*, 279–288. [CrossRef]
13. Ayton, S.; Lei, P.; Bush, A.I. Metallostasis in Alzheimer's disease. *Free Radic. Biol. Med.* **2013**, *62*, 76–89. [CrossRef]
14. Gupta, V.B.; Anitha, S.; Hegde, M.L.; Zecca, L.; Garruto, R.M.; Ravid, R.; Shankar, S.K.; Stein, R.; Shanmugavelu, P.; Rao, K.J. Aluminium in Alzheimer's disease: Are we still at a crossroad? *Cell Mol. Life Sci.* **2005**, *62*, 143–158. [CrossRef]
15. Lucaroni, F.; Ambrosone, C.; Paradiso, F.; Messinese, M.; Di Domenicantonio, R.; Alessandroni, C.; Cicero, C.E.; Cerutti, F.; Di Gaspare, F.; Morciano, L.; et al. Metals Dyshomeostasis in Alzheimer's Disease: A Systematic Review. *Biomed. Prev.* **2017**, *2*, 112. [CrossRef]
16. Cicero, C.E.; Mostile, G.; Vasta, R.; Rapisarda, V.; Santo Signorelli, S.; Ferrante, M.; Zappia, M.; Nicoletti, A. Metals and neurodegenerative diseases. A systematic review. *Environ. Res.* **2017**, *159*, 82–94. [CrossRef] [PubMed]
17. Ferreira, P.C.; Tonani, K.A.; Julião, F.C.; Cupo, P.; Domingo, J.L.; Segura-Muñoz, S.I. Aluminum concentrations in water of elderly people's houses and retirement homes and its relation with elderly health. *Bull. Environ. Contam. Toxicol.* **2009**, *83*, 565–569. [CrossRef] [PubMed]
18. Zapatero, M.D.; Garcia de Jalon, A.; Pascual, F.; Calvo, M.L.; Escanero, J.; Marro, A. Serum aluminum levels in Alzheimer's disease and other senile dementias. *Biol. Trace Elem. Res.* **1995**, *47*, 235–240. [CrossRef] [PubMed]
19. Smorgon, C.; Mari, E.; Atti, A.R.; Dalla Nora, E.; Zamboni, P.F.; Calzoni, F.; Passaro, A.; Fellin, R. Trace elements and cognitive impairment: An elderly cohort study. *Arch. Gerontol. Geriatr.* **2004**, *9*, 393–402. [CrossRef] [PubMed]
20. González-Domínguez, R.; García-Barrera, T.; Gómez-Ariza, J.L. Characterization of metal profiles in serum during the progression of Alzheimer's disease. *Metallomics* **2014**, *6*, 292–300. [CrossRef]

21. Graham, S.F.; Nasaruddin, M.B.; Carey, M.; Holscher, C.; McGuinness, B.; Kehoe, P.G.; Love, S.; Passmore, P.; Elliott, C.T.; Meharg, A.A.; et al. Age-Associated changes of brain copper, iron and zinc in alzheimer's disease and dementia with lewy bodies. *J. Alzheimers Dis.* **2014**, *42*, 1407–1413. [CrossRef]
22. Rembach, A.; Doecke, J.D.; Roberts, B.R.; Watt, A.D.; Faux, N.G.; Volitakis, I.; Pertile, K.K.; Rumble, R.L.; Trounson, B.O.; Fowler, C.J.; et al. Longitudinal analysis of serum copper and ceruloplasmin in Alzheimer's disease. *J. Alzheimers Dis.* **2013**, *34*, 171–182. [CrossRef]
23. Akatsu, H.; Hori, A.; Yamamoto, T.; Yoshida, M.; Mimuro, M.; Hashizume, Y.; Tooyama, I.; Yezdimer, E.M. Transition metal abnormalities in progressive dementias. *Biometals* **2012**, *25*, 337–350. [CrossRef]
24. Magaki, S.; Raghavan, R.; Mueller, C.; Oberg, K.C.; Vinters, H.V.; Kirsch, W.M. Iron, copper, and iron regulatory protein 2 in alzheimer's disease and related dementias. *Neurosci. Lett.* **2007**, *418*, 72–76. [CrossRef]
25. Xu, J.; Begley, P.; Church, S.J.; Patassini, S.; McHarg, S.; Kureishy, N.; Hollywood, K.A.; Waldvogel, H.J.; Liu, H.; Zhang, S.; et al. Elevation of brain glucose and polyol-pathway intermediates with accompanying brain-copper deficiency in patients with alzheimer's disease: Metabolic basis for dementia. *Sci. Rep.* **2016**, *6*, 27524. [CrossRef] [PubMed]
26. Smith, M.A.; Zhu, X.; Tabaton, M.; Liu, G.; McKeel Jr, D.W.; Cohen, M.L.; Wang, X.; Siedlak, S.L.; Dwyer, B.E.; Hayashi, T.; et al. Increased iron and free radical generation in preclinical Alzheimer disease and mild cognitive impairment. *J. Alzheimers Dis.* **2010**, *19*, 363–372. [CrossRef] [PubMed]
27. Szabo, S.T.; Harry, G.J.; Hayden, K.M.; Szabo, D.T.; Birnbaum, L. Comparison of metal levels between postmortem brain and ventricular fluid in alzheimer's disease and nondemented elderly controls. *Toxicol. Sci.* **2015**, *150*, 292–300. [CrossRef] [PubMed]
28. Emard, J.F.; Andre, P.; Thouez, J.P.; Mathieu, J.; Boily, C.; Beaudry, M.; Cholette, A.; Robitaille, Y.; Bouchard, R.; Daoud, N.; et al. Geographical distribution of Alzheimer's disease cases at birth and the geochemical profile of Saguenay-lac-Saint-Jean/Québec, Canada (image project). *Water Air Soil Pollut.* **1994**, *72*, 251–264. [CrossRef]
29. Shen, X.L.; Yu, J.H.; Zhang, D.F.; Xie, J.X.; Jiang, H. Positive relationship between mortality from Alzheimer's disease and soil metal concentration in mainland China. *J. Alzheimers Dis.* **2014**, *42*, 893–900. [CrossRef] [PubMed]
30. Jomova, K.; Valko, M. Advances in metal-induced oxidative stress and human disease. *Toxicology* **2011**, *283*, 65–87. [CrossRef]
31. Barnham, K.J.; Bush, A.I. Metals in Alzheimer's and Parkinson's diseases. *Curr. Opin. Chem. Biol.* **2008**, *12*, 222–228. [CrossRef]
32. Mohmand, J.; Eqani, S.A.M.A.S.; Fasola, M.; Alamdar, A.; Mustafa, I.; Ali, N.; Liu, L.; Peng, S.; Shen, H. Human exposure to toxic metals via contaminated dust: Bio-accumulation trends and their potential risk estimation. *Chemosphere* **2015**, *132*, 142–151. [CrossRef]
33. Antoniadis, V.; Shaheen, S.M.; Boersch, J.; Frohne, T.; Du Laing, G.; Rinklebe, J. Bioavailability and risk assessment of potentially toxic elements in garden edible vegetables and soils around a highly contaminated former mining area in Germany. *J. Environ. Manag.* **2017**, *186*, 192–200. [CrossRef]
34. Cabral Pinto, M.M.S.; Marinho-Reis, P.; Almeida, A.; Pinto, E.; Neves, O.; Inácio, M.; Gerardo, B.; Freitas, S.; Simões, M.R.; Dinis, P.A.; et al. Links between cognitive status and trace element levels in hair for an environmentally exposed population: A case study in the surroundings of the estarreja industrial area. *Int. J. Environ. Res. Public Health* **2019**, *16*, 4560. [CrossRef]
35. Cabral Pinto, M.M.S.; Marinho-Reis, A.P.; Almeida, A.; Ordens, C.M.; Silva, M.M.; Freitas, S.; Simões, M.R.; Moreira, P.I.; Dinis, P.A.; Diniz, M.L.; et al. Human predisposition to cognitive impairment and its relation with environmental exposure to potentially toxic elements. *Environ. Geochem. Health* **2018**, *40*, 1767–1784. [CrossRef] [PubMed]
36. Júlvez, J.; Paus, T.; Bellinger, D.; Eskenazi, B.; Tiemeier, H.; Pearce, N.; Ritz, B.; White, T.; Ramchandani, P.; Gispert, J.D.; et al. Environment and brain development: Challenges in the global context. *Neuroepidemiology* **2016**, *46*, 79–82. [CrossRef]
37. Genuis, S.J.; Kelln, K.L. Toxicant exposure and bioaccumulation: A common and potentially reversible cause of cognitive dysfunction and dementia. *Behav. Neurol.* **2015**, *620143*. [CrossRef] [PubMed]
38. Haut, M.W.; Hartzell, J.W.; Moram, M.T. Toxins in the Central Nervous System. In *Textbook of Clinical Neuropsychology*, 2nd ed.; Morgan, J.E., Ricker, J.H., Eds.; Routledge: New York, NY, USA, 2018; pp. 587–602.

39. Singer, R. Neurotoxicity in Neuropsychology. In *The Little Black Book of Neuropsychology: A Syndrome-Based Approach*; Schoenberg, M.R., Scott, J.G., Eds.; Springer: New York, NY, USA, 2014; pp. 813–838.
40. White, R.F.; Krengel, M.; Grashow, R. Neurotoxicology. In *Clinical Neuropsychology: A Pocket Handbook for Assessment*, 3rd ed.; Parsons, M.W., Hammeke, T.A., Eds.; American Psychological Association: Washington, DC, USA, 2014; pp. 338–362.
41. He, K. Trace elements in nails as biomarkers in clinical research. *Eur. J. Clin. Investig.* **2011**, *41*, 98–102. [CrossRef]
42. Clarkson, T.W.; Friberg, L.; Nordberg, G.F.; Sager, P.R. *Biological Monitoring of Toxic Metals*; Springer Science & Business Media: New York, NY, USA, 2012.
43. Sureda, A.; Bibiloni, M.M.; Julibert, A.; Aparicio-Ugarriza, R.; Blé, G.P.L.; Pons, A.; Gonzalez-Gross, M.; Tur, J.A. Trace element contents in toenails are related to regular physical activity in older adults. *PLoS ONE* **2017**, *12*. [CrossRef]
44. WHO. Human Biomonitoring: Facts and Figures. Available online: http://www.euro.who.int/__data/assets/pdf_file/0020/276311/Human-biomonitoring-facts-figures-en.pdf (accessed on 9 March 2020).
45. Cabral-Pinto, M.M.; Inácio, M.; Neves, O.; Almeida, A.A.; Pinto, E.; Oliveiros, B.; da Silva, E.A.F. Human health risk assessment due to agricultural activities and crop consumption in the surroundings of an industrial area. *Expos. Health* **2019**, 1–12. [CrossRef]
46. Yesavage, J.A.; Brink, T.L.; Rose, T.L.; Lum, O.; Huang, V.; Adey, M.; Leirer, O. Development and validation of a geriatric depression screening scale: A preliminary report. *J. Psychiatr Res.* **1983**, *17*, 37–49. [CrossRef]
47. Pocinho, M.T.S.; Farate, C.; Dias, C.A.; Lee, T.T.; Yesavage, J.A. Clinical and psychometric validation of the geriatric depression scale (GDS) for portuguese elders. *Clin. Gerontol.* **2009**, *32*, 223–236. [CrossRef]
48. Simões, M.R.; Prieto, G.; Pinho, M.S.; Firmino, H. Geriatric Depression Scale (GDS-30). In *Escalas e Testes na Demência [Scales and Tests in Dementia]*, 3rd ed.; Simões, M.R., Isabel Santana e Grupo de Estudos de Envelhecimento Cerebral e Demência, Eds.; Novartis: Lisboa, Portugal, 2015; pp. 128–133.
49. Folstein, M.; Folstein, S.; McHugh, P. Mini-mental state: A practical method for grading the cognitive state of patients for the clinician. *J. Psychiatr. Res.* **1975**, *12*, 189–198. [CrossRef]
50. Guerreiro, M.; Silva, A.P.; Botelho, M.A.; Leitão, O.; Castro-Caldas, A.; Garcia, C. Adaptação à população portuguesa da tradução do "Mini Mental State Examination" (MMSE). *Rev. Port. Neurol.* **1994**, *1*, 9.
51. Nasreddine, Z.S.; Phillips, N.A.; Bédirian, V.; Charbonneau, S.; Whitehead, V.; Collin, I.; Cummings, J.L.; Chertkow, H. The Montreal Cognitive Assessment, MoCA: A brief screening tool for Mild Cognitive Impairment. *J. Am. Geriatr. Soc.* **2005**, *534*, 695–699. [CrossRef] [PubMed]
52. Simões, M.R.; Freitas, S.; Santana, I.; Firmino, H.; Martins, C.; Nasreddine, Z.; Vilar, M. *Montreal Cognitive Assessment (MoCA): Versão Portuguesa [Montreal Cognitive Assessment (MoCA): Portuguese Version]*; Serviço de Avaliação Psicológica da Faculdade de Psicologia e de Ciências da Educação da Universidade de Coimbra: Coimbra, Portugal, 2008.
53. Freitas, S.; Simões, M.R.; Alves, L.; Santana, I. The relevance of sociodemographic and health variables on MMSE normative data. *Appl. Neuropsychol. Adult* **2015**, *22*, 311–319. [CrossRef] [PubMed]
54. Freitas, S.; Simões, M.R.; Alves, L.; Santana, I. Montreal cognitive assessment: Influence of sociodemographic and health variables. *Arch. Clin. Neuropsychol.* **2012**, *27*, 165–175. [CrossRef] [PubMed]
55. Anderson, T.M.; Sachdev, P.S.; Brodaty, H.; Trollor, J.; Andrews, G. Effects of sociodemographic and health variables on minimental state exam scores in older Australians. *Am. J. Geriatr. Psychiatry* **2007**, *15*, 467–476. [CrossRef]
56. Bravo, G.; Hébert, R. Age and education specific reference values for the mini-mental and modified mini-mental state examination derived from a non-demented elderly population. *Int. J. Geriatr. Psychiatry* **1997**, *12*, 1008–1018. [CrossRef]
57. Matallana, D.; Santacruz, C.; Cano, C.; Reyes, P.; Samper-Ternent, R.; Markides, K.S.; Ottenbacher, K.J.; Reyes-Ortiz, C.A. The relationship between educational level and Mini-Mental State Examination domains among older Mexican Americans. *J. Geriatr. Psych. Neur.* **2011**, *24*, 9–18. [CrossRef]
58. Costa, C.; Jesus-Rydin, C. Site investigation on heavy metals contaminated ground in Estarreja—Portugal. *Eng. Geol.* **2001**, *60*, 39–47. [CrossRef]
59. Inácio, M.; Neves, O.; Pereira, V.; da Silva, E.F. Levels of selected potential harmful elements (PHEs) in soils and vegetables used in diet of the population living in the surroundings of the estarreja chemical complex (Portugal). *J. Appl. Geochem.* **2014**, *44*, 38–44. [CrossRef]

60. Patinha, C.; Reis, A.P.; Dias, A.C.; Abduljelil, A.A.; Noack, Y.; Robert, S.; Cave, M.; da Silva, E.F. The mobility and human oral bioaccessibility of Zn and Pb in urban dusts of Estarreja (N Portugal). *Environ. Geochem. Health* **2015**, *37*, 115–131. [CrossRef]
61. Leitão, T.B.E. Metodologia Para A Reabilitação De Aquíferos Poluídos. Ph.D. Thesis, University of Lisbon, Lisbon, Portugal, 1996.
62. Van der Weijden, C.; Pacheco, F.A.L. Hydrogeochemistry in the Vouga River basin (central Portugal): Pollution and chemical weathering. *J. Appl. Geochem.* **2006**, *21*, 580–613. [CrossRef]
63. Ordens, C.M. Estudo Da Contaminacão Do Aquífero Superior Na Região De Estarreja. Master's Thesis, University of Coimbra, Coimbra, Portugal, 2007.
64. Abreu, M.M.; Tavares, M.T.; Batista, M.J. Potential use of Erica andevalensis and Erica australis in phytoremediation of sulphide mine environments: São Domingos, Portugal. *J. Geochem. Explor.* **2008**, *96*, 210–222. [CrossRef]
65. Oliveira, J.S.; Farinha, J.; Matos, J.X.; Ávila, P.; Rosa, C.; Machado, M.J.C.; Daniel, F.S.; Martins, L.; Leite, M.R.M. Diagnóstico ambiental das principais áreas mineiras degradadas do país [Environmental diagnosis of the main degraded mining areas of the country]. *Boletim Minas* **2002**, *39*, 67–85.
66. Pérez-López, R.; Álvarez-Valero, A.M.; Nieto, J.M.; Sáez, R.; Matos, J.X. Use of sequential extraction procedure for assessing the environmental impact at regional scale of the São Domingos Mine (Iberian Pyrite Belt). *Appl. Geochem.* **2008**, *23*, 3452–3463. [CrossRef]
67. Freitas, S.; Simões, M.R.; Alves, L.; Santana, I. Montreal Cognitive Assessment (MoCA): Normative study for the Portuguese population. *J. Clin. Exp. Neuropsychol.* **2011**, *33*, 989–996. [CrossRef]
68. Bass, D.A.; Hickok, D.; Quig, D.; Urek, K. Trace element analysis in hair: Factors determining accuracy, precision, and reliability. *Altern Med. Rev.* **2001**, *6*, 472–481.
69. IBM Corp. *IBM SPSS Statistics for Windows*; Version 22.0; IBM Corp: Armonk, NY, USA, 2013.
70. Freitas, S.; Simões, M.R.; Alves, L.; Santana, I. Mini Mental State Examination (MMSE): Normative study for the Portuguese population in a community stratified sample. *Appl. Neuropsychol. Adult* **2015**, *22*, 311–319. [CrossRef]
71. Freitas, S.; Simões, M.R.; Alves, L.; Santana, I. Montreal cognitive assessment: Validation study for mild cognitive impairment and Alzheimer disease. *Alzheimer Dis. Assoc. Disord.* **2013**, *27*, 37–43. [CrossRef]
72. Freitas, S.; Simões, M.R.; Alves, L.; Duro, D.; Santana, I. Montreal Cognitive Assessment (MoCA): Validation study for frontotemporal dementia. *J. Geriatr. Psychiatry Neurol.* **2012**, *25*, 146–154. [CrossRef]
73. Freitas, S.; Simões, M.R.; Alves, L.; Vicente, M.; Santana, I. Montreal Cognitive Assessment (MoCA): Validation study for vascular dementia. *J. Int. Neuropsychol. Soc.* **2012**, *18*, 1031–1040. [CrossRef]
74. Rodushkin, I.; Axelsson, M.D. Application of double focusing sector field ICP-MS for multielemental characterization of human hair and nails. Part II. A study of the inhabitants of northern Sweden. *Sci. Total Environ.* **2000**, *262*, 21–36. [CrossRef]
75. Li, Y.; Zou, X.; Lv, J.; Yang, L.; Li, H.; Wang, W. Trace elements in fingernails of healthy chinese centenarians. *Biol. Trace Elem. Res.* **2012**, *145*, 158–165. [CrossRef] [PubMed]
76. Cohen, J. *Statistical Power Analysis for the Behavioural Sciences*, 2nd ed.; Academic Press: New York, NY, USA, 1988.
77. Cohen, J. A power primer. *Psychol. Bull.* **1992**, *112*, 155–159. [CrossRef] [PubMed]
78. Moraes, C.; Pinto, J.A.; Lopes, M.A.; Litvoc, J.; Bottino, C.M. Impact of sociodemographic and health variables on mini-mental state examination in a community-based sample of older people. *Eur. Arch. Psychiatry Clin. Neurosci.* **2010**, *260*, 535–542. [CrossRef] [PubMed]
79. Gonçalves-Pereira, M.; Cardoso, A.; Verdelho, A.; da Silva, J.A.; de Almeida, M.C.; Fernandes, A.; Raminhos, C.; Ferri, C.P.; Prina, M.; Prince, M. The prevalence of dementia in a Portuguese community sample: A 10/66 Dementia Research Group study. *BMC Geriatr.* **2017**, *17*, 261. [CrossRef] [PubMed]
80. Hou, Y.; Dan, X.; Babbar, M.; Wei, Y.; Hasselbalch, S.G.; Croteau, D.L.; Bohr, V.A. Ageing as a risk factor for neurodegenerative disease. *Nat. Rev. Neurol.* **2019**, *15*, 565–581. [CrossRef]
81. Larson, E.B.; Yaffe, K.; Langa, K.M. New insights into the dementia epidemic. *N. Engl. J. Med.* **2013**, *369*, 2275–2277. [CrossRef]
82. Kumar, S.; Trivedi, A.V. A review on role of nickel in the biological system. *Int. J. Curr. Microbiol. Appl. Sci.* **2016**, *5*, 719–727. [CrossRef]

83. Song, X.; Kenston, S.S.F.; Kong, L.; Zhao, J. Molecular mechanisms of nickel induced neurotoxicity and chemoprevention. *Toxicology* **2017**, *392*, 47–54. [CrossRef]
84. Chen, Q.Y.; Brocato, J.; Laulicht, F.; Costa, M. Mechanisms of Nickel Carcinogenesis. In *Essential and Non-Essential Metals. Molecular and Integrative Toxicology*; Mudipalli, A., Zelikoff, J., Eds.; Springer International Publishing AG: New York, NY, USA, 2017; pp. 181–197. ISBN 978-3-319-55446-4.
85. Zambelli, B.; Uversky, V.N.; Ciurli, S. Nickel impact on human health: An intrinsic disorder perspective. *Biochim. Biophys. Acta Proteins Proteom.* **2016**, *1864*, 1714–1731. [CrossRef]
86. Poonkothai, M.; Vijayavathi, B.S. Nickel as an essential element and a toxicant. *Int. J. Eng. Sci. Technol.* **2012**, *1*, 285–288.
87. Zhao, J.; Shi, X.; Castranova, V.; Ding, M. Occupational toxicology of nickel and nickel compounds. *J. Environ. Pathol. Toxicol. Oncol.* **2009**, *28*, 177–208. [CrossRef] [PubMed]
88. Genchi, G.; Carocci, A.; Lauria, G.; Sinicropi, M.S.; Catalano, A. Nickel: Human health and environmental toxicology. *Int. J. Environ. Res. Public Health* **2020**, *17*, 679. [CrossRef] [PubMed]
89. Ijomone, O.M.; Okori, S.O.; Ijomone, O.K.; Ebokaiwe, A.P. Sub-acute nickel exposure impairs behavior, alters neuronal microarchitecture, and induces oxidative stress in rats' brain. *Drug Chem. Toxicol.* **2018**, *41*, 377–384. [CrossRef] [PubMed]
90. Ijomone, O.M.; Olatunji, S.Y.; Owolabi, J.O.; Naicker, T.; Aschner, M. Nickel-induced neurodegeneration in the hippocampus, striatum and cortex; an ultrastructural insight, and the role of caspase-3 and α-synuclein. *J. Trace Elem. Med. Biol.* **2018**, *50*, 16–23. [CrossRef]
91. Weekley, C.M.; Harris, H.H. Which form is that? The importance of selenium speciation and metabolism in the prevention and treatment of disease. *Chem. Soc. Rev.* **2013**, *42*, 8870–8894. [CrossRef]
92. Vinceti, M.; Grill, P.; Malagoli, C.; Filippini, T.; Storani, S.; Malavolti, M.; Michalke, B. Selenium speciation in human serum and its implications for epidemiologic research: A cross-sectional study. *J. Trace Elem. Med. Biol.* **2015**, *31*, 1–10. [CrossRef]
93. Marschall, T.A.; Bornhorst, J.; Kuehnelt, D.; Schwerdtle, T. Differing cytotoxicity and bioavailability of selenite, methylselenocysteine, selenomethionine, selenosugar 1 and trimethylselenonium ion and their underlying metabolic transformations in human cells. *Mol. Nutr. Food Res.* **2016**, *60*, 2622–2632. [CrossRef]
94. Labunskyy, V.M.; Hatfield, D.L.; Gladyshev, V.N. Selenoproteins: Molecular pathways and physiological roles. *Physiol. Rev.* **2014**, *94*, 739–777. [CrossRef]
95. Vinceti, M.; Filippini, T.; Wise, L.A. Environmental selenium and human health: An update. *Curr. Environ. Health Rep.* **2018**, *5*, 464–485. [CrossRef]
96. Rayman, M.P. Selenium intake, status, and health: A complex relationship. *Hormones* **2020**, *19*, 9–14. [CrossRef]
97. Varikasuvu, S.R.; Prasad, S.; Kothapalli, J.; Manne, M. Brain selenium in Alzheimer's disease (BRAIN SEAD study): A systematic review and meta-analysis. *Biol. Trace Elem. Res.* **2019**, *189*, 361–369. [CrossRef] [PubMed]
98. Cardoso, B.R.; Roberts, B.R.; Bush, A.I.; Hare, D.J. Selenium, selenoproteins and neurodegenerative diseases. *Metallomics* **2015**, *7*, 1213–1228. [CrossRef] [PubMed]
99. Robberecht, H.; De Bruyne, T.; Davioud-Charvet, E.; Mackrill, J.; Hermans, N. Selenium status in elderly people: Longevity and age-related diseases. *Curr. Pharm. Des.* **2019**, *25*, 1694–1706. [CrossRef] [PubMed]
100. Berr, C.; Balansard, B.; Arnaud, J.; Roussel, A.M.; Alpérovitch, A.; EVA Study Group. Cognitive decline is associated with systemic oxidative stress: The EVA study. *J. Am. Geriat. Soc.* **2000**, *48*, 1285–1291. [CrossRef] [PubMed]
101. Garcia, T.; Esparza, J.L.; Nogués, M.R.; Romeu, M.; Domingo, J.L.; Gómez, M. Oxidative stress status and RNA expression in hippocampus of an animal model of Alzheimer's disease after chronic exposure to aluminum. *Hippocampus* **2010**, *20*, 218–225. [CrossRef]
102. McGeer, P.L.; Rogers, J.; McGeer, E.G. Inflammation, antiinflammatory agents, and Alzheimer's disease: The last 22 years. *J. Alzheimers Dis.* **2016**, *54*, 853–857. [CrossRef]
103. Lin, T.; Liu, G.A.; Perez, E.; Rainer, R.D.; Febo, M.; Cruz-Almeida, Y.; Ebner, N.C. Systemic inflammation mediates age-related cognitive deficits. *Front Aging Neurosci.* **2018**, *10*, 236. [CrossRef]
104. Savaskan, N.E.; Bräuer, A.U.; Kühbacher, M.; Eyüpoglu, I.Y.; Kyriakopoulos, A.; Ninnemann, O.; Behne, D.; Nitsch, R. Selenium deficiency increases susceptibility to glutamate-induced excitotoxicity. *FASEB J.* **2003**, *17*, 112–114. [CrossRef]

105. Prabhu, K.S.; Zamamiri-Davis, F.; Stewart, J.B.; Thompson, J.T.; Sordillo, L.M.; Reddy, C.C. Selenium deficiency increases the expression of inducible nitric oxide synthase in RAW 264.7 macrophages: Role of nuclear factor-κB in up-regulation. *Biochem. J.* **2002**, *366*, 203–209. [CrossRef]
106. Schweizer, U.; Schomburg, L. Selenium, selenoproteins and brain function. In *Selenium*; Hatfield, D.L., Berry, M.J., Gladyshev, V.N., Eds.; Springer: Boston, MA, USA, 2006; pp. 233–248. ISBN 978-0-387-33827-9.
107. Pitts, M.W.; Kremer, P.M.; Hashimoto, A.C.; Torres, D.J.; Byrns, C.N.; Williams, C.S.; Berry, M.J. Competition between the brain and testes under selenium-compromised conditions: Insight into sex differences in selenium metabolism and risk of neurodevelopmental disease. *J. Neurosci.* **2015**, *35*, 15326–15338. [CrossRef]
108. Berr, C. Cognitive impairment and oxidative stress in the elderly: Results of epidemiological studies. *Biofactors* **2000**, *13*, 205–209. [CrossRef] [PubMed]
109. Berr, C.; Arnaud, J.; Akbaraly, T.N. Selenium and cognitive impairment: A brief-review based on results from the EVA study. *Biofactors* **2012**, *38*, 139–144. [CrossRef] [PubMed]
110. Cardoso, B.R.; Ong, T.P.; Jacob-Filho, W.; Jaluul, O.; Freitas, M.I.Á.; Cozzolino, S.M.F. Nutritional status of selenium in Alzheimer's disease patients. *Br. J. Nutr.* **2010**, *103*, 803–806. [CrossRef] [PubMed]
111. Gao, S.; Jin, Y.; Hall, K.S.; Liang, C.; Unverzagt, F.W.; Ji, R.; Murrell, J.R.; Cao, J.; Shen, J.; Ma, F.; et al. Selenium level and cognitive function in rural elderly Chinese. *Am. J. Epidemiol.* **2007**, *165*, 955–965. [CrossRef]
112. Reddy, V.S.; Bukke, S.; Dutt, N.; Rana, P.; Pandey, A.K. A systematic review and meta-analysis of the circulatory, erythrocellular and CSF selenium levels in Alzheimer's disease: A metal meta-analysis (AMMA study-I). *J. Trace Elem. Med. Biol.* **2017**, *42*, 68–75. [CrossRef]
113. Gao, S.; Jin, Y.; Hall, K.S.; Liang, C.; Unverzagt, F.W.; Ma, F.; Cheng, Y.; Shen, J.; Cao, J.; Matesan, J.; et al. Selenium level is associated with apoE ε4 in rural elderly Chinese. *Public Health Nutr.* **2009**, *12*, 2371–2376. [CrossRef]
114. Vinceti, M.; Chiari, A.; Eichmüller, M.; Rothman, K.J.; Filippini, T.; Malagoli, C.; Weuve, J.; Tondelli, M.; Zamboni, G.; Nichelli, F.; et al. A selenium species in cerebrospinal fluid predicts conversion to Alzheimer's dementia in persons with mild cognitive impairment. *Alzheimers Res. Ther.* **2017**, *9*, 100. [CrossRef]
115. McClain, C.J.; McClain, M.; Barve, S.; Boosalis, M.G. Trace metals and the elderly. *Clin. Geriatr. Med.* **2002**, *18*, 801–818. [CrossRef]
116. Cardoso, B.R.; Bandeira, V.S.; Jacob-Filho, W.; Cozzolino, S.M.F. Selenium status in elderly: Relation to cognitive decline. *J. Trace Elem. Med. Biol.* **2014**, *28*, 422–426. [CrossRef]
117. Forte, G.; Deiana, M.; Pasella, S.; Baralla, A.; Occhineri, P.; Mura, I.; Madeddu, R.; Muresu, E.; Sotgia, S.; Zinellu, A.; et al. Metals in plasma of nonagenarians and centenarians living in a key area of longevity. *Exp. Gerontol.* **2014**, *60*, 197–206. [CrossRef]
118. Alis, R.; Santos-Lozano, A.; Sanchis-Gomar, F.; Pareja-Galeano, H.; Fiuza-Luces, C.; Garatachea, N.; Lucia, A.; Emanuele, E. Trace elements levels in centenarian 'dodgers'. *J. Trace Elem. Med. Biol.* **2016**, *35*, 103–106. [CrossRef] [PubMed]
119. Xu, J.W.; Shi, X.M.; Yin, Z.X.; Liu, Y.Z.; Zhai, Y.; Zeng, Y. Investigation and analysis of plasma trace elements of oldest elderly in longevity areas in China. *Zhonghua Yu Fang Yi Xue Za Zhi* **2010**, *44*, 119–122. [PubMed]
120. Cai, Z.; Zhang, J.; Li, H. Selenium, aging and aging-related diseases. *Aging Clin. Exp. Res.* **2018**, 1–13. [CrossRef] [PubMed]
121. INE. Tábuas de Mortalidade em Portugal [Mortality Tables in Portugal]. Available online: https://www.ine.pt/xportal/xmain?xpid=INE&xpgid=ine_destaques&DESTAQUESdest_boui=354096866&DESTAQUESmodo=2&xlang=pt (accessed on 2 April 2020).

# 11

# Assessment of Clayey Peloid Formulations Prior to Clinical use in Equine Rehabilitation

**Carla Marina Bastos [1,2,*], Fernando Rocha [1], Ângela Cerqueira [1], Denise Terroso [1], Cristina Sequeira [1] and Paula Tilley [3]**

1 GeoBioTec Research Centre, Department of Geosciences, University of Aveiro, 3810-193 Aveiro, Portugal; tavares.rocha@ua.pt (F.R.); angelamcerqueira@ua.pt (Â.C.); laraterroso@ua.pt (D.T.); csequeira@ua.pt (C.S.)
2 Exatronic, Lda, 3800-373 Aveiro, Portugal
3 CIISA – Centre for Interdisciplinary Research in animal Health, Faculty of Veterinary Medicine, University of Lisbon, 1300-477 Lisbon, Portugal; paulatilley@fmv.ulisboa.pt
* Correspondence: a4872@ua.pt.

**Abstract:** Clays are natural ingredients used to prepare therapeutic cataplasms suitable for topical application. The knowledge about these formulations and their preparations to be applied on humans and animals has been orally transmitted since ancient times. Several empirical methods using clays have demonstrated fast and effective results in the reduction of the inflammatory response and the formation of edemas in horse limbs. The use of traditional and alternative medicine, such as pelotherapy, is now becoming more popular in veterinarian medical practice, alone or combined with other therapies in horse muscle and tendon rehabilitation. This study characterizes the use of commercial equine clays and an old therapeutic clay cataplasm formulation, using acetic acid, to treat tendon injuries in horses. This work might contribute to a major database characterization of clays used empirically on equine health, the potential of dermal absorption, the risks of exposure to some toxic elements, and safety assessment for these formulations. The present study was carried out to characterize the suitability of four commercial equine clays (Group II) and a protocoled healing mixture: "clay acetic acid cataplasm", (Group III), to treat tendon injuries in horses. In this mixture, three conventional "green" clays (Group I) without any mineralogical specificity were used and blended with acetic acid. The mineralogical composition was determined through X-ray powder diffraction and X-ray fluorescence data. To determine the performance of the samples, cooling kinetics, oil absorption, expandability, and specific surface area were measured. According to the mineralogical composition, Group I was mainly composed of carbonates and silicates, while Group II was much richer in silicates with the main clay minerals kaolinite and illite. Group II exhibited the highest values for As, Pb, Cr, Ni, and Zn, considered potentially toxic. Both groups showed low cation exchange capacities and exchanged mainly $Ca^{2+}$, with the exception of VET.1 and VET.7, which also highlight $Na^{+}$, and VET.5 and VET.6, which have $K^{+}$ as an exchangeable main cation. The addition of acetic acid (Group III) does not reveal any significant chemical changes. The results confirm that both clay groups are adequate for the therapeutic propose. They have good plastic properties (skin adherence), good oil absorptive capabilities (cleaning), and exchange an essential physiological element, calcium. Group II has prior industrial preparation, which is probably why it showed better results. Group I presented lower heat retention capacity and higher abrasiveness, which could be improved using cosmetic additives. The clinical benefit of the "clay acetic acid cataplasm" (Group III) could be the systemic anti-inflammatory effect established by the acetic acid.

**Keywords:** ethnoveterinary; pelotherapy; healing clays; quality control; equine limb injuries

## 1. Introduction

It is well documented that clayey formulations have been important resources for human and animal health care, because of their therapeutic and curative properties, since the first records in history [1,2].

Several traditional veterinary practices use zootherapeutic resources in the health care of domestic animals, the medicinal value of which maintains its relevance in ethnoveterinary medicine (EVM), the scientific term for traditional animal health care [3]. Clay minerals and their healing powers in wild animals are well documented by the practice of eating clay (geophagy) for detoxification of the body and alleviation of gastrointestinal infections and are now being rediscovered [4]. Kaolin and smectitic clays are commonly used in animal nutrition as growth promoters and supplements for the treatment of gastrointestinal disturbances. The introduction of kaolin clay, as feed additive, to treat foals with "heat" diarrhea, caused by disturbances in the intestinal osmotic balance of the young horses succeeds well as an absorbent and as an anticaking agent, alleviating the severity and duration of foal heat diarrhea [5].

The veterinary industry responds to the equine market with specific clayey products, promoting them by their pharmacological effects. The clayey products tailored for lameness injury prevention are relevant indicators for the evaluation of the therapeutic impact of pelotherapy in equine health and product procurement.

The clays' efficacy for lameness or other musculoskeletal injuries on horses is free of regulatory compliance. Most of these clayey products are designed accordingly with requirements and specifications supported by specialized equine technicians.

The use of pelotherapy as a therapeutic modality is scientifically little explored in equine health, despite its recognition as a valid non-invasive therapeutic option.

There are a few veterinarian databases, designed to search for relevant studies and clinical trials reported by researchers, such as PubMed and IVIS Quick-Links. The AVMA Animal Health Studies Database (www.avma.org/findvetstudies) allows submission and search of studies for health care issues in dogs, cats, horses, or other animals. Using "horse" and "equine" as a keyword we did not find any issue related to the use of clays in equine rehabilitation.

Clay minerals are widely used in pharmaceutical formulations as excipients and because of their biological activities [6] and are used in cosmetics because of their physical and physical–chemical properties such as adsorption capacity, specific surface area, swelling capacity, and reactivity to acids [7].

Williams and Haydel (2010) made the distinction between "healing clays" and "antibacterial clays", which may cure several diseases only by their unique physical and chemical properties (e.g., high absorbance, surface area, heat capacity, exchange capacity, etc.) or by killing pathogenic bacteria [8].

The absorptive capabilities of clays have been explored in a variety of cosmetic and pharmaceutical formulations and as a contributor to the healing of diseases, as well as for their cation exchange capacity and extremely fine particle size, which are important properties for removing oils, secretions, toxins, and contaminants from the skin. Cation exchange experiments showed that the antibacterial component of the clay can be moved, implying the presence of exchangeable cations in the antibacterial process [8]. Studies made on a natural clay from the Colombian Amazon and compared to the standard reference of smectite and kaolinite showed chemical interactions that are detrimental to bacteria by absorbing nutrients (e.g., Mg, P) and by toxic metal supply (e.g., Al) [9].

Humans and equine athletes share acute and chronic tendon injuries as the most common orthopedic affections, having similar structural (reparation) and functional (regeneration) recovery process. [10].

Ca, P, K, and S play a pivotal role in both the growth and the degeneration of the collagenous bone–cartilage interface of articulating joints demonstrated on equine osteoarthritic lesions (metacarpophalangeal joint) by detecting variations of elemental presence, using Synchrotron radiation micro X-ray fluorescence analysis [11].

The conservative veterinary therapy protocols include the same considerations as the human medicine protocols for orthopedic affections: cold applications, pressure-supporting bandages, controlled exercise, medicines to be injected, electrotherapy sessions, electromagnetic stimulation, ultrasound and laser therapy, or in an extreme clinical recommendation, surgical therapy [10]. Intra-articular (IA) administration of drugs in the treatment of musculoskeletal injuries has the objective of directing the drug delivery to the affected tissues and is commonly used by veterinarians and by Medical Physical Rehabilitation specialists. The use of corticosteroids or nonsteroidal anti-inflammatory drugs is common in racehorses and has become a problem for veterinarians due to the fact that the medicine could be masking a possible musculoskeletal condition and may contribute to injuries during competition [10,12]. Although there are a significant number of nonsteroidal anti-inflammatory drug (NSAID) formulations designed for the treatment of muscle and tendon traumatic conditions in human beings, when compared with the same problem in equine clinical practice, these formulations are more commonly used in horses. In vitro studies to evaluate and compare the penetration of diclofenac, a common NSAID designed for human application, revealed a significantly lower penetration through horse skin [13].

Complementary and alternative medicine (CAM) gained good acceptance in human medicine, mainly in the treatment of musculoskeletal pathologies and is now getting some popularity in veterinary medicine [14], therefore, regenerative therapies in horses may have applications for future human medicine and vice versa [10]. The initial interest and positive opinion on complementary alternative veterinary medicine (CAVM) started amongst horse owners. Most of them applied CAM therapies without the previous knowledge of their veterinarian, mainly to avoid possible conflict and fearing that their veterinarian might not want to continue providing veterinary care for their horse [14]. Some of this CAVM was supported by traditional Chinese veterinary medicine (TCVM), using acupuncture physiology to treat pain [15].

There is a lack of dissemination of traditional therapeutic procedures or rehabilitation programs using pelotherapy by key users (e.g., equine owners, equine trainers, equine veterinary, and industry) in the research field.

Equine rehabilitation programs must be designed with the previous identification of the risk factors that could predispose to musculoskeletal injury, considering the phases of healing, the rehabilitation goals, and the techniques used for acute injuries in horses [16].

The use of clays in these rehabilitation programs should fulfill the requirements regarding their safety and stability and should preferably be subjected to pre-market approval.

In this work, we characterize the mineralogical composition and technical performance of equine peloids used in prevention and rehabilitation programs. The main goals are contribution to the establishment of veterinarian clay therapeutic criteria, disclosure of the protocoled healing mixture: "clay acetic cataplasm", and to contribute to ethnoveterinary scientific databases.

## 2. Materials and Methods

### *2.1. Data Preparation*

For this study, we selected seven commercial clays routinely used in veterinary medicine and suggested by the CIISA—Center for Interdisciplinary Research in Animal Health (University of Lisbon, Portugal) for the treatment of equine musculoskeletal injuries, namely front- and hind-limb tendon and ligament injuries. VET.1, VET.5, VET.6, and VET.7 are four industrial pasty clays sold in the market as an equine clay treatment to be applied in a thick layer against the lay of the hair, after intensive exercise or a competition. VET.2, VET.3, and VET.4 are natural "green" clays, sold for human application and with no specific usage recommendations.

The protocol performed by CIISA for the treatment of horse musculoskeletal limb injuries proposes a 1:10 acetic acid (AA) and piped water solution with the necessary proportion of dried "green" clay. The CIISA protocoled solution was prepared with food acetic acid, pH 2.7 at 25 °C resulting in a solution (1:10) with pH of 2.9 at 25 °C. This protocol solution is only applied on VET.2, VET.3, and VET.4. This resulting cataplasm must ensure adhesive proprieties when applied to the injured area. After this, the injured area is wrapped in cling film, which acts as a thermal adjuvant, prolonging the therapeutic effect of the clays. The animal is then supervised by the clinical therapist, who decides when the clay effect is assured. We compared the commercial clays' results with the protocoled healing clay and acetic acid mixture.

Clays were distributed in three groups (Table 1), Group I and Group II according to their commercial purpose, and Group III for the protocoled healing mixture. All samples were dried at 50 °C, with no previous treatment, and maintained in closed containers at room temperature. For the preparation of the protocoled mixture, 10 g of the Group I clays was dispersed in 10 mL of acetic acid solution (1:10) and left to rest for 24 hours. These samples, VET.2AA, VET.3AA, and VET.4AA, were also dried at 50 °C. The pH value of the samples was measured with a HANNA HI 9126 pH meter, previously calibrated with standards (Titisol standard solutions) at pH 4 and pH 7 with an accuracy of ±0.05.

**Table 1.** Sample identification.

| Group | Samples | Type | Commercial Purpose |
|---|---|---|---|
| I | VET.2 | Powder | Human dermal application |
| | VET.3 | Powder | Human dermal application |
| | VET.4 | Powder | Human dermal application |
| II | VET.1 | Paste | Equine dermal application |
| | VET.5 | Paste | Equine dermal application |
| | VET.6 | Paste | Equine dermal application |
| | VET.7 | Paste | Equine dermal application |
| III [1] | VET.2AA | Paste | Protocoled healing mixture |
| | VET.3AA | Paste | Protocoled healing mixture |
| | VET.4AA | Paste | Protocoled healing mixture |

[1] CIISA—Center for Interdisciplinary Research in Animal Health protocol.

### 2.2. *Mineralogical, Chemical, and Technological Analysis*

The mineralogical analysis was carried out by X-ray diffraction (XRD) analysis, using a Philips/Panalytical X'Pert-Pro MPD, Kα Cu ($\alpha$ = 1.5405 Å) radiation, with 0.02° $2\theta$ $s^{-1}$ steps in goniometer speed. For the preparation of preferentially oriented aggregates of the clay (<2 μm fraction), a suspension was placed on a glass slide and air dried. XRD scans were run on this air-dried glass slide, and afterward a glycerol saturation and a final heat treatment at 500 °C were carried out [17]. The semi-quantitative identification of the principal clay minerals was obtained by measuring peak areas of the basal reflections, considering the full width at half maximum and then weighted by empirically estimated factors [17,18].

The particle size distribution of these clays was determined by an X-ray beam particle size analyzer (Micromeritics Sedigraph III Plus). The samples were dried and washed with distilled water, resting for 24 hours to ensure the separation between all particles. Then, the samples were sieved (106 μm) and dried. The dried sample was gently disaggregated, and the uniformity was ensured by quartile distribution. After that, 80 mL of sodium hexametaphosphate was added to 5.8 g of each sample and left to stand for 8 hours with magnetic stirring. At the end, the sample was sieved again (106 μm) and submitted to ultrasound for 40 s before the equipment measurement step.

The chemical composition of the commercial clays was assessed by X-ray fluorescence (XRF) using a Panalytical AX-IOS PW 4400/40. Loss on ignition (LOI) was also assessed by heating 1 g of the sample at 1000 °C for 1 hour in a furnace.

Abrasiveness was measured with an Einlehner AT-100 apparatus [19,20], and Atterberg limits were assessed using Casagrande Shell to obtain the liquid limit and using molding rolls in a glass plate for the plastic limit [20,21]. The plasticity index was calculated in accordance with the Portuguese standard, NP 143-1969. The expansion index test was performed by the standard LNEC E200-1967, Portuguese edition for ASTM (2008) to measure the swelling capacity of the samples when absorbing distilled water [20].

Samples were heated to 60 °C and the heat diffusiveness was assessed by a dual-channel thermometer, Lutron TM-9064. The range of time values was measured between 60 and 29 °C. Linseed oil was used to measure the oil absorption capacity of the clays. Fifteen grams of dry clay was weighed together with an amount of linseed oil. In a glass plate, linseed oil was slowly added, drop by drop, until it was possible to achieve the consistency necessary to obtain a solid roll of clay. The remaining oil and the clay roll were weighted for the oil absorption calculation.

The cation exchange capacity (CEC) was estimated by the ammonium acetate method, and the exchangeable cations ($Na^+$, $K^+$, $Mg^{2+}$, and $Ca^{2+}$) were determined by an atomic absorption spectrophotometer [22]. Specific surface area (SSA) was estimated by BET analysis—Gemini II 2370.

## 3. Results

### *3.1. Mineralogical and Chemical Characterization*

#### 3.1.1. Grain Size Distribution and Mineralogical Composition

The results from the particle size distribution of the samples are shown in Figure 1. Group I samples contained more than 55% of fine fraction content, with an average diameter of ~3 µm. In Group II, VET.1, VET.6, and VET.7 had around 65%, 68%, and 79%, respectively, of particles sized between 2 and 100 µm, and different average diameters, VET.1 and VET.6 had a $D_{50}$ of 0.708 and 0.373 µm, and VET.7 had a $D_{50}$ of 1.292 µm. For the VET.5 sample, the granulometry size distribution was as in Group I samples, 66% of fine fraction content and an average diameter of ~3 µm.

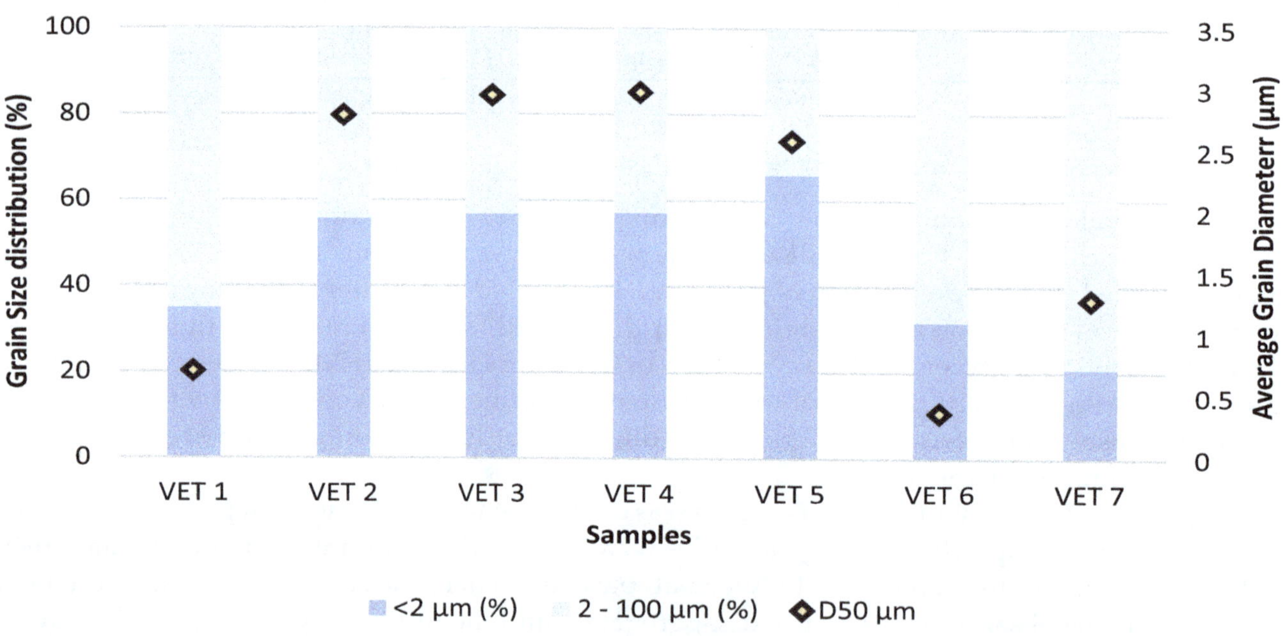


**Figure 1.** Grain size distribution of the Group I and Group II samples.

The mineralogical composition of the Group I and Group II clays is reported in Table 2. All samples were polymineralic (Figure 2) and exhibited differences in mineralogical composition. Considering the average for each group, we can classify Group I (n = 3) mineralogically as being composed by carbonates (calcite and dolomite) and silicates (quartz and phyllosilicates/clay minerals), while Group II is much richer in silicates, with a pronounced increase in phyllosilicates/clay minerals. The main clay minerals (Figure 3) are kaolinite (25%) followed by illite (7%) in Group I, and illite (68%) followed by kaolinite (2%) in Group II, except for VET.7 (67% kaolinite and 28% illite).

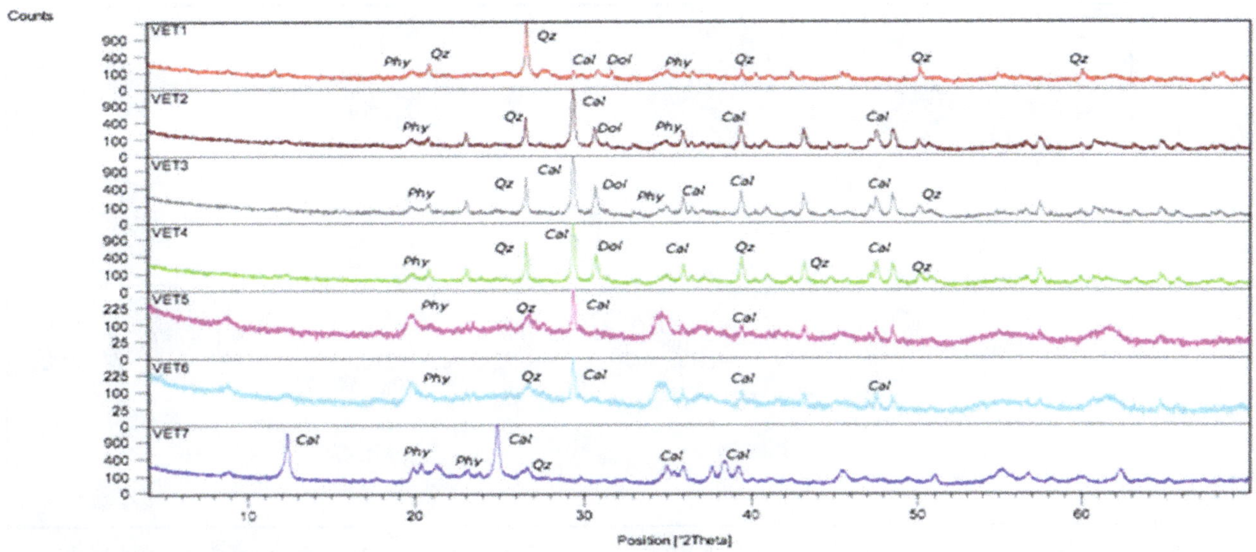


**Figure 2.** X-ray diffraction (Cal: calcite; Qz: quartz; Dol: dolomite; Phy: phyllosilicates).

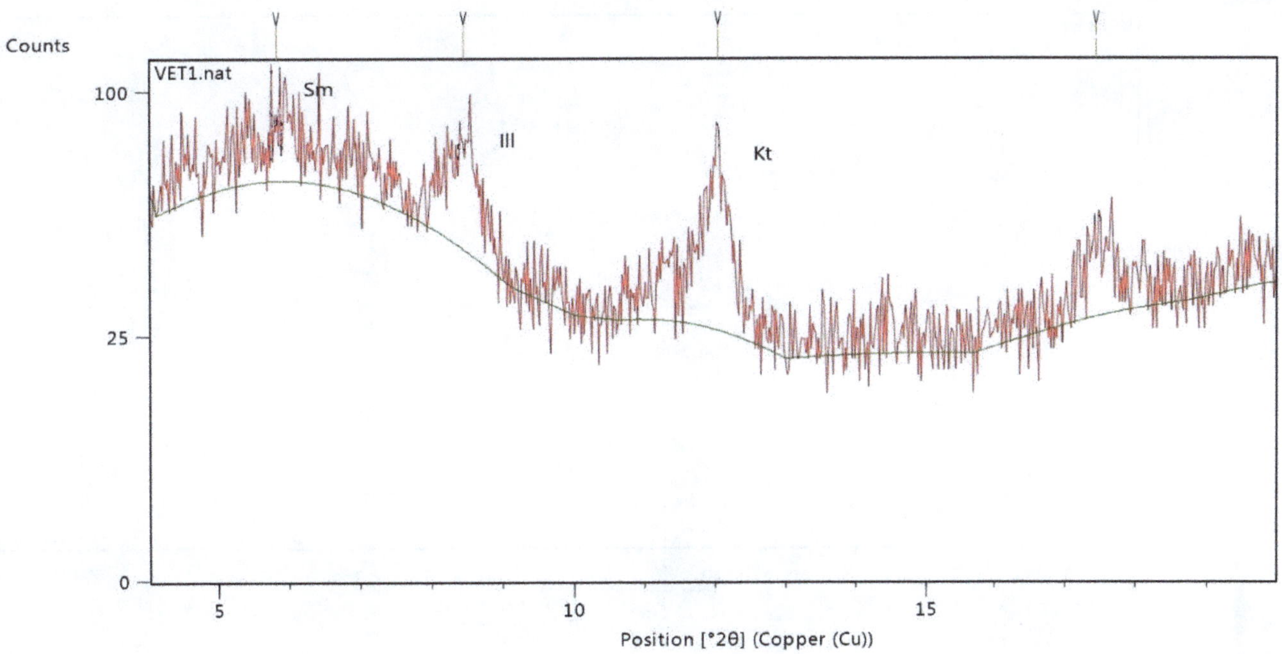


**(a)**

**Figure 3.** *Cont.*

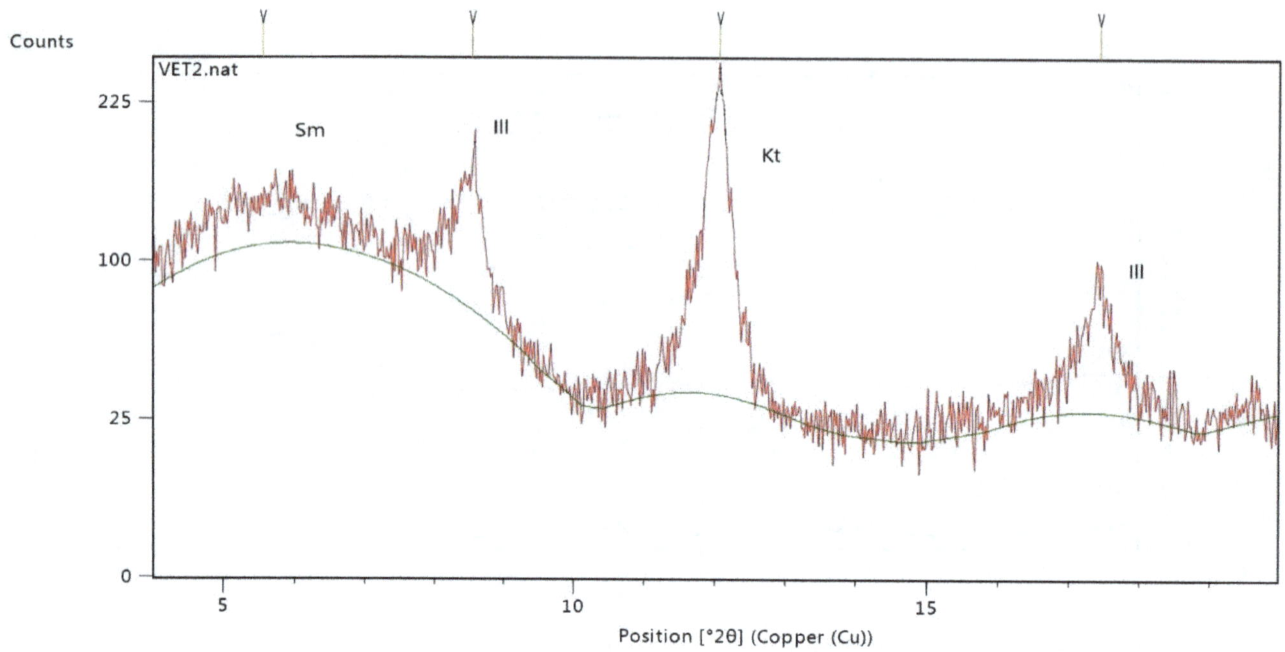


(b)

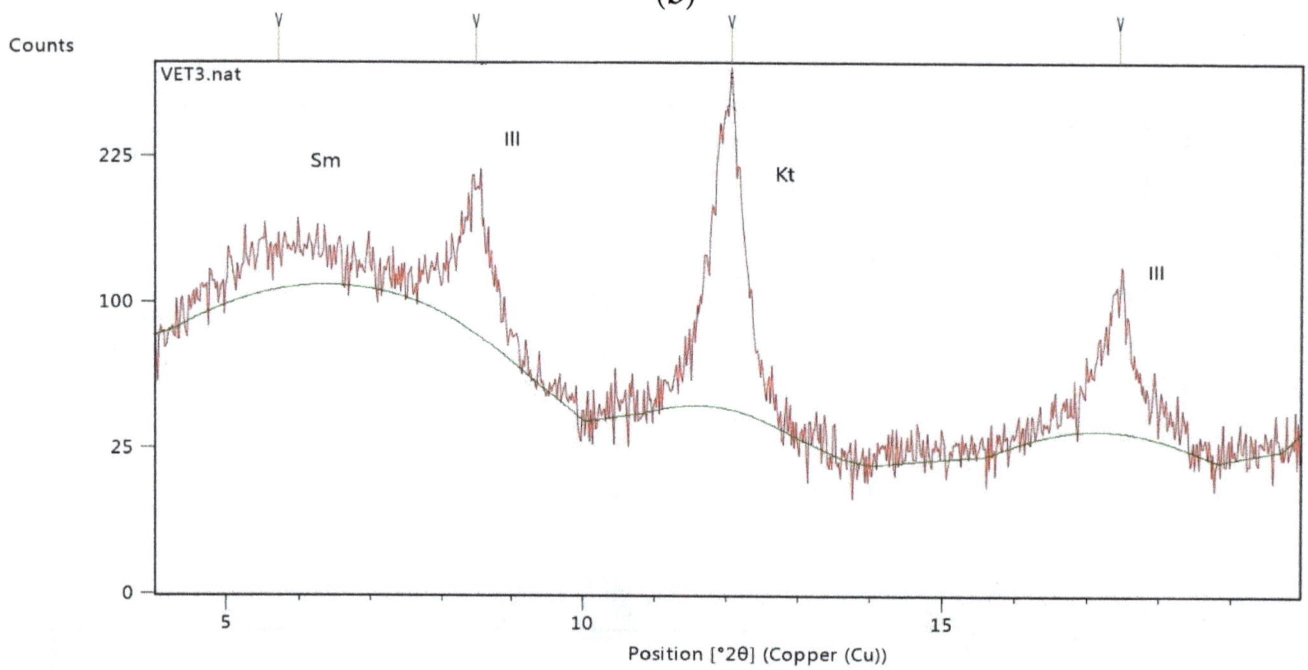


(c)

**Figure 3.** *Cont.*

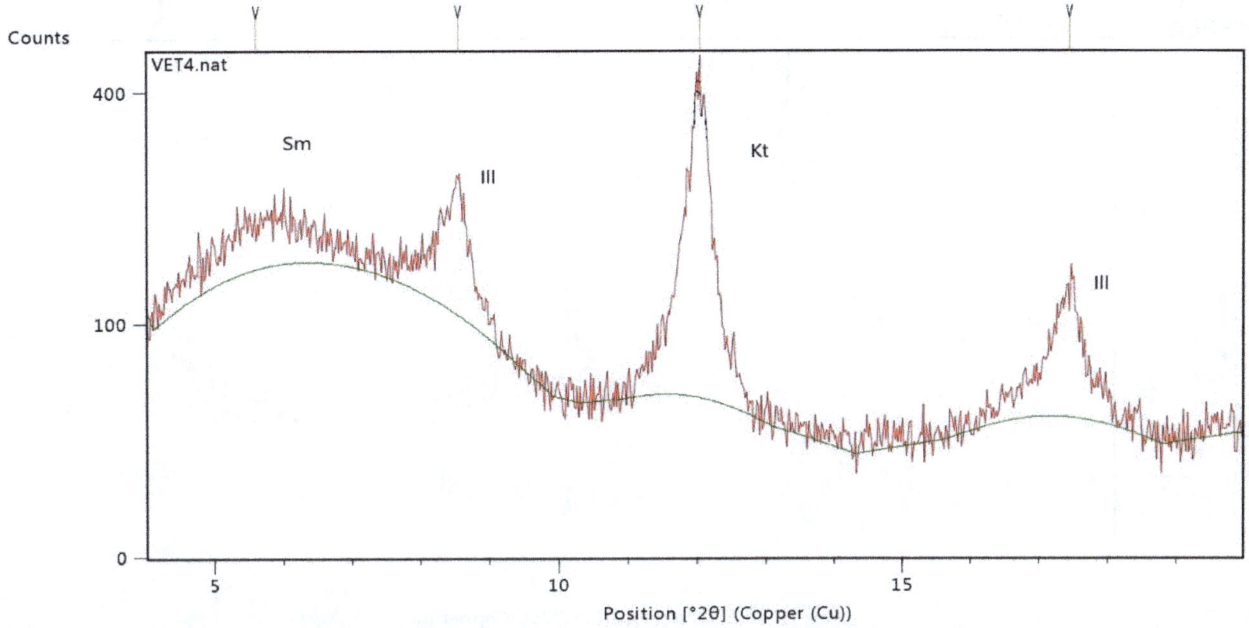


(d)

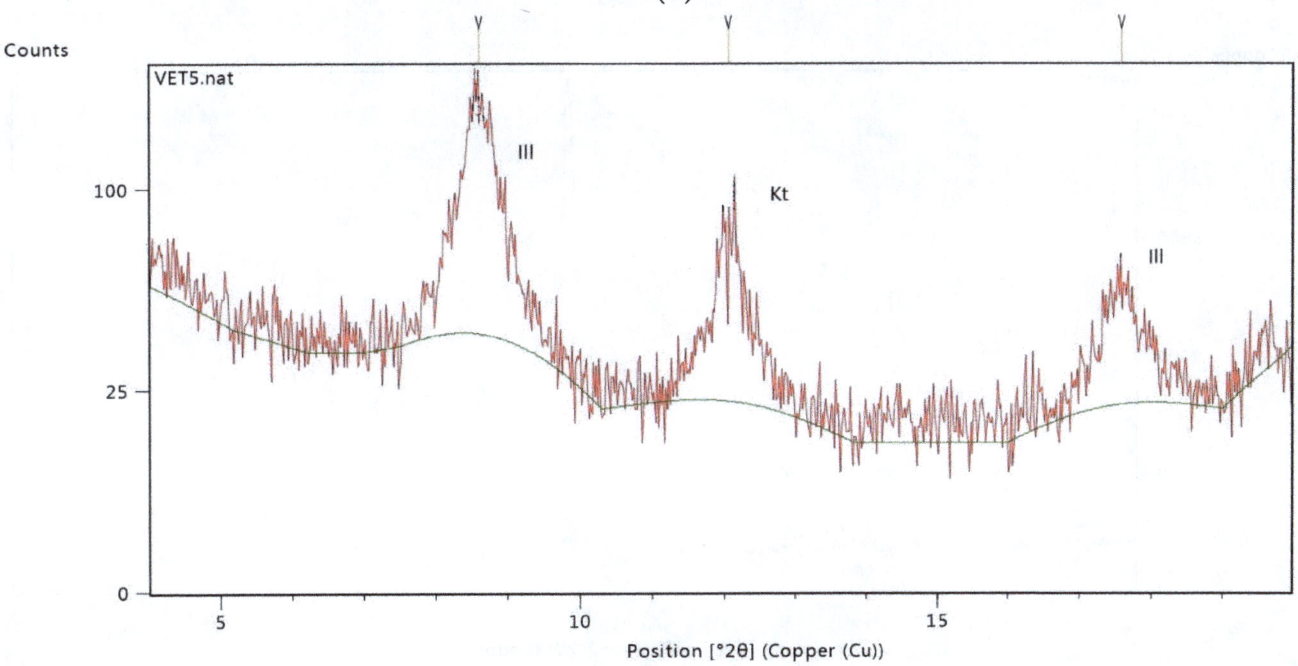


(e)

**Figure 3.** *Cont.*

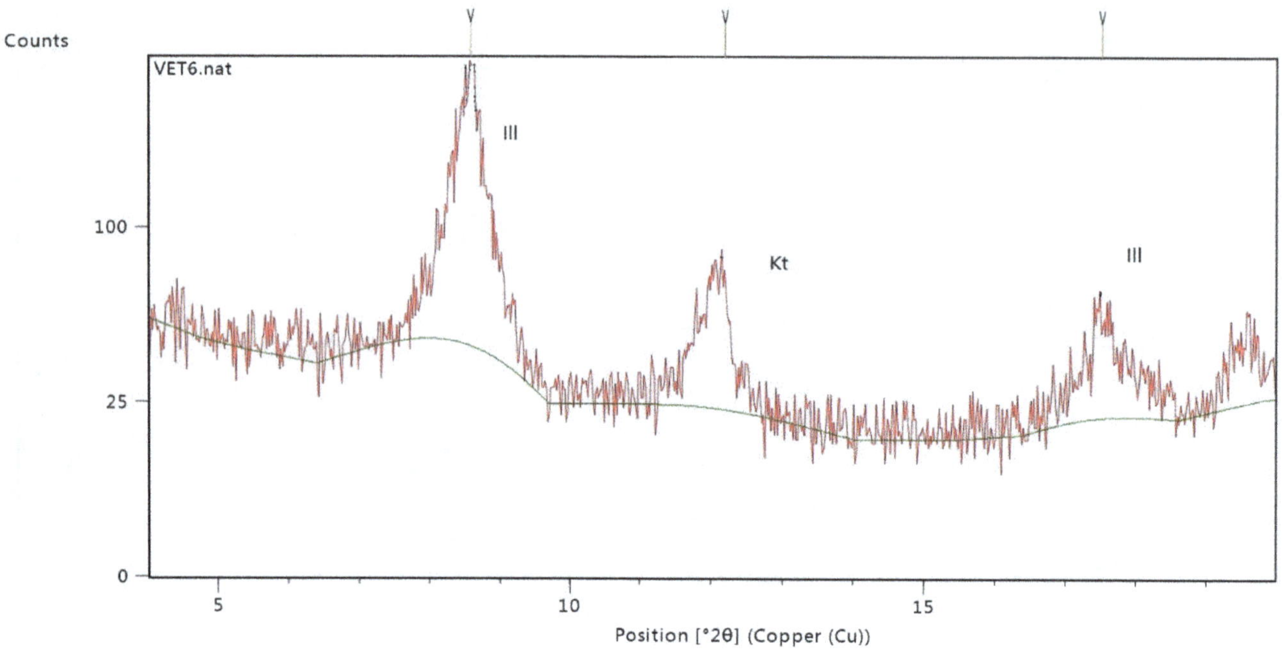


(f)

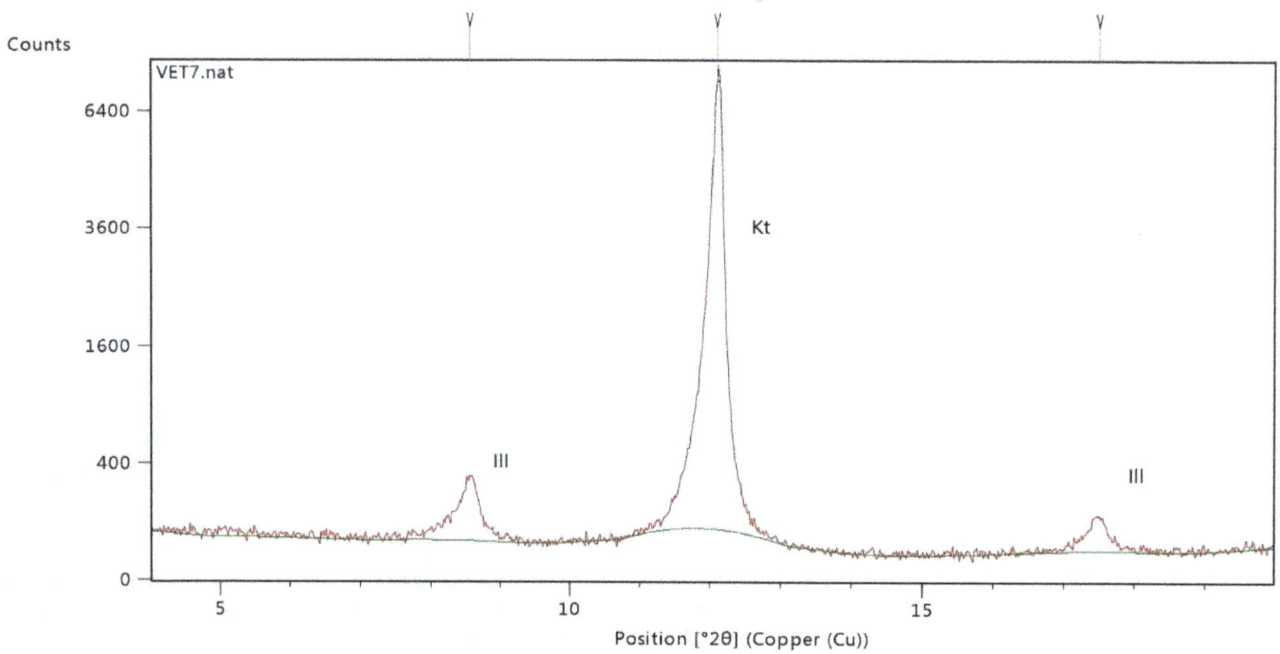


(g)

**Figure 3.** X-ray diffraction patterns of air-dried oriented aggregates (**a**) VET.1; (**b**) VET.2; (**c**) VET.3; (**d**) VET.4; (**e**) VET.5; (**f**) VET.6; (**g**) VET.7. (Ill: illite; Sm: smectite; Kt: kaolinite).

**Table 2.** Mineralogical composition (%).

| Group | Samples | Total Sample (%) | | | | Clay Minerals (%) | | |
|---|---|---|---|---|---|---|---|---|
| | | Qz | Hal | Calc | Dol | Sm | Kt | Ill |
| I | VET.2 | 6 | - | 47 | 13 | <1 | 29 | 5 |
| | VET.3 | 9 | - | 43 | 21 | <1 | 21 | 6 |
| | VET.4 | 10 | - | 39 | 17 | <1 | 24 | 10 |
| II | VET.1 | 44 | 2 | - | - | 1 | 3 | 50 |
| | VET.5 | 8 | - | 15 | - | - | 4 | 73 |
| | VET.6 | 3 | - | 16 | - | - | 1 | 80 |
| | VET.7 | 5 | - | - | - | - | 67 | 28 |
| III | VET.2AA | 7 | - | 52 | 16 | <1 | 19 | 6 |
| | VET.3AA | 14 | - | 56 | 10 | <1 | 17 | 3 |
| | VET.4AA | 11 | - | 58 | 11 | <1 | 17 | 3 |

Qz = quartz; Hal = halite; Calc = calcite; Dol = dolomite; Sm = smectite; Kt = kaolinite; Ill = illite.

### 3.1.2. Chemical Composition

The content of major and minor chemical elements is shown in Table 3. Differences in chemical composition were in accordance with those detected in the mineralogical composition; Group I (the more carbonated) was richer in CaO (27%) while Group II was richer in $SiO_2$ (44%) as well as in $Al_2O_3$, $Fe_2O_3$, and $K_2O$.

**Table 3.** Major and minor element composition of the samples. LOI = loss on ignition.

| Samples | | Group I | | | Group III | | | Group II | | | |
|---|---|---|---|---|---|---|---|---|---|---|---|
| | | VET.2 | VET.3 | VET.4 | VET.2 AA | VET.3 AA | VET.4 AA | VET.1 | VET.5 | VET.6 | VET.7 |
| Major elements (wt.%) | $SiO_2$ | 24.424 | 26.272 | 24.924 | 25.457 | 25.009 | 25.499 | 46.078 | 41.723 | 41.110 | 46.206 |
| | $Al_2O_3$ | 10.508 | 11.147 | 10.488 | 11.062 | 10.906 | 11.120 | 15.417 | 18.195 | 17.980 | 35.098 |
| | $Fe_2O_3$ | 2.597 | 2.957 | 2.664 | 2.885 | 2.721 | 2.916 | 6.185 | 6.423 | 6.354 | 0.778 |
| | MgO | 3.364 | 2.742 | 3.423 | 2.334 | 2.535 | 2.415 | 2.626 | 3.129 | 3.271 | 0.427 |
| | CaO | 26.859 | 25.976 | 27.391 | 28.636 | 27.071 | 28.811 | 4.233 | 4.901 | 5.841 | 0.049 |
| | $Na_2O$ | 0.096 | 0.083 | 0.082 | 0.093 | 0.084 | 0.088 | 3.292 | 0.146 | 0.147 | 0.781 |
| | $K_2O$ | 1.568 | 1.649 | 1.585 | 1.793 | 1.692 | 1.788 | 2.976 | 6.111 | 6.184 | 1.831 |
| | $TiO_2$ | 0.328 | 0.363 | 0.342 | 0.352 | 0.348 | 0.366 | 0.824 | 0.629 | 0.649 | 0.023 |
| | $P_2O_5$ | 0.046 | 0.048 | 0.041 | 0.052 | 0.047 | 0.046 | 0.196 | 0.166 | 0.233 | 0.144 |
| | $SO_3$ | 1.295 | 1.265 | 1.219 | 1.081 | 0.998 | 1.068 | 2.239 | 0.323 | 0.029 | 0.100 |
| | LOI | 28.65 | 27.20 | 27.65 | 25.96 | 28.270 | 25.550 | 13.83 | 17.85 | 17.80 | 14.3 |
| Minor elements (ppm) | As * | • | • | • | • | • | • | 17 | 24.6 | 21.5 | 8.2 |
| | Cd * | • | • | • | • | • | • | • | • | • | • |
| | Pb * | 13.7 | 14.4 | 12.8 | 16.1 | 14.8 | 16.1 | 31.6 | 32.2 | 31.8 | 21.6 |
| | Cr | 51.9 | 57.4 | 53.4 | 54.6 | 52.3 | 54.8 | 130 | 68.5 | 66.7 | 4.2 |
| | Cu | 15.4 | 8.1 | 15.0 | 8.4 | 11.6 | 11.6 | 14 | 22.2 | 24.4 | 29.5 |
| | Ni | 16.0 | 19.7 | 16.0 | 17.7 | 15.4 | 16.1 | 33.6 | 29.0 | 29.3 | 5.0 |
| | Zn | 24.3 | 28.2 | 26.9 | 24.3 | 23.0 | 23.2 | 95.2 | 120 | 140 | 22.3 |
| | Ba | 110 | 120 | 150 | 160 | 150 | 110 | 250 | 280 | 200 | 150 |
| | Co | • | 7.5 | 4.9 | 5.5 | 6.1 | 4.8 | 14.7 | 10.2 | 11.7 | • |
| | Sr | 230 | 220 | 200 | 240 | 220 | 240 | 210 | 180 | 280 | 160 |
| | V | 59.7 | 74.5 | 72.4 | 57.2 | 56.4 | 57.7 | 100 | 72.2 | 78.1 | 5.4 |
| | Sb * | • | • | • | • | • | • | • | • | • | • |
| | Sc | 19.8 | 18.6 | 17.1 | 20.6 | 17.9 | 21.0 | 10.9 | 10.1 | 11.2 | • |

• Not determined; * Potentially toxic elements.

Considering the chemical elements that are potentially toxic and not allowed in care products (Regulation (EC) 1223/2009), the Group II exhibited the highest values for As, Pb, Cr, Ni, and Zn when compared with Group I, for human usage.

Group II shows high levels of Pb and As, however, it is difficult to estimate equine exposure and the health risks associated.

### 3.2. *Physical and Technological Characterization*

Both groups show low cation exchange capacities and exchange mainly $Ca^{2+}$, with the exception of VET.1 and VET.7, which highlight also $Na^{+}$; and VET.5 and VET.6, which have $K^{+}$, as an exchangeable main cation (Tables 4 and 5). They have good plasticity, which is necessary to ensure adhesiveness to the skin. For Group I and VET.1, an abrasiveness action is expected when in contact with the skin surface, but the impact depends on the skin condition of the animal and horsehair protection.

**Table 4.** Main physical and technological properties of the studied samples.

| Group | Samples | P.I. (%) | A.I. ($g/m^2$) | C.E.C. (meq/100) | E.C. (mg/L) | | | |
|---|---|---|---|---|---|---|---|---|
| | | | | | Na | Mg | K | Ca |
| I | VET.2 | 25 | 142.85 | 7 | 1.67 | 25.28 | 9.63 | 730.81 |
| | VET.3 | 26 | 236.77 | 7 | 1.79 | 27.76 | 7.45 | 734.50 |
| | VET.4 | 25 | 140.21 | 7 | 1.36 | 26.53 | 7.11 | 699.98 |
| II | VET.1 | n.d. | 353.17 | 19 | 240.68 | 63.90 | 51.38 | 568.39 |
| | VET.5 | 34 | 26.45 | 10 | 2.56 | 27.10 | 110.51 | 724.47 |
| | VET.6 | 29 | 31.75 | 12 | 3.16 | 16.76 | 159.74 | 724.28 |
| | VET.7 | 18 | 6.61 | 3 | 315.12 | 5.22 | 8.19 | 11.74 |
| III | VET.2AA | n.d. | n.d. | 6 | 1.38 | 23.18 | 8.96 | 680.83 |
| | VET.3AA | n.d. | n.d. | 7 | 1.37 | 20.77 | 8.29 | 572.56 |
| | VET.4AA | n.d. | n.d. | 5 | 1.50 | 21.46 | 8.79 | 584.11 |

n.d.—not determined; P.I.—plasticity index; A.I.—abrasivity index; C.E.C.—cation exchange capacity; E.C.—exchange cations.

**Table 5.** Main physical and technological properties of the studied samples (Cont.).

| Group | Samples | C.K. (min) | O.A. (%) | pH | Exp. (%) | S.S.A. ($m^2/g$) |
|---|---|---|---|---|---|---|
| I | VET.2 | 13.6 | 29 | 7.0 | 19.5 | 22.50 |
| | VET.3 | 19.0 | 30 | 7.3 | 12.3 | 22.58 |
| | VET.4 | 19.4 | 31 | 7.7 | 14.8 | 22.11 |
| II | VET.1 | 18.4 | 43 | 6.8 | 17.8 | 13.75 |
| | VET.5 | 37.8 | 37 | 7.3 | 13.6 | 42.55 |
| | VET.6 | 30.1 | 37 | 7.7 | 10.9 | 44.71 |
| | VET.7 | 30.3 | 63 | 8.6 | 3.1 | 5.09 |
| III | VET.2AA | n.d. | n.d. | 7.6 | n.d. | n.d. |
| | VET.3AA | n.d. | n.d. | 7.8 | n.d. | n.d. |
| | VET.4AA | n.d. | n.d. | 7.8 | n.d. | n.d. |

n.d.—not determined; C.K.—cooling kinetics; O.A.—oil absorption; Exp.—expandability, S.S.A.—specific surface area.

## 4. Discussion

These two groups of clays recommended by the CIISA—Center for Interdisciplinary Research in Animal Health, University of Lisbon, for the treatment of horse musculoskeletal injuries, have different compositional and textural characteristics.

Considering application and topical use characteristics (Table 6), taking as reference recommended values published by several authors [8,19,20,22–34], the samples studied show good plastic properties, which are necessary for skin adherence; good oil absorptive capabilities, which is important to clean the skin from impurities or wound secretions; and Group II has a good heat retention capacity, important when it is necessary to heat the cataplasm, to active the blood circulation. The abrasiveness of Group I clays, which can cause unnecessary rubbing on the animal´s injured skin, should be smoothed.

**Table 6.** Veterinary clay group characterizations.

| Properties | VET.1 | VET.2 | VET.3 | VET.4 | VET.5 | VET.6 | VET.7 |
|---|---|---|---|---|---|---|---|
| Adhesiveness [17,18,20,22,29,31] | ▲ | ▲ | ▲ | ▲ | ▲ | ▲ | • |
| Abrasiveness [17,18,20–22,29,31,32] | ▼ | ▼ | ▼ | ▼ | ▲ | ▲ | ▲ |
| Hazardous elements [20,23,28,31,32] | ▼ | ▲ | ▲ | ▲ | ▼ | ▼ | ▼ |
| Essential elements [20,21,30–32] | ▲ $Ca^{2+}$, $Na^{+}$ | ▲ $Ca^{2+}$ | ▲ $Ca^{2+}$ | ▲ $Ca^{2+}$ | ▲ $Ca^{2+}$, $K^{+}$ | ▲ $Ca^{2+}$, $K^{+}$ | ▲ $Na^{+}$ |
| Oil Absorption [18,20–22,31,32] | ▲ | ▲ | ▲ | ▲ | ▲ | ▲ | ▲ |
| Heat Retention [18,20,21,27,31,32] | ▲ | • | • | • | ▲ | ▲ | ▲ |
| Antibacterial performance [6,24–26] | ▼ | ▲ VET.2AA | ▲ VET.3AA | ▲ VET.4AA | ▼ | ▼ | ▼ |

▲ Advisable; • Advisable with limitations; ▼ Needs vigilance.

Calcium is an essential element that is important for the growth and regeneration of collagenous bone-cartilage and this may explain the traditionally used "green" carbonated clay for the acetic acid cataplasm formulation. Despite As and Pb being technically avoidable above 0.5 and 2.0 mg/kg [25], respectively, the fact that they are above the risk limit that is internationally accepted for pharmaceutical formulations and cosmetics applied to human beings, and that several factors should be considered in transdermal penetration for humans and for horses, the same formulation may have different efficacies and safety profiles when used in species for which they were not developed [13]. Group II are clays that were industrially developed specifically for equine usage.

The addition of acetic acid (Group III) does not reveal any significant chemical changes when compared with Group I, apart from the pH that becomes more alkaline (closer to 8). All samples have a pH around 7 and 8. Organic acids (e.g., acetic acid) are usually used in food as natural preservatives and antibacterial agents. The manipulation of clay minerals to eliminate clinical and environmental bacteria is very common and the investigation of the antibacterial properties of natural clays has taken a new approach [8,26–28], where the possibility of incorporating bactericidal properties into clays may also be activated by the use of an acid solution in the treatment [27]. Some studies reveal the strong antibacterial effect of acetic acid combined with silver nanoparticles (AgNPS), where the release of $Ag^{+}$ responsible for the antibacterial activity increased by the addition of acetic acid [28].

The obtained data, when analyzed in comparison to reference values, allowed us to consider that both groups are adequate for therapeutic proposes, such as the treatment of horse musculoskeletal injuries, but Group II shows the best characteristics. Group II, having a prior industrial preparation, is technologically adapted for use in horses. The use of additives and preservatives in their preparation may be the reason why they are less abrasive and toxic.

The establishment of a health database considering ethnoveterinary medicine studies may enrich the equine health databases, useful to institutions, veterinarians, animal owners, and also providing guidelines for the investigation of new therapies and for scientific evidence findings.

## 5. Conclusions

The potential benefit of using therapeutic clays in equine lameness injuries is to minimize the side-effects associated with oral and intra-articular administration of anti-inflammatory medicines and to sustain a local release of therapeutic elements. This study can also be considered as a contribution to a major database of clays used for animal topical application, to the awakening of traditional and ancestral methodologies in healing clay preparations, and also to the knowledge about the contribution of these clays in the rehabilitation programs developed by veterinarians.

Through this study we tried to address the knowledge gaps about the mineralogical and chemical composition of clays used for equine peloid preparation and assessed their technical performance. This research assured that the studied clays fulfill the safety and stability requirements for these rehabilitation programs, thus, enabling them to be submitted for pre-market technical and legal approval process.

The main limitation of this study was the impossibility to correlate CIISA therapeutic formulation results with results obtained with specific equine commercial clays. Further work is needed to compare the antibacterial effectiveness of this protocol with other mud cataplasm protocols applied to horses and humans. This would be important and necessary for the establishment of healing criteria for veterinary clays.

Most of the thermal spas around the world recommend their own mud baths or local mud cataplasm applications, as they recognize therapeutic results through their anti-inflammatory, analgesic, and antiseptic effects on musculoskeletal and dermatologic pathologies, which are increasingly supported by clinical trials. The efficacy of the candidate clays to be used in veterinary pelotherapy should be evaluated and compared with human pelotherapy results and supported by clinical trials.

The safety and regulatory compliance of these products should also be a priority. The identification of unwanted trace elements or toxic substances based on the raw material source (natural or synthetic) should be a determinant for market surveillance of the appropriate limits expected in these natural products.

**Author Contributions:** Conceptualization, C.M.B.; methodology C.M.B. and F.R.; resources, C.M.B., Â.C., D.T. and C.S.; investigation, C.M.B., F.R. and P.T.; project administration, C.M.B.; writing—original draft preparation, C.M.B.; writing—review and editing, C.M.B. and F.R.; supervision, F.R. All authors have read and agreed to the published version of the manuscript.

**Acknowledgments:** We are grateful to CIISA—Center for Interdisciplinary Research in Animal Health, University of Lisbon, for the samples used for experiments.

## References

1. Gomes, C. Healing and edible clays: A review of basic concepts, benefits and risks. *Environ. Geochem. Health* **2018**, *40*, 1739–1765. [CrossRef]
2. Williams, B.W. Natural antibacterial clays: Historical uses and modern advances. *Clays Clay Miner.* **2019**, *67*, 7–24. [CrossRef]
3. González, J.A.; Amich, F.; Postigo-Mota, S.; Vallejo, J.R. The use of wild vertebrates in contemporary Spanish ethnoveterinary medicine. *J. Ethnopharmacol.* **2016**, *191*, 135–151. [CrossRef] [PubMed]
4. Slamova, R.; Trckova, M.; Vondruskova, H.; Zraly, Z.; Pavlik, I. Clay minerals in animal nutrition. *Appl. Clay Sci.* **2011**, *51*, 395–398. [CrossRef]
5. Pieszka, M.; Łuszczyński, J.; Hedrzak, M.; Goncharova, K.; Pierzynowski, S.G. The efficacy of kaolin clay in reducing the duration and severity of "heat" diarrhea in foals. *Turk. J. Vet. Anim. Sci.* **2016**, *40*, 323–328. [CrossRef]
6. López-Galindo, A.; Viseras, C.; Cerezo, P. Compositional, technical and safety specifications of clays to be used as pharmaceutical and cosmetic products. *Appl. Clay Sci.* **2007**, *36*, 51–63. [CrossRef]
7. Carretero, M.I.; Pozo, M. Clay and non-clay minerals in the pharmaceutical industry. Part I. Excipients and medical applications. *Appl. Clay Sci.* **2009**, *46*, 73–80. [CrossRef]
8. Williams, L.B.; Haydel, S.E. Evaluation of the medicinal use of clay minerals as antibacterial agents. *Int. Geol. Rev.* **2010**, *52*, 745–770. [CrossRef]
9. Londono, S.C.; Williams, L.B. Unraveling the antibacterial mode of action of a clay from the Colombian Amazon. *Environ. Geochem. Health* **2015**, *38*, 363–379. [CrossRef]

10. Spaas, J.H.; Guest, D.J.; Van de Walle, G.R. Tendon regeneration in human and equine athletes. *Sports Med.* **2012**, *42*, 871–890. [CrossRef]
11. Kaabar, W.; Gundogdu, O.; Tzaphlidou, M.; Janousch, M.; Attenburrow, D.; Bradley, D.A. Investigation of essential element distribution in the equine metacarpophalangeal joint using a synchrotron radiation micro x-ray fluorescence technique. In Proceedings of the National Physics Conference 2007—PERFIK 2007, Kuala Terengganu, Malaysia, 26–28 December 2007; Volume 25, pp. 18–24. [CrossRef]
12. Soma, L.R.; Uboh, C.E.; Maylin, G.M. The use of phenylbutazone in the horse. *J. Vet. Pharmacol. Ther.* **2012**, *35*, 1–12. [CrossRef] [PubMed]
13. Andreeta, A.; Verde, C.; Babusci, M.; Muller, R.; Simpson, M.I.; Landoni, M.F. Comparison of diclofenac diethylamine permeation across horse skin from five commercial medical human formulations. *J. Equine Vet. Sci.* **2011**, *31*, 502–505. [CrossRef]
14. Bergenstrahle, A.; Nielsen, B.D. Attitude and behavior of veterinarians surrounding the use of complementary and alternative veterinary medicine in the treatment of equine musculoskeletal pain. *J. Equine Vet. Sci.* **2016**, *45*, 87–97. [CrossRef]
15. Robinson, N.G. Making sense of the metaphor: How acupuncture works neurophysiologically. *J. Equine Vet. Sci.* **2009**, *29*, 642–644. [CrossRef]
16. Davies, L. Equine rehabilitation. In *Pain Management in Veterinary Practice*, 1st ed.; Egger, C.M., Love, L., Doherty, T., Eds.; John Wiley & Sons: Hoboken, NJ, USA, 2013; Volume 1, pp. 1–464. [CrossRef]
17. Oliveira, A.; Rocha, F.; Rodrigues, A.; Jouanneau, J.; Dias, J.; Weber, O.; Gomes, C. Clay minerals from the sedimentary cover from the Northwest Iberian shelf. *Prog. Oceanogr.* **2002**, *52*, 233–247. [CrossRef]
18. Galhano, C.; Rocha, F.; Gomes, C. Geostatistical analysis of the influence of textural, mineralogical and geochemical parameters on the geotechnical behaviour of the "Argilas de Aveiro" Formation (Portugal). *Clay Miner.* **1999**, *34*, 109–116. [CrossRef]
19. Quintela, A.; Costa, C.; Terroso, D.; Rocha, F. Abrasiveness index of dispersions of Portuguese clays using the Einlehner method: Influence of clay parameters. *Clay Miner.* **2014**, *49*, 27–34. [CrossRef]
20. Rebelo, M.; Viseras, C.; Galindo, A.L.; Rocha, F.; Silva, E.F. Rheological and thermal characterization of peloids made of selected Portuguese geological materials. *Appl. Clay Sci.* **2011**, *52*, 219–227. [CrossRef]
21. Quintela, A.; Costa, C.; Terroso, D.; Rocha, F. Liquid limit determination of clayey material by Casagrande method, fall cone test and EBS parameter. *Mater. Technol. Adv. Perform. Mat.* **2014**, *29*, 82–87. [CrossRef]
22. Rebelo, M.; Rocha, F.; Ferreira da Silva, E. Mineralogical and physicochemical characterization of selected Portuguese Mesozoic-Cenozoic muddy/clayey raw materials to be potentially used as healing clays. *Clay Miner.* **2010**, *45*, 229–240. [CrossRef]
23. Quintela, A.; Terroso, D.; Ferreira da Silva, E.; Rocha, F. Certification and quality criteria of peloids in use for therapeutic purposes. *Clay Miner.* **2012**, *47*, 441–451. [CrossRef]
24. Quintela, A.; Terroso, D.; Costa, C.; Sá, H.; Nunes, J.C.; Rocha, F. Characterization and evaluation of hydrothermally influenced clayey sediments from Caldeiras da Ribeira Grande fumarolic field (Azores Archipelago, Portugal) used for aesthetic and pelotherapy purposes. *Environ. Earth Sci.* **2015**, *73*, 2833–2842. [CrossRef]
25. Bund, B.V.L. Technically avoidable heavy metal contents in cosmetic products. *J. Consum. Prot. Food Saf.* **2017**, *12*, 51–53. [CrossRef]
26. Kalfa, A.; Rakovitsky, N.; Tavassi, M.; Ryskin, M.; Ben-Ari, J.; Etkin, H.; Shuali, U.; Nir, S. Removal of Escherichia coli and total bacteria from water by granulated micelle-clay complexes: Filter regeneration and modeling of filtration kinetics. *Appl. Clay Sci.* **2017**, *147*, 63–68. [CrossRef]
27. Santos, M.F.; Oliveira, C.M.; Tachinski, C.T.; Fernandes, M.P.; Pitch, C.T.; Angioletto, E.; Riella, H.G.; Fiori, M.A. Bactericidal properties of bentonite treated with Ag+ and acid. *Int. J. Miner. Process.* **2011**, *100*, 51–53. [CrossRef]
28. Sedira, S.; Ayachi, A.A.; Lakehal, S.; Fateh, M.; Achour, S. Silver nanoparticles in combination with acetic acid and zinc oxide quantum dots for antibacterial activities improvement—A comparative study. *Appl. Surf. Sci.* **2014**, *311*, 659–665. [CrossRef]
29. García-Villén, F.; Sánchez-Espejo, R.; Carazo, E.; Borrego-Sánchez, A.; Aguzzi, C.; Cerezo, P.; Viseras, C. Characterization of Andalusian peats for skin health care formulations. *Appl. Clay Sci.* **2017**, *160*, 201–205. [CrossRef]

30. Bocca, B.; Pino, A.; Alimonti, A.; Forte, G. Toxic metals contained in cosmetics: A status report. *Regul. Toxicol. Pharmacol.* **2014**, *68*, 447–467. [CrossRef]
31. Pozo, M.; Carretero, M.I.; Maraver, F.; Pozo, E.; Gómez, I.; Armijo, F.; Rubí, J.A.M. Composition and physico-chemical properties of peloids used in Spanish spas: A comparative study. *Appl. Clay Sci.* **2013**, *83–84*, 270–279. [CrossRef]
32. Fernández-González, M.V.; Martín-García, J.M.; Delgado, G.; Párraga, J.; Delgado, R. A study of the chemical, mineralogical and physicochemical properties of peloids prepared with two medicinal mineral waters from Lanjarón Spa (Granada, Spain). *Appl. Clay Sci.* **2013**, *80–81*, 107–116. [CrossRef]
33. Costa, C.; Fortes, A.; Rocha, F.; Cerqueira, A.; Santos, D.; Amaral, M.H. Characterization of Portuguese gypsums as raw materials for Dermocosmetics. *Clay Miner.* **2019**, *54*, 277–281. [CrossRef]
34. Cerqueira, A.; Costa, C.; Sequeira, C.; Terroso, D.; Rocha, F. Assessment of clayey materials from Santa Maria Island (Azores, Portugal) for peloids preparation. *Clay Miner.* **2019**, *54*, 299–307. [CrossRef]

12

# Preliminary Assessment of Chemical Elements in Sediments and Larvae of Gomphidae (Odonata) from the Blyde River of the Olifants River System, South Africa

**Abraham Addo-Bediako * and Karabo Malakane**

Department of Biodiversity, University of Limpopo, Private Bag X1106, Sovenga 0727, South Africa; karabo.malakane@ul.ac.za

* Correspondence: abe.addo-bediako@ul.ac.za.

**Abstract:** Benthic macroinvertebrates and sediments can act as good indicators of environmental quality. The aim of this study was to assess the accumulation of chemical elements in the Gomphidae (Odonata) collected in the Blyde River. Seven sites were sampled for river sediments assessment and five sites for larvae (naiads) of Gomphidae bioaccumulation analysis. The tissue samples were analysed using inductively coupled plasma optical emission spectrometry (ICP-OES). The results showed high levels of all of the tested elements except Cd in the sediment. The mean concentrations of As, Cu and Cr exceeded the standard guideline values, whereas Pb and Zn were below the standard guideline values. In the insect body tissue, the concentrations of most elements were higher than in the sediments. The elements with the highest concentrations were Mn, Zn, Cu, and As. The bioaccumulation factor (BF) showed a tendency for bioaccumulation for almost all of the selected elements in the insect. The BF value was high for Cu, Mn, Sb, and Zn (BF > 1). The high concentrations of elements in the insect body tissue may pose a risk to fish that consume them, and subsequently to humans when fish from the river are consumed. It is therefore important to monitor the river to reduce pollution to prevent health risks in humans, especially in communities that rely on the river for water and food.

**Keywords:** bioaccumulation; Gomphidae; heavy metals; naiads; metalloids; pollution; sediments

## 1. Introduction

Globally, rivers and streams are threatened by anthropogenic pollution, such as toxic elements, due to intensive land-use and inadequate environmental management practices [1–3]. Though most elements occur naturally in the biogeochemical cycle, many are released into inland waters as industrial, mining, agricultural, and domestic effluents, and may be harmful to aquatic systems [4]. River sediments serve as a habitat for various benthic macroinvertebrates and can serve as a sink for elements such as heavy metals. The burrowing activity of some benthic organisms leads to their chronic exposure to sediments contaminated with chemical elements [5].

Some elements are essential micronutrients for living organisms, while some (e.g., Cd, Cr and Pb) are toxic to living organisms, even at low concentrations. The toxicity of elements in aquatic ecosystems is complex and dependent on their bioavailability. Due to their prevalence and toxicity, heavy metal contamination in aquatic ecosystems poses a serious environmental threat [6–8]. This may lead to a decline in freshwater ecosystem functioning and biodiversity [9]. The available elements in the environment (sediment and water) can be assimilated into living tissues through direct uptake and the food chain, and if accumulated at unacceptable concentrations can affect the aquatic biota [10].

When the contaminants are incorporated into the food chain, it poses a toxicity risk to the organisms that consume them: fish, fish-eating birds, mammals and humans [11].

Many benthic organisms represent a link for the transfer of elements from the sediments to upper trophic levels. Macroinvertebrates play a major ecological role in conveying energy from lower trophic levels upwards. They serve as food for many predatory organisms in the water including fish, which are a vital food for many rural communities, especially low-income groups [12]. Humans who regularly consume contaminated fish are at risk to genotoxic, carcinogenic, and non-carcinogenic health impairment from long-term exposure to toxic contaminants [13,14]. Thus, it has become increasingly important to assess the levels of chemical elements in the body tissues of aquatic organisms as an indicator of metal and metalloid pollution in aquatic systems and to determine whether the food (e.g., fish) from impacted river systems are suitable for human consumption [15].

The Blyde River is one of the main tributaries of the Olifants River System. The river serves as a source of drinking water and food (fish) to the rural communities living in the catchment. The larvae (naiads) of dragonflies (Gomphidae, order Odonata) were selected for the study. They are good ecological indicators and reflect the quality of aquatic systems [16,17]. The larvae are important predators in aquatic ecosystems and prey on benthic and planktonic invertebrates [18] and also serve as food for many fish species. The aim of the study was to assess the concentration of chemical elements (bioaccumulation) in the larvae of Gomphidae and to predict the potential risk of transfer of toxic elements to fish species.

## 2. Materials and Methods

### *2.1. Study Area*

The Blyde River rises on the western slopes of the north-south trending Drakensberg Mountains and flows northwards towards the escarpment edge where it is dammed. From the dam, the Blyde River cascades down a steep series of rapids to its lower reaches, where the river again flows northwards to join the Olifants River at the town of Hoedspruit in Limpopo Province [19]. The Blyde River sub-catchment is approximately 2000 km$^2$ in size. Geologically, the northern part of the sub-catchment is made up of crystalline gneissic and granitic rocks of the Basement Complex, underlying the catchment [19]. The sub-catchment lies partly on the escarpment and, as a result, experiences considerably higher rainfall than the other sub-catchments in the Olifants River Basin, with mean annual precipitation sometimes exceeding 1000 mm [19]. During the last decade, there has been an increase in human activities in the area, especially agriculture, which are likely to cause environmental pollution in the freshwater systems.

The river is subjected to various sources of anthropogenic pollution, including domestic waste (S1 and S2), agricultural runoff (S3 and S5), and industrial waste (Site 4), while S6 and S7 are nature reserves (Table 1). The sampling sites were spread along the Blyde River until near the confluence with the Olifants River. The study sites ranged between 24°30′59.46″ S 30°47′56.14″ E and 24°15′30.38″ S 30°50′13.22″ E (Figure 1).

**Table 1.** Location, description of activities, vegetation cover and substrate type (%).

| Site | Activity | Vegetation Cover | Cobbles | Sand | Silt | Mud |
|---|---|---|---|---|---|---|
| S1 | Domestic | 70% (mainly shrubs and trees) | 50 | 20 | 20 | 10 |
| S2 | Domestic/agriculture | 60% (mainly shrubs, grass, and a few trees) | 30 | 30 | 20 | 20 |
| S3 | Agriculture (mainly mangoes and citrus) | 90% (mainly trees and shrubs) | 40 | 30 | 20 | 10 |
| S4 | Industries (mainly local furniture manufacturing, automotive services and fruit processing factories) | 20% (mainly shrubs and grass) | 30 | 20 | 30 | 20 |
| S5 | Agriculture (mainly mangoes and citrus) | 70% (mainly trees and shrubs) | 20 | 20 | 30 | 30 |
| S6 | Nature reserve (little human activity) | 80% (mainly trees, shrubs and grass) | 20 | 20 | 30 | 30 |
| S7 | Nature reserve (little human activity) | 50% (mainly shrubs and grass, and a few trees) | 30 | 30 | 20 | 20 |

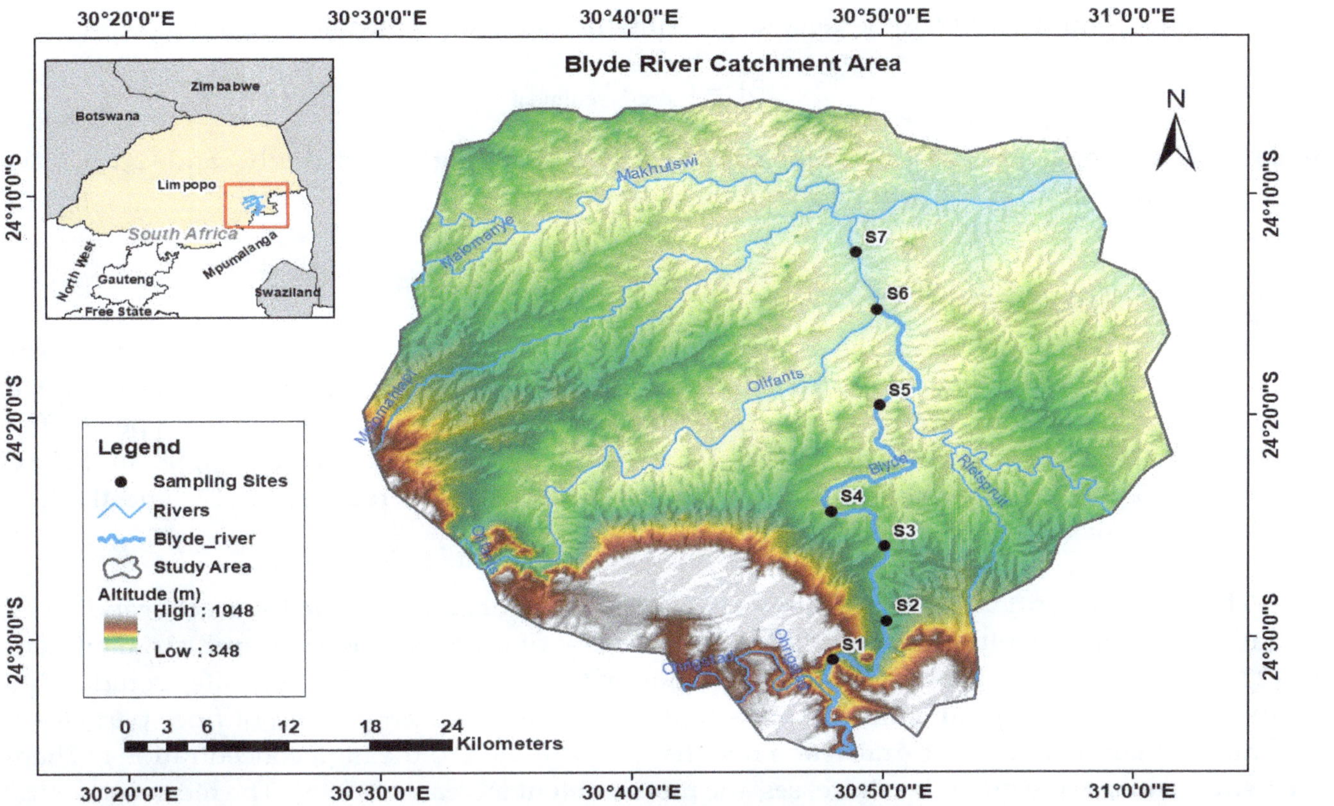


**Figure 1.** Map of the study area, showing the locations of the seven sampling sites of the Blyde River.

## 2.2. *Sampling and Analysis*

Sediment samples were collected at seven sites along the Blyde River during the months of February, April, July and October, in 2018. The samples were collected in acid pre-treated polyethylene bottles. The sediment was frozen prior to chemical analysis. Gomphidae larvae were sampled using a 30 by 30 cm SASS net with a 500 µm mesh size [20]. The samples collected at S3 and S7 were not sufficient for chemical analysis. Sediments and macroinvertebrate samples were then analysed for elements at an accredited (ISO 17025) chemical laboratory (WATERLAB (PTY) LTD, Pretoria, South Africa). The samples were put in acid-washed polypropylene pre-weighed vials and dried at 60 °C for 24 h, and a mixture of $HNO_3$ and HCl was added. Subsequently, the samples were digested in an oven [21]. The digested samples were cooled at room temperature, filtered using filter papers, and collected in beakers. The following metals and metalloids were then analysed in batches with blanks using inductively coupled plasma–optical emission spectrometry (ICP-OES; Perkin Elmer, Optima 2100 DV, Pretoria, South Africa): Arsenic (As), Antimony (Sb), Cadmium (Cd), Chromium (Cr), Copper (Cu), Lead (Pb), Manganese (Mn), Nickel (Ni) and Zinc (Zn). The analytical accuracy was determined using certified standards (De Bruyn Spectroscopic Solutions 500 MUL20 - 50 STD2) and recoveries were within 10% of certified values. The detection limits were: As—0.001 mg/kg, Cd—0.0001 mg/kg, Cr—0.001 mg/kg, Cu—0.001 mg/kg, Mn—0.0025 mg/kg, Ni—0.001 mg/kg, Pb—0.001 mg/kg, Sb—0.001 mg/kg, and Zn—0.001mg/kg.

## 2.3. *Statistical Analysis*

The mean and standard deviation of four samples at each site from the respective concentrations of the elements in the sediments were calculated. Analysis of variance (ANOVA) was performed using SPSS to determine whether there were significant differences among the different sites for the concentrations of the elements. Pearson's correlation matrix was used to identify the relationship between the metals. The ability of benthic macroinvertebrates to accumulate chemical elements was quantified through the bioaccumulation factors (BF) according to Klavinš et al. [22]

The bioaccumulation factor is calculated using the following formula:

$$BF = C_{org}/C_{sediment}$$

where $C_{org}$ is the element mass fraction in the organism (mg $kg^{-1}$ dry weight) and $C_{sediment}$ is the element concentration of the sediment (mg $kg^{-1}$ dry weight).

## 3. Results and Discussion

The mean concentrations of the elements in the sediment samples at the different sites are shown in Table 2. The concentrations of As, Cu and Sb varied significantly among the different sites ($p < 0.05$). The variations in the concentrations of the elements among sites could be due to the type of effluents washed into the river from the catchment. The highest concentrations of As, Cu, Sb, and Zn were recorded at S3. The highest concentrations of Cr, Mn and Ni were recorded at S5, and the highest concentrations of Cd at S6.

The high concentrations of most of the chemical elements may be due to direct or indirect land surface runoff of agricultural fields at S3 and the release of urban sewage and industrial effluents at S5 [23,24]. Furthermore, the grain-size distribution of the sediments at different sites could have also contributed to the type and concentrations of the elements. The proportion of finer particles at S5 was higher than that of coarse grains and may have contributed to the high concentration of chemical elements. Thus, as the grain size decreases, the metal content increases [25,26]. The mean concentration of As was greater than the CCME [27] guideline value of 13 mg $kg^{-1}$, dw at all the sites. The high As concentration at S3 might have been coming from pesticides and fertilizers used in agricultural fields [28,29]. The mean concentrations of Cr exceeded the guideline value of 37.3 mg $kg^{-1}$, dw at all the sites. Chromium and its salts are used in pigments and paints, in fungicides, and in chrome alloy and chromium metal production [30]. In this study, the main source of Cr in the sediment was mainly from agricultural activities. The concentration of Cu exceeded the guideline value of 37.3 mg $kg^{-1}$, dw at all the sites except S7. The high concentration of Cu in the study sites could be attributed to agricultural activities (pesticides, herbicides and fungicides) and to municipal wastewater and discharges from the catchment.

**Table 2.** Concentrations (mg $kg^{-1}$) of chemical elements at different sites in the Blyde River sediment samples.

| Element | S1 | | S2 | | S3 | | S4 | | S5 | | S6 | | S7 | | SQG |
|---|---|---|---|---|---|---|---|---|---|---|---|---|---|---|---|
| | AVE | ± SD | AVE | ±SD | AVE | ±SD | AVE | ±SD | AVE | ±SD | AVE | ±SD | AVE | ±SD | |
| As | 29.04 | 19.6 | 57.2 | 59.2 | 107.57 | 49.3 | 51.03 | 40.5 | 44.88 | 46.6 | 50.79 | 46.0 | 6.23 | 3.6 | 5.9 |
| Cd | ND | - | 0.04 | 0.05 | 0.09 | 0.1 | 0.01 | 0.02 | 0.11 | 0.18 | 0.41 | 0.7 | ND | - | 0.6 |
| Cr | 56.33 | 15.5 | 48.9 | 16.8 | 98.24 | 42.5 | 80.44 | 50.3 | 108.0 | 73.8 | 41.5 | 12.8 | 76.1 | 449 | 37.3 |
| Cu | 36.74 | 20.4 | 82.0 | 90.2 | 274.34 | 148.3 | 73.99 | 50.8 | 63.46 | 52.8 | 63.62 | 62.1 | 15.23 | 8.9 | 35.7 |
| Mn | 494.6 | 69.1 | 748.7 | 530.4 | 1175 | 490.5 | 949.8 | 635 | 1298.8 | 776 | 685.31 | 263 | 984.3 | 404 | - |
| Ni | 137.4 | 118 | 126.9 | 111.2 | 166.4 | 104.8 | 115.1 | 101 | 281.69 | 329 | 109.9 | 104 | 288.1 | 301 | - |
| Pb | 4.94 | 0.57 | 7.23 | 1.68 | 16.13 | 4.1 | 7.18 | 1.83 | 7.36 | 1.1 | 7.49 | 2.1 | 6.57 | 0.75 | 35 |
| Sb | 1.48 | 1.1 | 8.19 | 7.71 | 24.74 | 6.7 | 6.3 | 5.6 | 7.24 | 7.0 | 7.11 | 5.4 | 0.4 | 0.69 | - |
| Zn | 29.19 | 24.2 | 30.1 | 22.19 | 75.68 | 62.2 | 48.26 | 25.0 | 38.58 | 23.6 | 42.83 | 38.7 | 45.58 | 40.9 | 123 |

AVE: Average; SD: standard deviation; ND—not detected. SQG: Sediment quality guideline (CCME).

The correlation matrix showed a very strong relationship between Cr and Ni (0.868), Cu and Zn (0.897), and Pb and Zn (0.766), at a significance level of 0.01. There was a strong relationship between As and Cr (0.635), Cd and Cu (0.760), Cr and Mn (0.679), Ni and Mn (0.750), and Cd and Zn (0.727) at a significance level of 0.05 (Table 3). These results indicated that these elements originated from similar pollution sources. The absence of a correlation among some of the elements suggests that they are not controlled by a single factor [31]. The high concentrations of these elements in the sediments may pose an ecological risk to the aquatic biota, especially bottom-dwelling organisms. The concentration of Cd was very low in the river. The relatively low levels of the elements at the downstream sites (S6 and S7) is attributed to the nature conservation practices at these two sites. This is an indication that the conservation practice is having a positive impact on the downstream of the river.

**Table 3.** The correlation coefficients between chemical elements of the sediments in the Blyde River.

| Element | Sb | As | Cd | Cr | Cu | Pb | Mn | Ni | Zn |
|---|---|---|---|---|---|---|---|---|---|
| Sb | 1 | 0.217 | 0.103 | 0.099 | −0.271 | −0.389 | 0.574 | 0.452 | −0.289 |
| As | | 1 | 0.111 | 0.635 | 0.259 | −0.342 | 0.597 | 0.386 | −0.161 |
| Cd | | | 1 | −0.452 | 0.760 | 0.368 | −0.110 | −0.552 | 0.727 |
| Cr | | | | 1 | −0.086 | −0.500 | 0.679 | 0.868 | −0.392 |
| Cu | | | | | 1 | 0.586 | −0.202 | −0.386 | 0.897 |
| Pb | | | | | | 1 | −0.531 | −0.628 | 0.766 |
| Mn | | | | | | | 1 | 0.750 | −0.467 |
| Ni | | | | | | | | 1 | −0.556 |
| Zn | | | | | | | | | 1 |

The tissue of Gomphidae (Odonata) was analysed for these chemical elements; As, Cd, Cr, Cu, Mn, Ni, Pb, Sb and Zn. Aquatic insects can accumulate pollutants such as heavy metals from stream sediments and from food [32,33]. There were significant differences in the concentrations of Mn, Ni, Pb and Zn recorded in the body tissues of the insect larvae ($p < 0.05$). The concentrations of the elements in the body tissues varied among the sites, with the highest concentrations of all the elements with the exception of Mn and Ni were at S1. The concentrations of most of the elements in the aquatic insect were about five to 10 times those of the sediments. The larvae bioaccumulated lower concentrations at the downstream site, S6 (Table 4). The highest bioaccumulation of elements was at S1, instead of S3 or S5, which had the highest concentrations of most of the elements in the sediments and could partly be due to the local bioavailability of these elements.

**Table 4.** Concentration of chemical elements (Mean ± S.E) in the tissue of Gomphidae (Odonata) larvae at different sites of the Blyde River (S.E: standard deviation).

| Element | S1 | S2 | S4 | S5 | S6 |
|---|---|---|---|---|---|
| As | 32.26 ± 0.0 | 19.81 ± 2.4 | 12.32 ± 0.0 | 16.59 ± 3.5 | 7.3 ± 3.3 |
| Cd | 0.56 ± 0.0 | 0.28 ± 0.03 | 0.17 ± 0.0 | 0.09 ± 0.03 | 0.25 ± 0.25 |
| Cr | 13.81 ± 0.0 | 4.59 ± 1.78 | 2.05 ± 0.0 | 5.55 ± 2.2 | 1.82 ± 0.53 |
| Cu | 187.13 ± 0.0 | 101.07 ± 28.1 | 78.18 ± 0.0 | 61.1 ± 29.2 | 52.9 ± 26.6 |
| Mn | 3173 ± 0.0 | 2106 ± 395 | 3068 ±0.0 | 3637 ± 1038 | 563.2 ± 33.6 |
| Ni | 11.13 ± 0.0 | 8.17 ± 2.79 | 9.99 ± 0.0 | 29.47 ± 10.3 | 6.23 ± 5.7 |
| Pb | 1.9 ± 0.0 | 0.55 ± 0.1 | 0.33 ± 0.0 | 1.11 ± 0.6 | 0.38 ± 0.05 |
| Sb | 3.54 ± 0.0 | 0.97 ± 0.05 | 1.46 ± 0.07 | 2.18 ± 0.95 | 1.03 ± 0.17 |
| Zn | 362.2 ± 0.0 | 168.2 ± 3.9 | 183.8 ± 0.0 | 108.7 ± 57.6 | 102.3 ± 41.2 |

Most of the elements detected in high concentrations in the sediments and in the insect larvae are widely used in several fertilizers as a source of micronutrients. The larvae of Odonata are known to tolerate heavy metals [34]. The concentrations of Cd, Cu, Zn and Mn were found in higher concentrations (>50%) in the insect tissue than in the sediment. Meanwhile, the concentrations of As, Cr, Ni, Pb and Sb were higher in the sediments than in the tissue of the larvae (Figure 2). The elements in high concentrations in the sediments, such as Mn, Cu and Zn, were highly bioaccumulated in the insects. In this study, the transfer of Cr, Ni and Sb into the body tissue of the insect larvae was relatively less efficient, whereas Cu, Mn and Zn showed relatively high transfer efficiency. In aquatic insects, the concentrations of Cd, Ni, Cr, As, Pb, Cu, Ti, Zn and Mn change with size, life cycle stages, and different bioaccumulation patterns [35]. For example, Caddisflies have been found to accumulate Pb, regulate Zn and Cu, while Stoneflies accumulate Pb and regulate Zn [36].

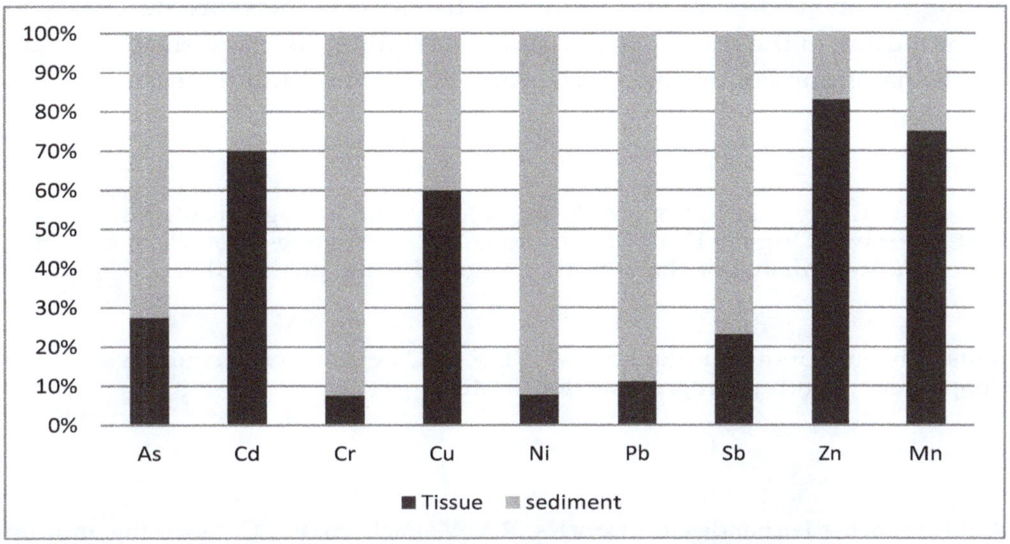


**Figure 2.** Composition of chemical elements in the sediments and the tissue of Gomphidae larvae.

The bioaccumulation factor (BF) of the elements in the insect larvae of Gomphidae from the Blyde River are shown in Figure 3. The BF value was >1 for Cu, Mn, Sb and Zn, thus these elements may be transferred to fish, and then to humans who consume fish from the river. The BF was high at the upstream sites, S1 and S2, indicating a high bioavailability of the elements for the insect larvae, whereas the lowest BF was at S6 (downstream site), with relatively low concentrations of the elements in the sediments. The results show that the larvae of Gomphidae accumulate chemical elements from the environment and they can be used to detect metal and metalloid pollution in aquatic environments [37].

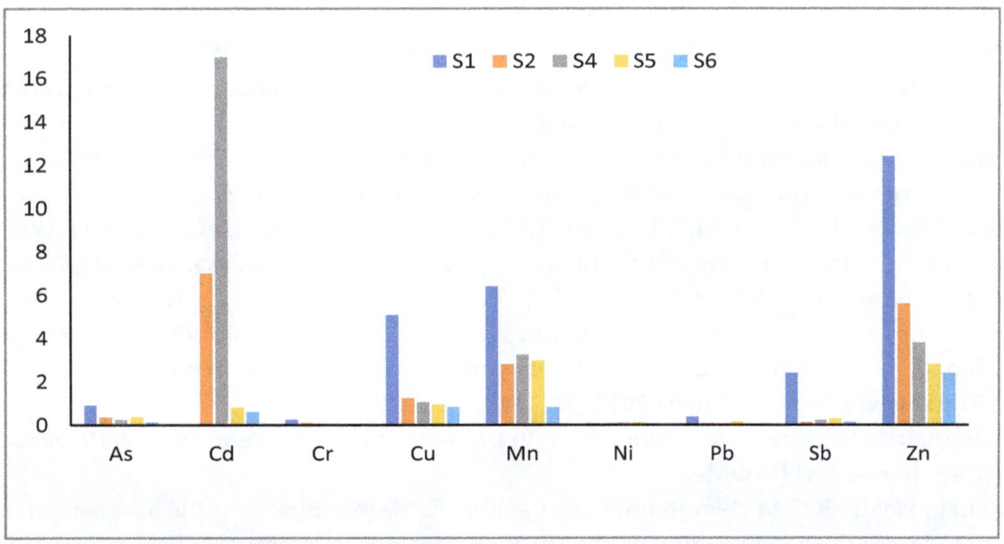


**Figure 3.** Bioaccumulation factor (BF) for larvae of Gomphidae samples from the Blyde River (ratio of concentrations of chemical elements in the larval tissue and in the sediment).

## 4. Conclusions

The metal and metalloid analysis of the river sediments showed variations in their concentrations among the sites. The effects of these elements may have consequences not only on aquatic insects, but also on higher trophic levels, such as fish and humans. In the insect body tissue, the concentrations of most of the chemical elements were higher than in the sediments, meaning that the insects accumulated the elements from the sediments. The study suggests that the concentrations of many of the elements studied are too high in the sediment and the larval tissue; it is therefore necessary to monitor and control chemical pollution in the river. Further study is required to assess the level of accumulation in the different functional groups of macroinvertebrates and to determine the transfer of toxic elements through the food chain.

**Author Contributions:** Conceptualization, A.A.-B., K.M.; Methodology, K.M. Formal Analysis, A.A.-B.; Investigation, K.M.; Writing—Original Draft Preparation, A.A.-B.; Writing—Review and Editing, A.A.-B., K.M.; Funding Acquisition, A.A.-B. All authors have read and agreed to the published version of the manuscript.

**Acknowledgments:** The authors thank the Belgian VLIR-IUC (Vlaamse Interuniversitaire Raad—University Development Cooperation) Funding Programme, for funding.

## References

1. Ferreira, A.R.L.; Sanches Fernandes, L.F.; Cortes, R.M.V.; Pacheco, F.A.L. Assessing anthropogenic impacts on riverine ecosystems using nested partial least squares regression. *Sci. Total Environ.* **2017**, *583*, 466–477. [CrossRef]
2. Santos, R.M.B.; Sanches Fernandes, L.F.; Cortes, R.M.V.; Varandas, S.G.P.; Jesus, J.J.B.; Pacheco, F.A.L. Integrative assessment of river damming impacts on aquatic fauna in a portuguese reservoir. *Sci. Total Environ.* **2017**, *601–602*, 1108–1118. [CrossRef]
3. Addo-Bediako, A.; Matlou, K.; Makushu, E. Heavy metal concentrations in water and sediment of the Steelpoort River, Olifants River System, South Africa. *Afr. J. Aquat. Sci.* **2018**, *43*, 413–416. [CrossRef]
4. Dallas, H.F.; Day, J.A. *The Effect of Water Quality Variables on Aquatic Ecosystems: A Review*; Water Research Commission: Pretoria, South Africa, 2004.
5. Thomas, P.; Liber, K. An estimation of radiation doses to benthic invertebrates from sediments collected near a Canadian uranium mine. *Environ. Int.* **2001**, *27*, 341–353. [CrossRef]
6. Xu, J.; Chen, Y.; Zheng, L.; Liu, B.; Liu, J.; Wang, X. Assessment of heavy metal pollution in the sediment of the main tributaries of Dongting lake, China. *Water* **2018**, *1*, 1060. [CrossRef]
7. Addo-Bediako, A. Assessment of heavy metal pollution in the Blyde and Steelpoort Rivers Of The Olifants River System, South Africa. *Pol. J. Environ.* **2020**, *29*, 3023–3029. [CrossRef]
8. Venkateswarlu, V.; Venkatrayulu, C. Bioaccumulation of heavy metal lead (Pb) in different tissues of brackish water fish *Mugil cephalus* (Linnaeus, 1758). *J. Appl. Biol. Biotechnol.* **2020**, *8*, 1–5.
9. Bhaskar, A.S.; Beesley, L.; Burns, M.J.; Fletcher, T.D.; Hamel, P.; Oldham, C.E.; Roy, A.H. Will it rise or will it fall? Managing the complex effects of urbanization on base flow. *Freshw. Sci.* **2016**, *35*, 293–310. [CrossRef]
10. Chen, C.Y.; Stemberger, R.S.; Klaue, B.; Blum, J.D.; Pickhardt, P.C.; Folt, C.L. Accumulation of heavy metals in food web components across a gradient of lakes. *Limnol. Oceanogr.* **2000**, *45*, 1525–1536. [CrossRef]
11. Varol, M.; Sen, B. Assessment of nutrient and heavy metal contamination in surface water and sediments of the Upper Tigris River, Turkey. *Catena* **2012**, *92*, 1–10. [CrossRef]
12. Sayer, J.; Cassman, K.G. Agricultural innovation to protect the environment. *Proc. Natl. Acad. Sci. USA* **2013**, *110*, 8345–8348. [CrossRef] [PubMed]
13. Du Preez, H.H.; Heath, R.G.M.; Sandham, L.A.; Genthe, B. Methodology for the assessment of human health risks associated with the consumption of chemical contaminated freshwater fish in South Africa. *Water SA* **2003**, *29*, 69–90. [CrossRef]

14. Jooste, A.; Marr, S.M.; Addo-Bediako, A.; Luus-Powell, W.J. Sharptooth catfish shows its metal: A case study of metal contamination at two impoundments in the Olifants River System, South Africa. *Ecotoxicol. Environ. Saf.* **2015**, *112*, 96–104. [CrossRef]
15. Addo-Bediako, A.; Marr, S.; Jooste, A.; Luus-Powell, W.J. Are metals in the muscle tissue of Mozambique tilapia a threat to human health? A case study of two impoundments in the Olifants River, Limpopo, South Africa. *Ann. Limnol. Int. J. Lim.* **2014**, *50*, 201–210. [CrossRef]
16. Haro, R.J. All along the watchtower: Larval dragonflies are promising biological sentinels for monitoring methylmercury contamination. *Park Sci.* **2014**, *31*, 70–73.
17. Pryke, J.S.; Sammays, M.J.; De Saedleer, K. An ecological network is as good as a major protected area for conserving dragonflies. *Biol. Conserv.* **2015**, *191*, 537–545. [CrossRef]
18. Nasirian, H.; Irvine, K.N. Odonata larvae as a bioindicator of metal contamination in aquatic environments: Application to ecologically important wetlands in Iran. *Environ. Monit. Assess.* **2017**, *189*, 436. [CrossRef]
19. Department of Water Affairs and Forestry (DWAF). *Olifants Water Management Area: Internal Strategic Perspective*; DWAF: Cape Town, South Africa, 2004.
20. Dickens, C.W.; Graham, P. The South African Scoring System (SASS) version 5 rapid bio-assessment method for rivers. *Afr. J. Aquat. Sci.* **2002**, *27*, 1–10. [CrossRef]
21. Bervoets, L.; Blust, R. Metal concentrations in water, sediment and gudgeon (*Gobio gobio*) from a pollution gradient: Relationship with fish condition factor. *Environ. Pollut.* **2003**, *26*, 9–19. [CrossRef]
22. Klavinš, M.; Briede, A.; Parele, A.; Rodinov, V.; Klavina, I. Metal accumulation in sediments and benthic invertebrates in lakes of Latvia. *Chemosphere* **1998**, *36*, 3043–3053. [CrossRef]
23. Pandey, J.; Singh, R. Heavy metal in sediment of Ganga River: Up- and downstream urban influences. *Appl. Water Sci.* **2017**, *7*, 1669–1678. [CrossRef]
24. Kumar, A.; Jha, K.D. Assessment of heavy metal concentration in the sediments of Mahananda River in the Seemanchal zone of Northen Bihar, India. *J. Emerg. Technol.* **2019**, *6*, 876–892.
25. Dinis, P.; Armando, A.; Pratas, J. Sources of potentially toxic elements in sediments of the Mussulo Lagoon (Angola) and implications for human health. *Int. J. Environ. Res. Public Health.* **2020**, *17*, 2466. [CrossRef]
26. Yu, T.; Zhanga, Y.; Zhanga, Y. Distribution and bioavailability of heavy metals in different particle-size fractions of sediments in Taihu Lake, China. *Chem. Spec. Bioavailab.* **2012**, *24*, 205–215. [CrossRef]
27. CCME (Canadian Council of Ministers of the Environment). Canadian Water Quality Guidelines for the Protection of Aquatic Life. In *Canadian Council of Ministers of the Environment*; CCME: Winnipeg, MB, Canada, 2012.
28. Wei, X.; Gao, B.; Wang, P.; Zhou, H.D.; Lu, J. Pollution characteristics and health risk assessment of heavy metals in street dusts from different functional areas in Beijing, China. *Ecotoxicol. Environ. Saf.* **2015**, *112*, 186–192. [CrossRef]
29. Zhou, J.; Liang, J.N.; Hu, Y.M.; Zhang, W.T.; Liu, H.L.; You, L.Y.; Zhang, W.H.; Gao, M.; Zhou, J. Exposure risk of local residents to copper near the largest flash copper smelter in China. *Sci. Total Environ.* **2018**, *630*, 453–461. [CrossRef]
30. WHO (World Health Organization). *Guidelines for Drinking Water Quality*; World Health Organization: Geneva, Switzerland, 1996; Volume 2.
31. Bhuyan, M.S.; Bakar, M.A.; Rashed-Un-Nabi, M.; Senapathi, V.; Chung, S.Y.; Islam, M.D. Monitoring and assessment of heavy metal contamination in surface water and sediment of the Old Brahmaputra River, Bangladesh. *Appl. Water Sci.* **2019**, *9*, 125. [CrossRef]
32. Prommi, T.O.; Payakka, A. Monitoring cadmium concentrations in sediments and aquatic insects (Hydropsychidae: Trichoptera) in a stream near a zinc mining area. *Pol. J. Environ. Stud.* **2018**, *27*, 2237–2243. [CrossRef]
33. Corbi, J.J.; Froehlich, C.G.; Trivinho-Strixino, S.; Dos Santos, A. Bioaccumulation of metals in aquatic insects of streams located in areas with sugar cane cultivation. *Química Nova* **2010**, *3*, 644–648. [CrossRef]
34. Tollett, V.D.; Benvenutti, E.L.; Deer, L.A.; Rice, T.M. Differential toxicity to Cd, Pb, and Cu in dragonfly larvae (Insecta: Odonata). *Arch. Environ. Contam. Toxicol.* **2009**, *56*, 77–84. [CrossRef]
35. Cid, N.; Ibánez, C.; Palanques, A.; Prat, N. Patterns of metal bioaccumulation in two filter-feeding macroinvertebrates: Exposure distribution, inter-species differences and variability across developmental stages. *Sci. Total Environ.* **2010**, *408*, 2795–2806. [CrossRef]

36. Goodyear, K.L.; McNeillb, S. Bioaccumulation of heavy metals by aquatic macro-invertebrates of different feeding guilds: A review. *Sci. Total Environ.* **1999**, *229*, 1–19. [CrossRef]
37. Souto, R.M.G.; Juliano, J.; Corbi, J.J.; Jacobucci, J.B. Aquatic insects as bioindicators of heavy metals in sediments in Cerrado streams. *Limnetica* **2019**, *38*, 575–586. [CrossRef]

# 13

# Sources of Potentially Toxic Metals in Sediments of the Mussulo Lagoon (Angola) and Implications for Human Health

**Pedro Dinis [1,*], Amílcar Armando [2] and João Pratas [3]**

1 MARE - Marine and Environmental Sciences Centre, Department of Earth Sciences, University of Coimbra, 3030-790 Coimbra, Portugal
2 Universidade Metodista de Angola, Rua Nossa Senhora da Muxima 10, Caixa Postal 6739, Luanda, Angola; amilcarquizembe@hotmail.com
3 Geosciences Center, Department of Earth Sciences, University of Coimbra, Rua Sílvio Lima, Univ. Coimbra – Pólo II, 3030-790 Coimbra, Portugal; jpratas@uc.pt
* Correspondence: pdinis@dct.uc.pt

**Abstract:** The Mussulo lagoon is a coastal environment located near Luanda, one of the SW African cities that has been growing more rapidly during the last decades. Geochemical, mineralogical, and grain-size data obtained for the lagoon sediments are analyzed together, in order to establish the factors that control the distribution of some potentially toxic elements (PTEs). Sediments from northern location tend to be enriched in feldspar and, despite some variability in grain-size distributions, in fine-grained detrital minerals; southern lagoon sediments display very homogenous grain-size distribution and are enriched in minerals associated with salt precipitation (halite and gypsum). Multivariate statistics reveal a close link between some PTEs, namely Co, Hg, Ni, and Pb, for which an anthropogenic source can be postulated. On the other end, As seems to be associated with natural authigenic precipitation in southern lagoon sectors. Sediments enriched in clay also tend to yield more Fe, Mn, Zn, and Cu, but it is unclear whether their sources are natural or anthropogenic. Hazard indexes calculated for children are higher than 1 for As and Co, indicating potential non-carcinogenic risk. For the other elements, and for adults, there is no potential carcinogenic or non-carcinogenic risk.

**Keywords:** Mussulo lagoon; Sediment composition; Factors controlling sediment geochemistry; Human Health Risk Assessment

## 1. Introduction

Luanda is one the African cities that has been growing more rapidly during the early 21st century [1]. The tendency for rapid rises in the number of inhabitants started before, in particular during the civil war after the independence of Angola, in 1975, when the population left the rural areas and sought refuge in the main cities. Hence, the city, projected to hold ~500,000 inhabitants, grew dramatically during the last decades, holding more than 10 times that number today. The rise in population also saw a rise in problems at the level of basic sanitation, collection, transport, and treatment of municipal solid waste, as well as limitations in the regulation of potentially hazardous waste disposal. Besides its large population, the city of Luanda also comprises the biggest industrial park in the country, leading to additional risks of environmental pollution. Some of the potentially hazardous wastes produced in Luanda are actually disposed of in open sites and dragged into the rivers and the sea during periods of rainfall. Thus, significant concentrations of potentially toxic elements (PTEs) are likely transported to coastal environments and can be concentrated in low hydrodynamic settings, such as the Mussulo lagoon, which is located a few km to the south of Luanda city centre.

Knowing that sediments in coastal ecosystems serve as sinks for PTEs, numerous investigations focused on the relations between the concentrations of PTEs in sediments and living organism have been conducted (e.g., [2–5]). High concentration of PTEs is frequently attributed to anthropogenic inputs [6–11], and strong relations between the levels of PTEs in coastal sediments and human activities were proposed for many locations worldwide [7,12–18]. Some works even established links between the history of human occupation and the concentration of harmful elements in coeval depositional sequences [13,14,16,18]. However, sediment geochemistry is necessarily controlled by the geology of the source-area. Furthermore, significant enrichments relative to source-rocks can be promoted by exogenous transformations due to weathering [19–21] and sorting processes [22–24]. Because of this complexity, understanding the factors responsible for the concentration of PTEs is of major importance in environmental studies.

In the present research, a set of textural, mineralogical, and geochemical properties of the sediments of the Mussulo lagoon are joint-analyzed, in order to investigate the factors that control the concentration of elements usually considered to be harmful in the environment. An assessment of the carcinogenic and non-carcinogenic risks associated with exposure to these sediments is also presented.

## 2. Geological and Geomorphological Setting

In central-west Angola, the Mussulo spit (~30 km long and <2 km in width) separates the elongated Mussulo lagoon from the South Atlantic (Figure 1). The spit is attached to the mainland some 30 km downdrift of the Cuanza River mouth, in a shifting point of coastal direction from SSE-NNW, southward, to SSW-NNE, northward. The lagoon reaches a maximum width of ~6 km in a bay near its aperture to the ocean. Approximately 6 km to the north of the tip of the Mussulo spit occur smaller linear features, including the so-called "Island of Luanda". This was a narrow island (~12 km long and <500 in width) that, after human intervention in the first half of the 20th century and several reinforcements until present times, became permanently attached to the continent.

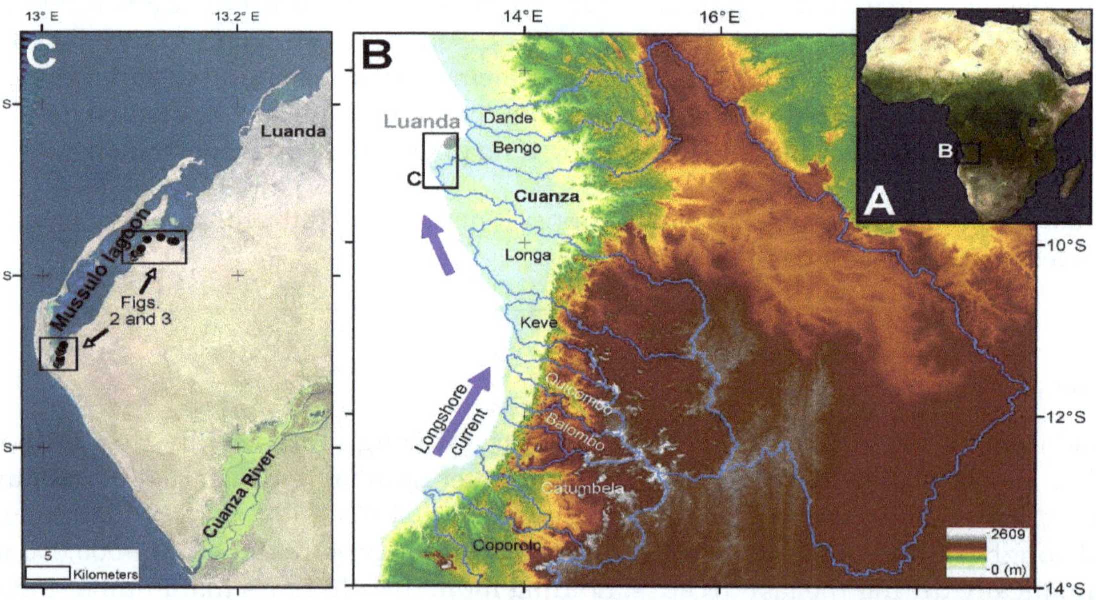


**Figure 1.** Geological setting of the lagoon of Mussulo. (**A**) The Angolan Atlantic margin in SW Africa and (**B**) orographic features of the regional drainage basins. Note that Cuanza is by far the biggest regional river. (**C**) The Mussulo spit and lagoon system extending until approximately the southern limit of Luanda urban area and location of the sampled sectors.

Climatic conditions in coastal Angola are influenced by the cold, northward-flowing, Benguela Current that is responsible for the drier conditions than are seen at similar latitudes inland. In central and northern Angola, the climate evolves in ~200 km from semi-arid (Bsh type of Koppen) in the littoral to tropical savannah (Aw type of Koppen), a transition clearly evidenced by an increase in rainfall from less than 500 mm of annual precipitation to more than ~1000 mm. Further inland rainfall becomes even higher, reaching ~1500 mm in the most elevated areas draining towards the Atlantic Ocean. Annual average temperatures are usually above 25 °C near the coast, decreasing inland with an elevation of slightly less than 20°. In Luanda region, almost all rain occurs from October to April, with March and April being the months with the highest rainfall. The dry months tend to be slightly cooler, but still with average temperatures above 20°. The semi-arid conditions and the limited connectivity of the lagoon with the open ocean are responsible for high water salinity, averaging 39 in the dry season.

Throughout the Atlantic margin of Central Angola, the basement is made mostly of silica-rich plutonic and metamorphic rocks that belong to the Congo Craton, which also includes mafic complexes in its north-eastern tip [25,26]. To the north, the West Congo Supergroup, with a volcano-sedimentary succession, is covered by siliciclastic and carbonate strata outcrops along a large area that extends parallel to the coastline [27]. These basement units are overlain by Meso-Cenozoic successions deposited in the Benguela, Cuanza, and West Congo basins in association with the opening of the South Atlantic [28,29], and in the hinterland basins of Congo [30,31] and Kalahari [32]. Locally, thick tholeiitic rocks occur close to the contact between the basement and the sedimentary succession of the Cuanza Basin [33]. The natural development of the Mussulo spit and Luanda Island is strongly linked with persisting northward littoral transport, controlled by the oblique wave incidence on the coast. In Angola, where sand spits grow in the downdrift side of the mouth of major regional rivers, it is assumed that they are mostly sourced by bedload material supplied by these rivers [34,35]. But a comparison of the composition of sediments, collected at Cuanza River mouth with those from the Mussulo spit and the Island of Luanda, point to an additional sediment contribution that resembles the bedloads transported from the southern regions by coastal drift, or derived from volcanic rocks that outcrop in Mezo-Cenozoic basins [36].

At the time of its foundation, around 1575, Luanda was just a small colonial settlement that was designed for a limited number of families and was placed in "Luanda Island". A year later, in search of better amenities, it moved to the mainland, and then evolved into the so-called village of "São Paulo de Loanda". Most accurate estimations from the mid-18th century onward indicate that the population oscillated between a few thousand, and may have even decreased slightly during some periods [37], reaching more than 100,000 during the first half of the 20th century [38]. During the last decades of colonial occupation, the population started to grow more rapidly, becoming approximately 500,000 at the time of independence [39], ~3 million by the end of the 20th century, and ~6 million in 2018 [1]. Growth with limited land planning and the development of diverse industrial facilities likely promoted environmental pollution in potentially toxic metals.

## 3. Materials and Methods

### 3.1. Sampling and Pre-Treatment

Samples of fine-grained sediments of the Mussulo bay were collected during October 2013 in areas exposed at low-tide periods. Sampling sites are clustered in outer (i.e., northern) and inner (i.e., southern) sectors of the Mussulo lagoon (Figure 1). All samples were air-dried in areas isolated of possible atmospheric contamination and sieved at 2 mm, before being analyzed in the laboratories of the Earth Sciences Department of University of Coimbra (DCT-UC).

### 3.2. Analytical Procedures

The grain-size distributions of the sampled sediments were determined in the Sedimentology Laboratory of DCT-UC using a laser diffraction granulometer Coulter LS 230 that is able to measure the proportion of particles ranging from 0.5–2000 μm. Each sample was measured at least twice, and averaged results of these runs were used in the grain-size characterization.

The mineralogy was determined by X-ray diffraction (XRD) using a Philips®® PW 3710 equipment with CuKα radiation and the software APD-PW1877 (version 3.6 J). Bulk mineralogy was determined on randomly oriented grains in the range 2–60° 2θ. Clay mineralogy was determined on oriented aggregates after pipetting clay suspension (<2 μm) to glass slides. XRD was then applied on air dried slides (2–30° 2θ) and after solvation with ethylene-glycol and heating at 550 °C (2–15° 2θ). Semi-quantitative estimations of mineral proportions are based on the areas of characteristic reflections after confirming the presence of the mineral with other XRD peaks. In bulk samples the following reflections were adopted: ~7.6 Å for gypsum, 4.26 Å for quartz, 3.23 Å for K-feldspar, 3.18 Å for plagioclase, 3.03 Å for calcite, 2.89 Å for dolomite, 2.82 Å for halite, and 2.72 Å for pyrite. In the clay fractions the glycolated diffractograms were adopted using the reflection of ~15–17 Å for smectite, ~12–13 Å for mixed layer clays, 10 Å for mica illite, 7.6 Å for gypsum, and 7.1 Å for kaolinite.

Concentrations of chemical elements were obtained with approximately 0.5 g of the fraction <0.18 mm of each sample. Dried sediment samples were placed in Teflon vessels (Multiwave 3000, Anton Paar) with 9 mL of 37% HCl, and 3 mL of 70% $HNO_3$. The vessels were heated in a microwave apparatus within 10 min of ramp, and remained at about 180 °C for 15 min. The determination of total metal(loid) contents was performed using current analytical methods, including: Atomic Absorption Spectrometry (AAS, SOLAAR M Series equipment from Thermo Scientific, Madison, USA) for Ca, Cu, Co, Fe, Mg, Mn, Ni, and Zn with atomization source of flame. The same equipment using the grafite furnace mode was used to determine As, Co, and Pb, with ashing temperatures of 1100 °C, 1000 °C, and 800 °C, respectively; and atomization temperatures of 2600 °C, 2100 °C, and 1200 °C respectively. The observed detection limits were 0.05 mg/kg for As, Co and Hg, 0.3 mg/kg for Mg, 0.5 mg/kg for Pb and Zn, 1 mg/kg for Cu, and Ni, 3 mg/kg for Ca, Mn, and Fe. As a control, reference materials NIST 2709-San Joaquin Soil and RTC - CRM015 were used, and the recoveries obtained for the different elements showed values in the range of 86.2 and 100.7%, being highest for Fe and lowest for Ni.

### 3.3. Statistical Data Treatment

Conventional univariate and multivariate statistical analysis were performed using the software JMP Pro 14.0. In order to better evaluate the associations between textural and compositional parameters obtained for the present investigation, a correlation-based Principal Component Analysis (PCA) was performed. A centered log–ratio transformation (clr) [40] was previously applied to the compositional data to remove the non-negativity and constant-sum constraints on compositional data. When the concentration of geochemical variables was below the detection limit in some samples, to allow their inclusion in the PCA, it adopted the square root of this limit divided by two. Supplementary Material ST1 shows the compositional data obtained for the present study.

### 3.4. Human Health Risk Assessment

The non-carcinogenic and carcinogenic risks were estimated according to the United States Environmental Protection Agency (USEPA) methodology [41]. The human health risk assessment was calculated assuming that both children and adult groups are directly exposed to potentially toxic elements (HMTE) hosted by sediments. Chronic Daily Intake (CDI.; $mg{\cdot}kg^{-1}$ bw per day) was determined for exposition to toxic elements by ingestion ($CDI_{ingest}$), dermal contact ($CDI_{dermal}$), and inhalation ($CDI_{inhalation}$). The following equations were adopted:

$$CDI_{ingest} = \frac{C \times IR \times EF \times ED \times CF}{BW \times AT} \tag{1}$$

$$CDI_{dermal} = \frac{C \times SA \times AF \times ABS \times EF \times ED \times CF}{BW \times AT} \tag{2}$$

$$CDI_{inhalation} = \frac{C \times InhR \times ET \times EF \times ED}{PEF \times BW \times AT} \tag{3}$$

where C is the concentration of PTEs in sediment ($mg{\cdot}kg^{-1}$). An explanation for the remaining parameters is presented in Table 1.

**Table 1.** Values adopted for the parameters used in the determination of Chronic Daily Intake (CDI) based on USEPA (2011).

| Parameter | Adult | Children |
|---|---|---|
| IR (Ingestion Rate of sediment) | 100 $mg{\cdot}day^{-1}$ | 200 $mg{\cdot}day^{-1}$ |
| EF (Exposure Frequency) | 312 $days{\cdot}year^{-1}$ | 312 $days{\cdot}year^{-1}$ |
| ED (Exposure Duration) | 35 years | 6 years |
| BW (Body Weight) | 15 kg | 70 kg |
| AT (Averaging Time for non-carcinogenic risk | 365 days × 6 | 365 days × 35 |
| AT (Averaging Time for carcinogenic risk | 365 days × 70 | 365 days × 70 |
| CF (Conversing Factor) | 10–6 $mg{\cdot}day^{-1}$ | 10–6 $mg{\cdot}day^{-1}$ |
| SA (Skin Surface Area available for contact) | 6032 $cm^2$ | 2373 $cm^2$ |
| AF (Soil to skin Adherence Factor) | 0.07 mg $cm^{-2}$ | 0.2 mg $cm^{-2}$ |
| ABS (Absorption Factor) | 0.001 | 0.001 |
| InhR (Inhalation Rate) | 1.56 $m^3{\cdot}h^{-1}$ | 1.2 $m^3{\cdot}h^{-1}$ |
| ET (Exposure Time) | 8 $h{\cdot}day^{-1}$ | 4 $h{\cdot}day^{-1}$ |
| PEF (Particle Emission Factor) | 1.36 × 109 $m^3{\cdot}kg^{-1}$ | 1.36 × 109 $m^3{\cdot}kg^{-1}$ |

The human health non-carcinogenic risk caused by PTEs exposure is expressed as a hazard quotient (HQ) = CDI/RfD. The CDI is the average daily dose that a child or adult is exposed. The RfD is a reference dose, below which no adverse non-carcinogenic health effects should result from a lifetime of exposure. The HI is the chronic hazard index that is the sum of the hazard quotients for multiple exposure pathways. For HI values > 1, there is a chance that non-carcinogenic risk may occur; otherwise, the individuals are exposed to concentrations that do not present a hazard. The carcinogenic risks (CR) for As and Ni exposure of the studied groups were calculated according to the Exposure Factors Handbook [41] and using the Slope Factors according to the U.S. Department of Energy (USDE) [42].

## 4. Results and Discussion

### 4.1. Compositional Variability Within the Lagoon

#### 4.1.1. Grain-Size

The sampled sediments can be organized into four groups based on their grain-size distributions (Figure 2). Samples collected in inner lagoon locations (group A.; samples A1 to A8) are characterized by a clear predominance of sand-size particles (95–98%) and very low clay content (<1%), displaying coarse-skewed unimodal distributions with modal sizes in classes ranging 0.12–0.25 mm. In western locations of the northern sectors of the lagoon, finer-grained sediments (58% < sand < 77%) predominate, which tend to show relatively wide grain-size distributions, frequently with a main mode in the interval 63–177 μm and a secondary population with modal size in the interval 0.25–0.5 mm (group B.; samples A9 to A15). Sediments further north (group C.; samples A16 to A19) are characterized by fine-skewed unimodal distributions, which tend to become finer northwards with decreasing sand content (from 70% to 40%), increasing silt (from 27% to 51%) and clay (from 3% to 6%) content, and a modal-size evolving from 0.088–0.125 to 0.063–0.088 mm. Finally, in the outermost locations of the lagoon (group D.; samples A20 to A22), grain-size becomes coarser again, with higher sand content (79%–89%), but still with significant amounts of clay (up to 5%).

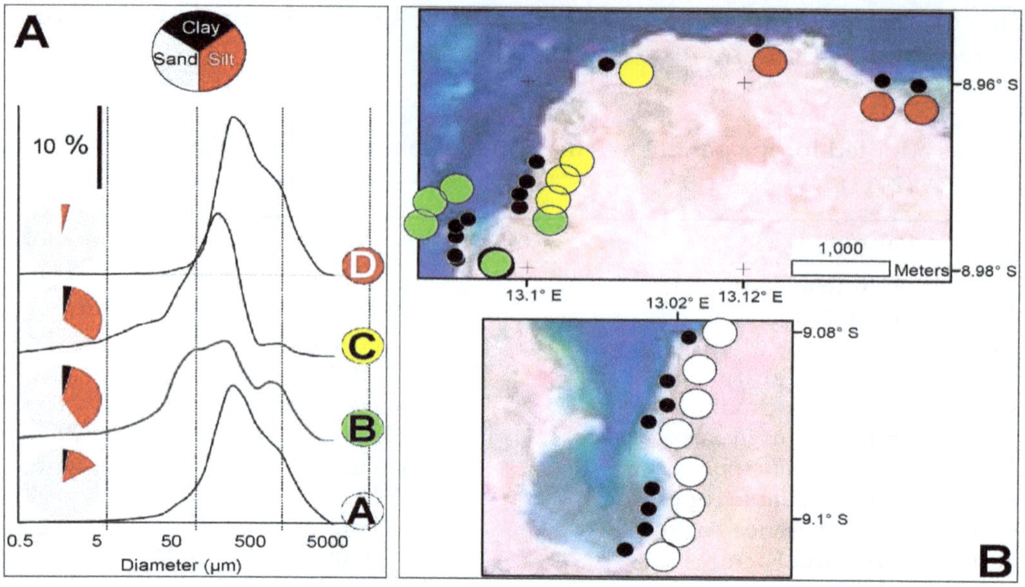


**Figure 2.** Grain-size of the Mussulo lagoon sediments. General features of different grain-size types (**A**) and their spatial distribution (**B**).

4.1.2. Mineralogy

Sediment mineralogy for bulk samples and the clay fraction are represented in Figure 3. The sediments of Mussolo lagoon are strongly enriched in quartz (65%–89%), followed by feldspars (2%–32%), phyllosilicates (1%–14%), and halite (<1%–13%). Gypsum, calcite, and dolomite, are usually present, but always in minor or trace amounts (<3%). Traces of pyrite are occasionally found and Mg-salts (e.g., carnalite, kainite, polyhalite) may also be present. Halite contents are higher in southern (3%–13%) than in northern locations (1%–4%), whilst feldspar tends to display an opposite spatial distribution (Figure 3). There are no clear geographic trends for quartz content or the remaining minerals.

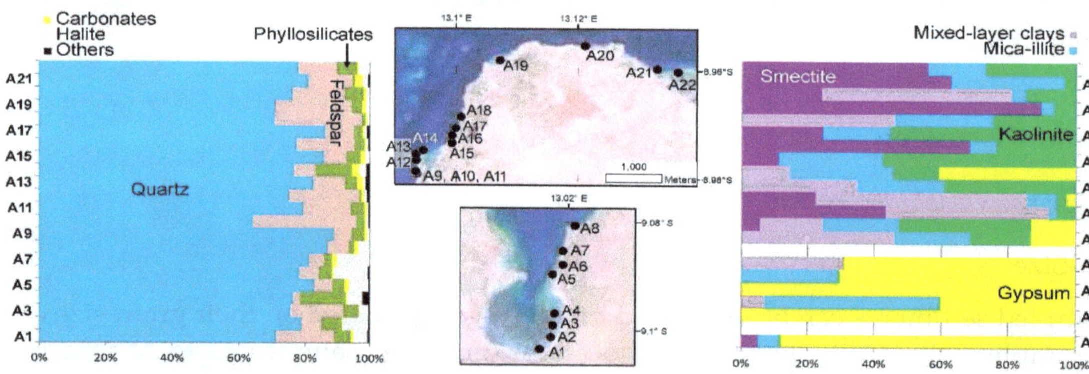


**Figure 3.** Bulk and clay mineral assemblages obtained for sediments of the Mussolo lagoon.

The composition of the clay fraction is highly variable and can be organized in three mineral assemblages with specific spatial distribution. Most sediments collected in southern realms of the lagoon (samples A1, A3, A5, and A6) are strongly enriched in gypsum (40%–100%), followed by mica-illite (0%–30%). A second assemblage is characterized by an enrichment in expansive clays (smectite and mixed-layer clays; 72%–100%), usually with secondary amounts of mica-illite (6%–26%) and kaolinite (3%–21%). This clay-mineral assemblage is characteristic of samples collected near the mouth of coastal streams in the northern sector of the lagoon (samples A11, A12, A20, and A21). Sediments from the northern sector of the lagoon can also yield a mix composition, with a variable abundance of kaolinite (13%–59%) and expansive clays (<55%), and more homogenous mica-illite (13%–25%).

The opposite behavior of feldspar and halite can be regarded as evidence of different orthochemical and detrital contributions. Because of the semi-arid climatic conditions and the high-water salinities, chemical precipitation in inner lagoon sectors likely occurs. The overall higher gypsum content in the clay fraction of the samples collected in the south can also be ascribed to authigenic formation in saline environments. Higher detrital supply, either with marine or continental sources, should occur in the northern part of the lagoon.

### 4.1.3. Geochemistry

Eleven chemical elements were selected for this research (As, Ca, Cu, Co, Fe, Hg, Mg, Mn, Ni, Pb, and Zn). As expected for sediments of a lagoon environment with high salinity, Ca (0.03%–1.98%) and Mg (0.12%–1.89%) are among the elements with highest measured concentration, and they tend to be more abundant in inner lagoon locations. Iron is more evenly distributed throughout the Mussulo lagoon (0.14%–1.20%), although, in general, with higher contents in northern samples. Samples enriched/depleted in Fe also yield high/low Cu (5.97–15.47 mg/kg), Mn (56.71–164.28 mg/kg) and Zn (9.59–56.48 mg/kg) contents (Figure 4).

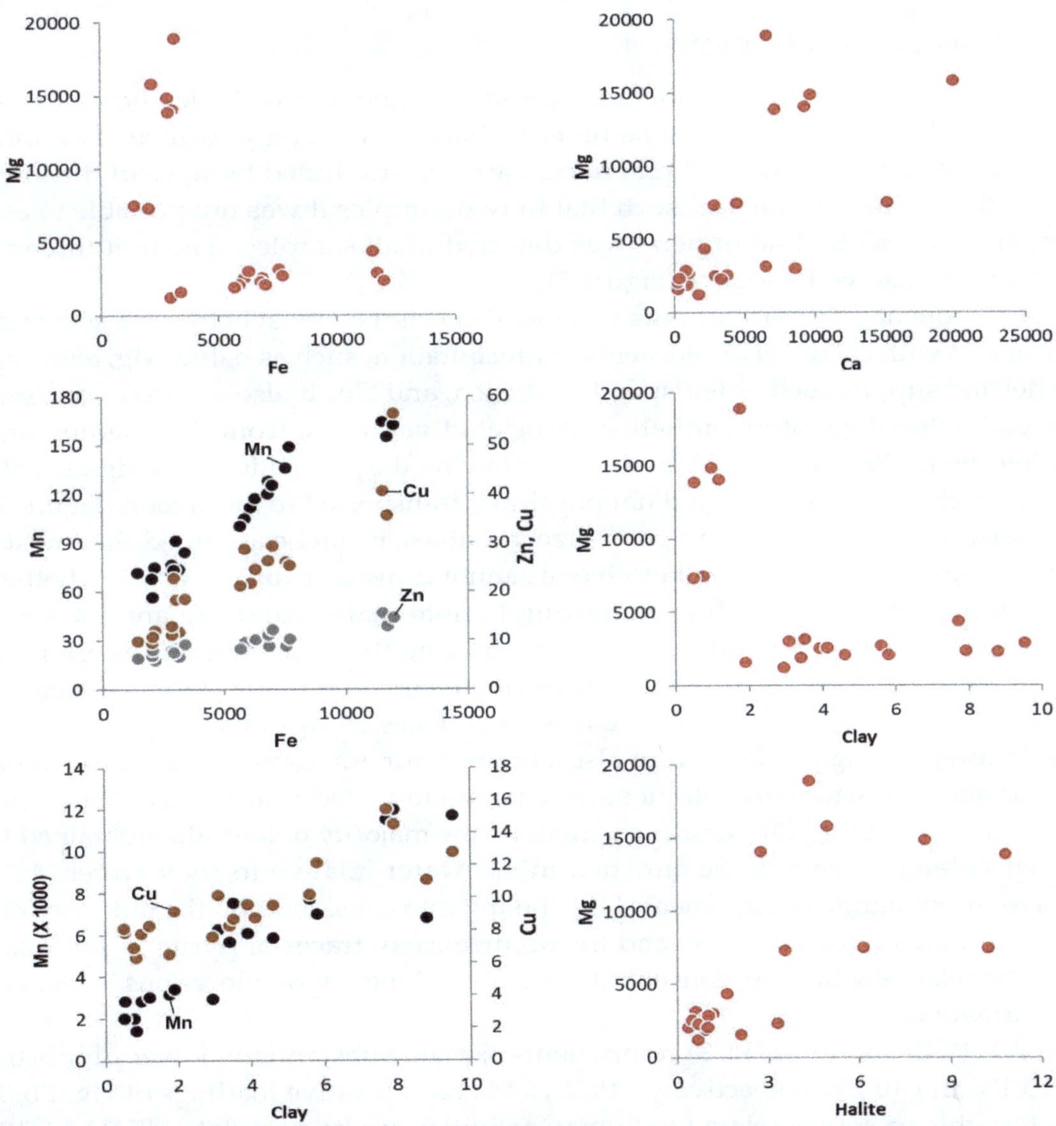


**Figure 4.** Ratios between chemical element concentrations and mineralogical and textural features of the sampled sediments. The plots Fe vs. Mg and clay (%) vs. Mg show two groups of samples, both displaying trends for increasing Mg with Fe and clay. Although with high scattering, Mg appears to correlate with Ca and halite. Samples enriched in clay also tend to yield higher Cu, Mn, Zn, and Fe.

Arsenic and Co were found in all samples, but in wide variable abundances (0.09–31.49 mg/kg and 0.78–26.04 mg/kg, respectively) and high contents can occur both in northern and southern locations. The other measured elements (Hg, Ni, and Pb) were detected only in part of the samples, Hg in eight of them (<0.12 mg/kg), Ni in 11 (<13.34 mg/kg), and Pb in 12 (<30.12 mg/kg). Sediments with relatively high amounts of at least two of these elements were collected both in the northern sector of the lagoon, namely near Luanda (samples A20, A21, and A22), and in some of the innermost southern locations (samples A2, A3, A6, and A8).

The plots of Mg vs. Fe and of Mg vs. grain-size variables reveal two distinct groups (Figure 4). Finer samples collected in the northern sector yield lower Mg and higher Fe; coarser samples collected in the southern sector yield higher Mg and lower Fe. However, if the two groups are isolated, Mg concentrations appear to be higher in finer sediments, which also tend to be enriched in Fe. The trend for higher Mg in sediments enriched in halite supports the possibility that Mg is strongly influenced by authigenic mineral formation, which prevails in southern lagoon settings with coarser sediments. On the other hand, the distribution of the two groups of samples in the bi-plots suggests that the presence of detrital fine-grained particles hosting Mg and Fe, along with other siderophile elements, is also influencing their concentrations.

### *4.2. Factors Controlling Sediment Composition*

A PCA using grain-size, bulk mineralogy, and geochemical variables helps to ascertain the compositional variability for the studied sediments. Variables strongly correlated or anti-correlated were not considered in the PCA. Clay mineralogy was also excluded because of the low amount of clay fraction in the studied sediments, such that in two samples it was not possible to establish clay assemblages, and the fact that no mineral was detected in all samples. The first three components explain 64.3% of the observed variance (Figure 5).

The first component (PC1, 41.7% of the variance) reflects a contrast between a set of variables that are probably linked with chemical or biochemical precipitation, such as halite, Mg, and Ca, and others linked with detrital supply, such as feldspar, Fe, Mn, Zn, and Cu. It also separates coarser sediments with high negative loadings of precipitation-associated variables, from finer sediments with high positive loadings of detrital-associated variables, reinforcing the possibility that a significant proportion of the coarser particles are authigenic and not physically transported to the lagoon (Figure 5). Based on the relations between geochemical and grain-size variables, it can be assumed that most Zn and Cu are hosted by fine-grained particles, although one cannot draw conclusions about whether they have natural or anthropogenic sources. It is interesting to note that Ca and Mg are not correlated with carbonate content, indicating that other minerals are hosting these elements. The correlation between halite and Mg suggest that Mg-bearing salts are being precipitated in the lagoon, as suggested before. XRD data is compatible with the possible occurrence of traces of Mg-salts.

Relatively low loadings of PC1 were also obtained for As, associating this element with the authighenic variables. Arsenic concentrations in non-contaminated near shore or estuarine sediments are in the order of 5–15 mg/kg [43], which encompasses the majority of the values obtained for Mussulo sediments. This element tends to be enriched in sea water relative to river water [44] and it can precipitate in reduced marine environments [45]. Taking into consideration the influence of authigenic salt-minerals on sediment composition and the occurrence of traces of pyrite in some samples, it is probable that the relatively high As concentration in some inner lagoon locations is related with local salt-mineral formation.

The second (PC2) and third (PC3) components explain substantially lower proportions of total variability (13.1% and 10.2%, respectively). PC2 yields high positive loadings of Hg, Pb, Ni, and Co, all elements for which a source related to human activities can be postulated [7–11,13–18]. The plot PC1 vs. PC2 (Figure 5) shows the links among this group of elements, suggesting a closer association of Pb and Ni with the variables with high loadings of PC1 (clay, Fe, Mn, Zn, and Cu), whilst Co is largely independent of PC1. Arsenic is not plotted with these elements, reinforcing the possibility of a natural origin associated with salt precipitation in the lagoon. Samples with high scores of this component are from both the northern and southern sectors.

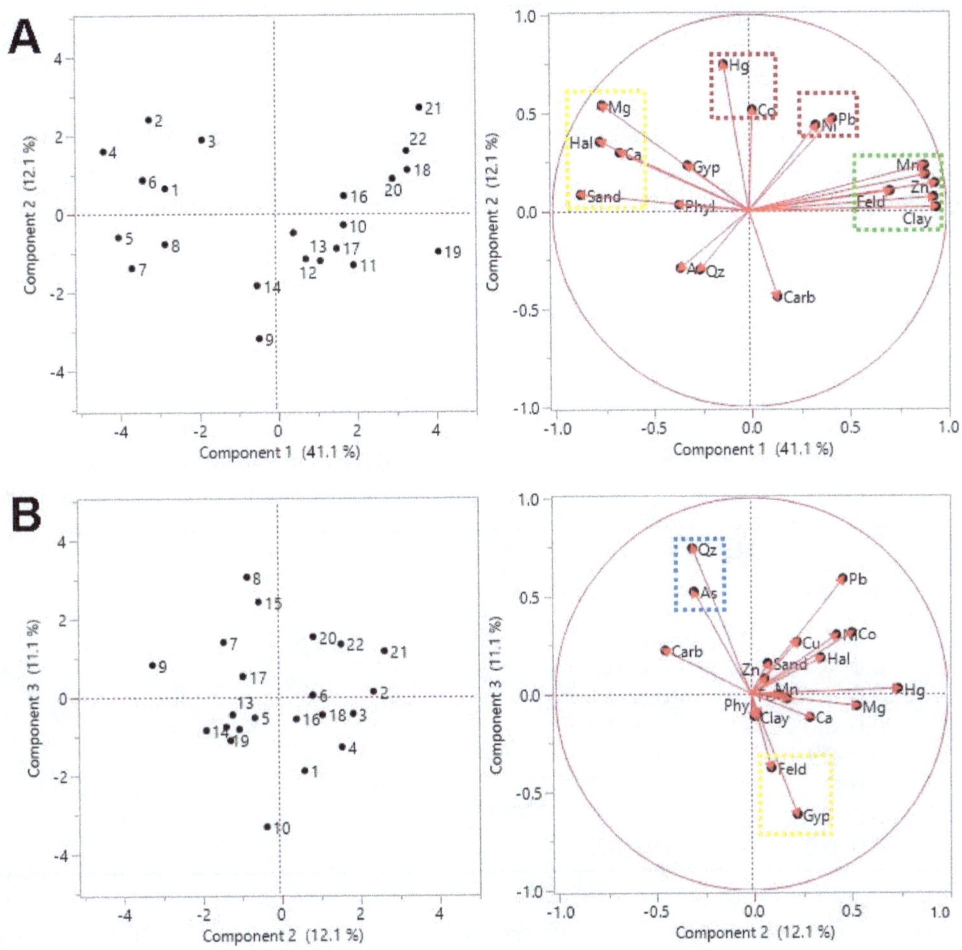


**Figure 5.** Maps of the principal components for geochemical, grain-size and mineralogical variables. (**A**) The vector loadings for the two main components of the Principal Component Analysis (PCA) define two perpendicular links, indicating two independent controls on the data. The first link connects elements assumed to be carried with fine-grained detrital minerals in opposition to elements that may result from precipitation in the lagoon. The second link connects a set of elements that may have anthropogenic sources. (**B**) The plot of PC2 vs. PC3 shows the opposition of quartz and feldspar, two detrital minerals whose relative enrichment in the lagoon sediments can be attributed to specific source areas. Gypsum seems to be associated with feldspar, which is compatible with a source in Meso-Cenozoic units of the neighboring Cuanza Basin.

PC3 shows an opposition between quartz and feldspar (Figure 5), which are the two most common minerals in the studied sediments, being both of detrital origin. Gypsum appears linked with feldspar. Differences between the sands of the Mussulo spit and at the mouth of the Cuanza river, which yield less feldspar along with alkali and alkaline earth metals, and more quartz and ultra-stable heavy minerals [36], support the possibility of different detrital sources in this costal environment. We propose here that sediment material enriched in feldspar and gypsum came from a proximal source, most likely the previous-cycle depositional units of the Cuanza Meso-Cenozoic basin that are enriched in these minerals; quartz-rich sediment is probably derived from more distant regions (i.e., the Cuanza River mouth and further up-drift). Except for As, which displays a close link with quartz, this component does not seem to have a major influence on the distribution of the studied PTEs.

In summary, an anthropogenic influence on sediment composition is probably responsible for occasional enrichments in Hg, Pb, Ni, and Co. Other elements found in higher abundances in fine grained sediments near Luanda, such as Cu and Zn, may also be partially human derived. On the other hand, natural factors seem to account for the observed distribution of As contents.

### *4.3. Health Implications of Sediment Composition*

High concentrations of PTEs in near-surface environment can threaten human health via sediment ingestion (geophagism), rare in adults but quite common in children, or by hand to-mouth intake, inhalation of dust particles, or dermal contact [46,47]. The Hazard Indices obtained from the health risk assessment for Cu, Hg, Mn, Ni, Pb, and Zn were below 1, pointing to no non-carcinogenic risk (Table 2). Regarding As and Co for children, HI reached 1.2 and 1.1, respectively, indicating potential non-carcinogenic risk due to sediment exposure. The proportion of cases with HI > 1 is 5% for As and 9% for Co. According to these results, it is highly recommendable that children spent less time playing in the lagoon sediments and avoid hand-to-mouth intake.

**Table 2.** Range of results obtained for Hazard Indexes (HI) and Cancer Risks (CR) due to sediment exposure for the different PTEs. Maximum results that are above the target values are in bold.

| | HI | | CR | |
|---|---|---|---|---|
| | Children | Adult | Children | Adult |
| As | 0.0–**1.2** | 0.0–0.1 | $9 \times 10^{-10}$ to $5 \times 10^{-5}$ | $3 \times 10^{-11}$ to $3 \times 10^{-5}$ |
| Mn | 0.0–0.2 | 0.0–0.0 | | |
| Co | 0.0–**1.1** | 0.0–0.1 | | |
| Cu | 0.0–0.0 | 0.0–0.0 | | |
| Hg | 0.0–0.1 | 0.0–0.0 | | |
| Ni | 0.0–0.0 | 0.0–0.0 | $1 \times 10^{-11}$ to $1 \times 10^{-9}$ | $4 \times 10^{-11}$ to $1 \times 10^{-10}$ |
| Pb | 0.0–0.1 | 0.0–0.0 | | |
| Zn | 0.0–0.0 | 0.0–0.0 | | |

According to the International Agency on Research of Cancer (IARC) [48], of the analyzed elements, only As and Ni poses significant carcinogenic risk. Cancer Risks values determined with the sampled sediments of the Mussulo lagoon (Table 2) are within the classes of acceptable carcinogenic risk for As ($1 \times 10^{-4}$ to $1 \times 10^{-6}$; [41]) and no risk for Ni ($<1 \times 10^{-6}$; [27]).

## 5. Conclusions

The geochemical composition of the sediments of the Mussulo lagoon is mainly determined by natural processes that influence the abundance of detrital components, which can be derived from different source areas, and authigenic components associated with mineral precipitation within the lagoon setting and surrounding regions. Human action may have a subsidiary role, contributing to the abundance of some PTEs, such as Co, Hg, Ni, and Pb, in parts of the lagoon sediments. Fine grained samples collected near Luanda yield relatively high concentrations of Zn and Cu, but whether their sources are natural or anthropogenic is uncertain. Arsenic is more abundant in inner (i.e., southern) locations of the lagoon, probably in association with natural processes due to precipitation from salt-water. The health risk assessments suggest no major carcinogenic risk due to sediment intake, but the concentrations of As and Co indicate potential non-carcinogenic risk for children.

The present research shows that health risks associated with element concentrations can emerge from enrichment due to both natural and anthropogenic processes. Although some PTEs preferentially hosted by fine grained particles (e.g., Co, Hg, Ni, Pb, Zn, and Cu) may be associated with human activities, in coastal settings of arid regions, the possibility of high concentrations of As due to natural precipitation should be fully considered in environmental assessments.

**Author Contributions:** Conceptualization: P.D.; Data curation: P.D., J.P.; Funding acquisition: J.P., P.D.; Methodology: P.D., J.P.; Sampling: A.A.; Formal analysis: A.A., P.D., J.P.; Investigation: P.D., J.P., A.A.; Visualization: P.D.; Writing original draft: P.D., J.P.; Writing, reviewing and editing: P.D., J.P. All authors have read and agreed to the published version of the manuscript.

**Acknowledgments:** The present work was supported by the FCT through the project UIDB/04292/2020 (MARE-Marine and Environmental Sciences Centre). The Guest Editor, Marina C. Pinto, provided valuable indications for the assessment of human health risks. Two anonymous reviewers are thanked for their constructive and pertinent comments, which helped to improve the article.

## References

1. UN (United Nations). The World's Cities in 2018: Data Booklet, Statistical Papers—United Nations (Ser. A), Population and Vital Statistics Report, UN, New York. 2018. Available online: https://doi.org/10.18356/c93f4dc6-en (accessed on 4 January 2020).
2. Cheggour, M.; Chafik, A.; Langston, W.J.; Burt, G.R.; Benbrahim, S.; Texier, H. Metals in sediments and the edible cockle Cerastoderma edule from two Moroccan Atlantic lagoons: Moulay Bou Selham and Sidi Moussa. *Environ. Pollut.* **2001**, *115*, 149–160. [CrossRef]
3. Fernandes, C.; Fontainhas-Fernandes, A.; Peixoto, F.; Salgado, M.A. Bioaccumulation of heavy metals in Liza saliens from the Esmoriz–Paramos coastal lagoon, Portugal. *Ecotoxicol. Environ. Saf.* **2007**, *66*, 426–431. [CrossRef]
4. Spada, L.; Annicchiarico, C.; Cardellicchio, N.; Giandomenico, S.; Di Leo, A. Mercury and methylmercury concentrations in Mediterranean seafood and surface sediments, intake evaluation and risk for consumers. *Int. J. Hyg. Environ. Health* **2012**, *215*, 418–426. [CrossRef]
5. María-Cervantes, A.; Jiménez-Cárceles, F.J.; Álvarez-Rogel, J. As, Cd, Cu, Mn, Pb, and Zn contents in sediments and mollusks (Hexaplex trunculus and Tapes decussatus) from coastal zones of a Mediterranean lagoon (Mar Menor, SE Spain) affected by mining wastes. *Water Air Soil Pollut.* **2009**, *200*, 289–304. [CrossRef]
6. Pinto, M.M.; Silva, M.M.; Neiva, A.M. Pollution of water and stream sediments associated with the Vale De Abrutiga Uranium Mine, Central Portugal. *Mine Water Environ.* **2004**, *23*, 66–75. [CrossRef]
7. Pekey, H. The distribution and sources of heavy metals in Izmit Bay surface sediments affected by a polluted stream. *Mar. Pollut. Bull.* **2006**, *52*, 1197–1208. [CrossRef]
8. Biasioli, M.; Grčman, H.; Kralj, T.; Madrid, F.; Díaz-Barrientos, E.; Ajmone-Marsan, F. Potentially toxic elements contamination in urban soils. *J. Environ. Qual.* **2007**, *36*, 70–79. [CrossRef]
9. Qingjie, G.; Jun, D.; Yunchuan, X.; Qingfei, W.; Liqiang, Y. Calculating pollution indices by heavy metals in ecological geochemistry assessment and a case study in parks of Beijing. *J. China Univ. Geosci.* **2008**, *19*, 230–241. [CrossRef]
10. Ye, S.; Zeng, G.; Wu, H.; Zhang, C.; Liang, J.; Dai, J.; Liu, Z.; Xiong, W.; Wan, J.; Xu, P.; et al. Co-occurrence and interactions of pollutants, and their impacts on soil remediation—A review. *Crit. Rev. Environ. Sci. Technol.* **2017**, *47*, 1528–1553. [CrossRef]
11. Cabral-Pinto, M.M.; Inácio, M.; Neves, O.; Almeida, A.A.; Pinto, E.; Oliveiros, B.; da Silva, E.A. Human health risk assessment due to agricultural activities and crop consumption in the surroundings of an industrial area. *Expo. Health* **2019**, 1–2. [CrossRef]
12. Mirlean, N.; Andrus, V.E.; Baisch, P.; Griep, G.; Casartelli, M.R. Arsenic pollution in Patos Lagoon estuarine sediments, Brazil. *Mar. Pollut. Bull.* **2003**, *46*, 1480–1484. [CrossRef]
13. Elbaz-Poulichet, F.; Dezileau, L.; Freydier, R.; Cossa, D.; Sabatier, P. A 3500-year record of Hg and Pb contamination in a Mediterranean sedimentary archive (The Pierre Blanche Lagoon, France). *Environ. Sci. Technol.* **2011**, *45*, 8642–8647. [CrossRef]
14. Covelli, S.; Langone, L.; Acquavita, A.; Piani, R.; Emili, A. Historical flux of mercury associated with mining and industrial sources in the Marano and Grado Lagoon (northern Adriatic Sea). *Estuar. Coast. Shelf Sci.* **2012**, *113*, 7–19. [CrossRef]
15. Fujita, M.; Ide, Y.; Sato, D.; Kench, P.S.; Kuwahara, Y.; Yokoki, H.; Kayanne, H. Heavy metal contamination of coastal lagoon sediments: Fongafale Islet, Funafuti Atoll, Tuvalu. *Chemosphere* **2014**, *95*, 628–634. [CrossRef]
16. Alyazichi, Y.M.; Jones, B.G.; McLean, E. Source identification and assessment of sediment contamination of trace metals in Kogarah Bay, NSW, Australia. *Environ. Monit. Assess.* **2015**, *187*, 20. [CrossRef]

17. Ke, X.; Gui, S.; Huang, H.; Zhang, H.; Wang, C.; Guo, W. Ecological risk assessment and source identification for heavy metals in surface sediment from the Liaohe River protected area, China. *Chemosphere* **2017**, *175*, 473–481. [CrossRef]
18. Mejjad, N.; Laissaoui, A.; El-Hammoumi, O.; Fekri, A.; Amsil, H.; El-Yahyaoui, A.; Benkdad, A. Geochemical, radiometric, and environmental approaches for the assessment of the intensity and chronology of metal contamination in the sediment cores from Oualidia lagoon (Morocco). *Environ. Sci. Pollut. Res.* **2018**, *25*, 22872–22888. [CrossRef]
19. Lottermoser, B.G. Natural enrichment of topsoils with chromium and other heavy metals, Port Macquarie, New South Wales, Australia. *Soil Res.* **1997**, *35*, 1165–1176. [CrossRef]
20. Kraepiel, A.M.; Dere, A.L.; Herndon, E.M.; Brantley, S.L. Natural and anthropogenic processes contributing to metal enrichment in surface soils of central Pennsylvania. *Biogeochemistry* **2015**, *123*, 265–283. [CrossRef]
21. Pinto, M.M.; Silva, M.M.; da Silva, E.A.; Dinis, P.A.; Rocha, F. Transfer processes of potentially toxic elements (PTE) from rocks to soils and the origin of PTE in soils: A case study on the island of Santiago (Cape Verde). *J. Geochem. Explor.* **2017**, *183*, 140–151. [CrossRef]
22. Komar, P.D. The entrainment, transport and sorting of heavy minerals by waves and currents. In *Heavy Minerals in Use*; Developments in Sedimentology Series; Mange, M.A., Wright, D.T., Eds.; Elsevier: Amsterdam, The Netherlands, 2007; Volume 58, pp. 3–48.
23. Garzanti, E.; Andò, S.; Vezzoli, G. Grain-size dependence of sediment composition and environmental bias in provenance studies. *Earth Planet. Sci. Lett.* **2009**, *277*, 422–432. [CrossRef]
24. Garzanti, E.; Dinis, P.; Vermeesch, P.; Andò, S.; Hahn, A.; Huvi, J.; Limonta, M.; Padoan, M.; Resentini, A.; Rittner, M.; et al. Sedimentary processes controlling ultralong cells of littoral transport: Placer formation and termination of the Orange sand highway in southern Angola. *Sedimentology* **2018**, *65*, 431–460. [CrossRef]
25. De Carvalho, H.; Tassinari, C.; Alves, P.H.; Guimarães, F.; Simões, M.C. Geochronological review of the Precambrian in western Angola: Links with Brazil. *J. Afr. Earth Sci.* **2000**, *31*, 383–402. [CrossRef]
26. Vaughan, A.P.; Pankhurst, R.J. Tectonic overview of the West Gondwana margin. *Gondwana Res.* **2008**, *13*, 150–162. [CrossRef]
27. Tack, L.; Wingate, M.T.; Liégeois, J.P.; Fernandez-Alonso, M.; Deblond, A. Early Neoproterozoic magmatism (1000–910 Ma) of the Zadinian and Mayumbian Groups (Bas-Congo): Onset of Rodinia rifting at the western edge of the Congo craton. *Precambrian Res.* **2001**, *110*, 277–306. [CrossRef]
28. Moulin, M.; Aslanian, D.; Unternehr, P. A new starting point for the South and Equatorial Atlantic Ocean. *Earth Sci. Rev.* **2010**, *98*, 1–37. [CrossRef]
29. Guiraud, M.; Buta-Neto, A.; Quesne, D. Segmentation and differential post-rift uplift at the Angola margin as recorded by the transform-rifted Benguela and oblique-to-orthogonal-rifted Kwanza basins. *Mar. Pet. Geol.* **2010**, *27*, 1040–1068. [CrossRef]
30. Daly, M.C.; Lawrence, S.R.; Diemu-Tshiband, K.; Matouana, B. Tectonic evolution of the Cuvette Centrale, Zaire. *J. Geol. Soc.* **1992**, *149*, 539–546. [CrossRef]
31. Kadima, E.; Delvaux, D.; Sebagenzi, S.N.; Tack, L.; Kabeya, S.M. Structure and geological history of the Congo Basin: An integrated interpretation of gravity, magnetic and reflection seismic data. *Basin Res.* **2011**, *23*, 499–527. [CrossRef]
32. Haddon, I.G.; McCarthy, T.S. The Mesozoic–Cenozoic interior sag basins of Central Africa: The late-cretaceous–Cenozoic Kalahari and Okavango basins. *J. Afr. Earth Sci.* **2005**, *43*, 316–333. [CrossRef]
33. Marzoli, A.; Melluso, L.; Morra, V.; Renne, P.R.; Sgrosso, I.; D'antonio, M.; Morais, L.D.; Morais, E.A.; Ricci, G. Geochronology and petrology of Cretaceous basaltic magmatism in the Kwanza basin (western Angola), and relationships with the Paraná-Etendeka continental flood basalt province. *J. Geodyn.* **1999**, *28*, 341–356. [CrossRef]
34. Abecasis, C.K. *As Formações Lagunares e Seus Problemas de Engenharia Litoral*; Junta de Investigação do Ultramar: Lisbon, Portugal, 1961; pp. 1–174.
35. Dinis, P.A.; Huvi, J.; Cascalho, J.; Garzanti, E.; Vermeesch, P.; Callapez, P. Sand-spits systems from Benguela region (SW Angola). An analysis of sediment sources and dispersal from textural and compositional data. *J. Afr. Earth Sci.* **2016**, *117*, 171–182. [CrossRef]
36. Garzanti, E.; Dinis, P.; Vermeesch, P.; Andò, S.; Hahn, A.; Huvi, J.; Limonta, M.; Padoan, M.; Resentini, A.; Rittner, M.; et al. Dynamic uplift, recycling, and climate control on the petrology of passive-margin sand (Angola). *Sediment. Geol.* **2018**, *375*, 86–104. [CrossRef]

37. Curto, J.C.; Gervais, R.R. A dinâmica demográfica de Luanda no contexto do tráfico de escravos do Atlântico Sul, 1781–1844. *Topoi (Rio de Janeiro)* **2002**, *3*, 85–138. [CrossRef]
38. Amaral, I.D. Subsídios para o estudo da evolução da população de Luanda. *Garcia Orta Rev. Junta Missões Geográficas Investig. Do Ultramar* **1959**, *7*, 211–226.
39. Amaral, I. Luanda e os seus "muceques", problemas de Geografia Urbana. *Finisterra* **1983**, *18*, 293–325. [CrossRef]
40. Aitchison, J. The statistical analysis of compositional data. *J. R. Stat. Soc. Ser. B (Methodological)* **1982**, *44*, 139–160. [CrossRef]
41. USEPA (United States Environmental Protection Agency). *Exposure Factors Handbook 2011 Edition (Final)*; United States Environmental Protection Agency: Washington, DC, USA, 2011. Available online: http://cfpub.epa.gov/ncea/risk/recordisplay.cfm?deid=236252 (accessed on 18 January 2020).
42. USDE (U.S. Department of Energy). *The Risk Assessment Information System (RAIS)*; U.S. Department of Energy's Oak Ridge Operations Office: Oak Ridge, TN, USA, 2013. Available online: https://rais.ornl.gov/ (accessed on 4 December 2019).
43. Neff, J.M. Ecotoxicology of arsenic in the marine environment. *Environ. Toxicol. Chem. Int. J.* **1997**, *16*, 917–927. [CrossRef]
44. Dong, W.Q.; Cui, Y.; Liu, X. Instances of soil and crop heavy metal contamination in China. *Soil Sediment Contam.* **2001**, *10*, 497–510. [CrossRef]
45. Kalia, K.; Khambholja, D.B. Arsenic contents and its biotransformation in the marine environment. In *Handbook of Arsenic Toxicology*; Flora, S.J.S., Ed.; Academic Press: San Diego, CA, USA, 2015; Chapter 28; pp. 675–700.
46. Viers, J.; Dupré, B.; Gaillardet, J. Chemical composition of suspended sediments in World Rivers: New insights from a new database. *Sci. Total Environ.* **2009**, *407*, 853–868. [CrossRef]
47. Sun, Y.; Zhou, Q.; Xie, X.; Liu, R. Spatial, sources and risk assessment of heavy metal contamination of urban soils in typical regions of Shenyang, China. *J. Hazard. Mater.* **2010**, *174*, 455–462. [CrossRef]
48. IARC (International Agency for Research on Cancer). List of Classifications 1–123. 2017. Available online: https://monographs.iarc.fr/agents-classified-by-the-iarc/ (accessed on 25 March 2019).

# Fine Particulate Matter and Gaseous Compounds in Kitchens and Outdoor Air of Different Dwellings

Célia Alves [1,*], Ana Vicente [1], Ana Rita Oliveira [1], Carla Candeias [2,*], Estela Vicente [1], Teresa Nunes [1], Mário Cerqueira [1], Margarita Evtyugina [1], Fernando Rocha [2] and Susana Marta Almeida [3]

1 Centre for Environmental and Marine Studies (CESAM), Department of Environment, University of Aveiro, 3810-193 Aveiro, Portugal; anavicente@ua.pt (A.V.); ritaanaoliveira@ua.pt (A.R.O.); estelaavicente@gmail.com (E.V.); tnunes@ua.pt (T.N.); cerqueira@ua.pt (M.C.); margarita@ua.pt (M.E.)
2 Geobiosciences, Geotechnologies and Geoengineering Research Centre (GeoBioTec), Department of Geosciences, University of Aveiro, 3810-193 Aveiro, Portugal; tavares.rocha@ua.pt
3 Centre for Nuclear Sciences and Technologies (C2TN), Instituto Superior Técnico, University of Lisbon, Estrada Nacional 10, 2695-066 Bobadela, Portugal; smarta@ctn.tecnico.ulisboa.pt
* Correspondence: celia.alves@ua.pt (C.A.); candeias@ua.pt (C.C)

**Abstract:** Passive diffusion tubes for volatile organic compounds (VOCs) and carbonyls and low volume particulate matter ($PM_{2.5}$) samplers were used simultaneously in kitchens and outdoor air of four dwellings. $PM_{2.5}$ filters were analysed for their carbonaceous content (organic and elemental carbon, OC and EC) by a thermo-optical technique and for polycyclic aromatic hydrocarbon (PAHs) and plasticisers by GC-MS. The morphology and chemical composition of selected $PM_{2.5}$ samples were characterised by SEM-EDS. The mean indoor $PM_{2.5}$ concentrations ranged from 14 $\mu g\ m^{-3}$ to 30 $\mu g\ m^{-3}$, while the outdoor levels varied from 18 $\mu g\ m^{-3}$ to 30 $\mu g\ m^{-3}$. Total carbon represented up to 40% of the $PM_{2.5}$ mass. In general, the indoor OC/EC ratios were higher than the outdoor values. Indoor-to-outdoor ratios higher than 1 were observed for VOCs, carbonyls and plasticisers. PAH levels were much higher in the outdoor air. The particulate material was mainly composed of soot aggregates, fly ashes and mineral particles. The hazard quotients associated with VOC inhalation suggested a low probability of non-cancer effects, while the cancer risk was found to be low, but not negligible. Residential exposure to PAHs was dominated by benzo[a]pyrene and has shown to pose an insignificant cancer risk.

**Keywords:** dwellings; indoor/outdoor; VOCs; carbonyls; $PM_{2.5}$; OC/EC; morphology; PAHs

## 1. Introduction

We spend most of our time in indoor environments. As building infrastructures are increasingly airtight to save energy, epidemiological studies need to understand the extent to which outdoor levels of air pollutants persist as a determining factor of exposure and, consequently, of health in the indoor environment. Many of the outdoor pollutants are also prevalent within homes, contributing to unhealthy air. Kitchens are spaces not only for preparing meals, but also for socialising with family and friends, and where children often do their homework, and most of us watch television. The kitchen is indeed the heart of the home. Places where meals are made and eaten are considered microenvironments with specific characteristics [1]. After the bedroom, this indoor area is probably the room people spend the most time in. Air quality in a kitchen is influenced by many factors, such as the method of meal preparation and ingredients used, the cooking style, the temperature of the cooking process, the volume of the room, the efficiency of the exhaust hood, and the number of persons using the space [1–3]. However, in developing countries wood stove emissions are the main cause of kitchen-related air pollution in many deprived homes [4–6]. Nearly 3 billion people use solid fuels, especially biomass and coal, for cooking and heating and this number will continue to rise in the next decade [5] (and references therein). Since biomass-burning cookstoves are a noteworthy source of carbonaceous aerosols and gaseous compounds, this source of pollution has received increasing attention because emissions greatly contribute to the global burden of disease [5,7–13]. Estimates by the World Health Organisation (WHO) indicate that exposure to air pollution from cooking with solid fuels contribute to more than four million annual premature deaths globally, half a million of which are children under the age of 5 who die of pneumonia [14]. Emissions from biomass-burning cookstoves encompass products of incomplete combustion, such as volatile organic compounds (VOCs) and particulate matter with an aerodynamic diameter less than or equal to 2.5 µm ($PM_{2.5}$), which can penetrate deeply into the alveolar sacs, where they can deposit and be absorbed, contributing to the entry of toxic substances into the bloodstream, including carcinogenic polycyclic aromatic hydrocarbons (PAHs) [15].

In the last two years, a substantial number of research articles has been published with the objective of documenting the indoor air quality in kitchens with low-efficiency biomass cookers, especially in underdeveloped countries. Some of these articles aimed at comparing emissions from traditional biomass stoves for household cooking with those from improved cookstoves [16–33]. In developed countries, in most cases, gas or electric stoves/cookers are used for meal preparation. Furthermore, a ventilation system mounted directly over the cooker/stove is an essential element in every kitchen to reduce the transport of odours and pollutants to neighbouring rooms. However, despite the pollutant levels in well-equipped modern kitchens are reportedly much lower, studies on this type of microenvironment are scarce and mostly focused on gaseous contaminants [34–38]. The WHO concluded that there is no convincing evidence of a difference in the hazardous nature of particulate matter from indoor sources as compared with those from outdoors and that the indoor levels are usually higher than the outdoor levels [39]. Continuous pressure to re-evaluate air quality standards stems from studies that have observed effects at low levels of particulate matter. These studies have suggested that, instead of mass concentration, some chemical components (e.g. carbonaceous compounds) may be a better metric for estimating the health risks [40]. A better understanding of indoor air pollutants, their levels and sources in specific microenvironments can help in adopting more efficient management strategies and mitigation measures to reduce health risks from exposure to $PM_{2.5}$ and associated toxic constituents.

This study is based on a multi-pollutant monitoring campaign carried out in four biomass-free kitchens, for which studies are comparatively much scarcer, in order to answer the following questions: Are there significant differences in pollutant levels between modern kitchens equipped with gas ranges or electric hobs? Do the observed levels and compounds depend on housing factors or outdoor air? Are the risks resulting from inhalation of pollutants (VOCs and $PM_{2.5}$-bond PAHs) routinely considered by international agencies of concern to health? Are these metrics sufficient to infer sources and effects or can the particle morphological analysis give us additional indications? The aim of this pilot study is not only to characterise air quality in a poorly studied microenvironment, such as kitchens, but also to draw lessons for conducting wider researches in the future.

## 2. Methodologies

### 2.1. *Sampling and Analysis*

A monitoring programme involving four kitchens with different characteristics (Table 1) and the respective outdoor air was conducted in the region of Aveiro, Portugal, in October and November 2017. Along with the neighbouring city of Ílhavo, Aveiro is part of an urban agglomeration that includes 120,000 inhabitants. Aveiro is located on the Atlantic coast, in the Central Region, at about 250 km to the North of Lisbon and 70 km to the South of Oporto. It is surrounded by beaches and by an extensive coastal lagoon.

**Table 1.** Characterisation of dwellings and sampling details.

| Characteristics | House 1 | House 2 | House 3 | House 4 |
|---|---|---|---|---|
| Location | Ílhavo | Aveiro | Aveiro | Aveiro |
| Area | Periurban/rural | Urban | Suburban | Urban |
| Type of dwelling | Detached house with lawn and extensive vegetable garden | City centre apartment with permanent occupancy | Detached house on the outskirts, with small garden, near a main road with intense traffic | Terraced house in a residential neighbourhood, near a main road with intense traffic |
| Kitchen area ($m^2$) | 24.5 | 16.5 | 15.4 | 20.2 |
| Number of permanent occupants | 3 | 3 | 3 | 4 |
| Number of daily occupancy hours | 15 | 24 | 15 | 17 |
| Smokers | No | No | No | No |
| Pets | 1 dog, 1 cat | No | No | No |
| Source of energy for cooking | Gas | Gas | Electricity | Electricity |
| Ventilation | Natural | Natural | Natural | Natural |
| Range hood | Under cabinet | Under cabinet | Under cabinet | Under cabinet |
| $PM_{2.5}$ sampling equipment | TCR Tecora | MiniVol | TCR Tecora | MiniVol |
| Number of samples | 10 (indoor)<br>10 (outdoor) | 7 (indoor)<br>7 (outdoor) | 10 (indoor)<br>10 (outdoor) | 7 (indoor)<br>7 (outdoor) |
| Sampling period | 16/10/2017 to 05/11/2017 | 16/10/2017 to 06/11/2017 | 06 to 26/11/2017 | 07 to 28/11/2017 |

Low volume samplers were used to collect particulate matter ($PM_{2.5}$) onto 47 mm diameter quartz filters. ECHO PM samplers (TCR Tecora, Cogliate, Italy) operating at 38.3 L min$^{-1}$ were deployed in two residences, in which ten pairs of $PM_{2.5}$ samples were collected for periods of 48 h. MiniVol™ TAS samplers (Airmetrics, Springfield, OR, USA) were used in the other two dwellings, in which seven pairs of $PM_{2.5}$ samples were collected for periods of 72 h at a flow of 5 L min$^{-1}$. In the kitchens, the samplers were positioned near the dining tables in a central location. Outside, the equipment was placed on the porch, terrace or balcony adjacent to the kitchens. VOCs and carbonyls were sampled in parallel, also indoors and outdoors, using Radiello® (Merck, Darmstadt, Germany) diffusive passive tubes (cartridges codes 145 and 165, respectively) in triplicate. Two consecutive samplings, each lasting 10 days, were performed at each site. VOCs were analysed by thermal desorption coupled to gas chromatography-mass at the Istituti Clinici Scientifici Maugeri (Pavia, Italy). The carbonyl-DNPH derivatives were analysed at the University of Aveiro by eluting with 2 mL of acetonitrile poured directly into the cartridge and stirring from time to time for 30 min. The extracts were filtered and then analysed in a high-performance liquid chromatography system (Jasco, Cremella, Italy) equipped with a PU- 980 pump, also from Jasco (Cremella, Italy), a manual injection valve (20 μL loop, Rheodyne, Rohnert Park, CA, USA), a Supelcosil LC-18 column (250×4.6 mm; 5 μm; Supelco, Darmstadt, Germany) and a Jasco MD-1510 diode array detector (Cremella, Italy). The elution was performed with an isocratic mixture of acetonitrile and water (60:40), with a flow rate of 1.5 mL $min^{-1}$. External calibration curves in six concentration levels were constructed from standard solutions.

The gravimetric quantification of $PM_{2.5}$ was performed on a RADWAG MYA 5/2Y/F (Radom, Poland) microbalance with an accuracy of 1 μg in a humidity (50 ± 5%) and temperature (20 ± 1 °C) controlled room. Filter weights were obtained from the average of six consecutive measurements with variations between them of less than 0.02%. The carbonaceous content (organic and elemental carbon, OC and EC) of $PM_{2.5}$ samples was analysed by a thermal-optical transmission technique. At least, two replicate analyses were performed for each filter. In each analytical run, two 9 mm punches are first heated in a non-oxidising atmosphere of $N_2$ in order to volatilise the carbonaceous organic compounds. After the first step of controlled heating, the remaining carbonaceous fraction is burnt in an oxygen-containing gas mixture. During anoxic heating, some OC is pyrolysed (PC), and quantified as EC in the second stage of heating. The minimise the bias in the OC/EC split, the blackening of the filter is continuously monitored by a laser beam and a photodetector, which allows reading the light transmittance. OC and EC are measured in the form of $CO_2$ by an infrared non-dispersive analyser.

For chemical and morphological characterisation, the filters with the lowest and highest concentration in each kitchen and the respective outdoor pairs were chosen. Two 5 mm diameter punches were cut from each of these filters. A Hitachi S-4100 scanning electron microscope (SEM) coupled to a Bruker Quantax 400 Energy Dispersive Spectrometer (EDS) (Bonsai Advanced, Madrid, Spain) was employed.

Since several punches were removed from each filter for analysis of the carbonaceous material and for morphological characterisation, the remaining area did not contain enough mass for the quantification of PAHs. Thus, for each site, the leftover area of the various filters was combined and extracted together to obtain an "average" of the concentrations. Each set of filters was extracted three times with dichloromethane (DCM) in an ultrasonic bath (25 mL for 15 min, each extraction, with 5 min stops between them). After each extraction, the 3 DCM organic extracts of each composite sample were combined, filtered through pre-cleaned cotton and concentrated to a volume of 0.5 mL using a Turbo Vap®II evaporation system (Biotage, Charlotte, NC, USA). The concentrated samples were transferred into vials and dried under a gentle nitrogen stream. The extracts were analysed in a gas chromatographer-mass spectrometer (GC-MS) from Agilent (Santa Clara, CA, USA) with single quadrupole. The chromatographic system (GC model 7890B, MS model 5977A) was equipped with a CombiPAL autosampler (Agilent, Santa Clara, CA, USA) and a TRB-5MS (60 m × 0.25 mm × 0.25 μm) column (Teknokroma, Barcelona, Spain). The quantitative analysis was performed by single ion monitoring (SIM). Blank filters were analysed in the same way to obtain blank-corrected results. Data were acquired in the electron impact (EI) mode (70 eV). The oven temperature programme was as follows: 60 °C (1 min), 60 to 150 °C (10 °C $min^{-1}$), 150 to 290 °C (5 °C $min^{-1}$), 290 °C (30 min) and using helium as carrier gas at 1.2 mL $min^{-1}$. The following mixture of deuterated internal standards (IS) was used to quantify PAHs: 1,4-dichlorobenzene-d4, naphtalene-d8, acenaphthene-d10, phenanthrene-d10, chrysene-d12, perylene-d12, fluorene-d10 and benzo[a]pyrene-d12 (Supelco). In the case of plasticisers, deuterated diethyl phthalate-3,4,5,6-d4 and bis (2-ethylhexyl) phthalate-3,4,5,6-d4 (Supelco) were used as IS. Calibrations were performed with authentic standards (Sigma-Aldrich, St. Louis, MO, USA) at eight different concentration levels.

### 2.2. Data Analysis

For the statistical treatment, SPSS (IBM Statistics Software V.25, Armonk, NY, USA) was used. The normality of the data was assessed by the Shapiro-Wilk test. The Mann-Witney non-parametric test was applied to obtain the statistically significant differences with a significance of 0.05 (Tables S1–S4). Uncertainties of measurements were estimated as 5/6 times the method detection limit, which is a common procedure adopted in factor analysis. On average, the absolute uncertainties for $PM_{2.5}$, OC and EC were 0.40, 0.14 and 0.13 μg $m^{-3}$, which correspond to relative errors of 1.4–2.9%, 1.8–4.4% and 2.0–5.8%, respectively. For organic compounds, depending on the PAH or plasticiser, uncertainties were estimated to be in the range from 1.2 to 25 pg $m^{-3}$, accounting for relative errors of 1.3–5.2%. In the case of volatile organic compounds, individual uncertainties were always < 0.1 μg $m^{-3}$ with relative errors ranging from 0.28 to 6.6%.

### *2.3. Health Risk Assessment*

To estimate the risk associated with inhalation of pollutants, the methodology proposed by the United States Protection Agency (USEPA), and extensively described in the literature (e.g.) [41], was followed. The assessment refers only to the period at home, since information on the time spent at work or other microenvironments or outdoors was not available. Given that several studies suggest that there are no significant differences between the levels in the various subcompartments of the residential dwelling (e.g.) [42,43], measurements in the kitchens were taken as representative of exposure at home. To account for the permanence in each household, time-adjusted concentrations ($E_i$) were calculated using the following equation:

$$E_i = \sum_j C_{ij} t_j \times \frac{EF}{NY} \times \frac{ED}{AL} \quad (1)$$

where $E_i$ is the time-weighted daily personal exposure to compound i (μg $m^{-3}$), $C_{ij}$ is the measured concentration of compound i (μg $m^{-3}$) in each household, $t_j$ is the time fraction spent at home, EF is the exposure frequency (350 days/year considering that people spend 15 days on vacation away from home), NY is the number of days per year (365 days/year), ED is the exposure duration (30 years), and AL is the average lifetime (70 years).

The inhalation unit risk (IUR) is the excess cancer risk resulting from continuous exposure to a unit increase of a compound via inhalation. IUR values listed in Table 2 are derived from previous studies by the USEPA for the general population with a default body weight of 70 kg and a default inhalation rate of 20 $m^3$ $day^{-1}$. The chronic inhalation cancer risk (CR) is the increased probability of developing cancer as a result of a specific exposure to a certain compound. CR is calculated using the following equation:

$$CR_i = E_i \times IUR_i \quad (2)$$

Cancer risks < 1 in a million are considered negligible, whereas values above $1.0 \times 10^{-4}$ are classically considered of concern.

The inhalation non-cancer risk is estimated as follows:

$$HQ_i = E_i / R_f C_i \quad (3)$$

where $HQ_i$ is the hazard quotient of compound i, and $R_fC_i$ is the chronic reference concentration of compound i in μg $m^{-3}$ (Table 2). The hazard index (HI) is the summation of non-cancer risks from multiple compounds. Values higher than 1 express a chance that non-carcinogenic effects may happen, whilst values below 1 indicate low or no risk of non-carcinogenic effects on humans.

**Table 2.** Toxicity parameters for VOCs provided by the Office of Environmental Health Hazard Assessment (OEHHA) and Integrated Risk Information System (IRIS) of USEPA (n.a.—not available).

| Compound. | IUR (μg $m^{-3}$) $^{-1}$ | $R_fC$ (μg $m^{-3}$) |
|---|---|---|
| Benzene | $2.2 \times 10^{-6}$ | 30 |
| Toluene | n.a. | $5 \times 10^3$ |
| Xylenes | n.a. | 100 |
| Ethylbenzene | $2.5 \times 10^{-6}$ | $1 \times 10^3$ |
| Styrene | $1.63 \times 10^{-7}$ | 900 |
| Tetrachloroethylene | $2.6 \times 10^{-7}$ | 40 |
| 1,4-Dichlorobenzene | $1.1 \times 10^{-5}$ | 800 |
| Formaldehyde | $1.3 \times 10^{-5}$ | 9 |
| Acetaldehyde | $2.7 \times 10^{-6}$ | 9 |

The carcinogenic risk due to exposure to PAHs is based on benzo[a]pyrene equivalent concentrations ($BaP_{eq}$). These are calculated multiplying the individual PAH concentrations by their toxic equivalent factor (TEF) [44]. The inhalation cancer unit risk of BaP is $1.11 \times 10^{-6}$ (ng $m^{-3}$) $^{-1}$. It is estimated from the cancer potency factor (CPF) using the following equation:

$$IUR = CPF \times 20\ m^3/(70\ kg \times 10^6) \tag{4}$$

where inhalation unit risk (IUR) represents the excess cancer risk associated with an exposure to a concentration of 1 µg $m^{-3}$, CPF (equal to 3.9 (mg/kg-day) $^{-1}$ for BaP) indicates the excess cancer risk for an exposure to 1 mg of a compound per kg of body weight (70 kg), 20 $m^3$ is the default inhalation rate per day, and $10^6$ is the conversion factor from mg to ng. The excess cancer risk for a receptor exposed to PAHs via the inhalation pathway can be estimated by equation 2, where $E_i$ represents the time-weighted daily personal exposure to the sum of $BaP_{eq}$ concentrations.

## 3. Results and Discussion

### *3.1. Carbonyls and Volatile Organic Compounds*

Formaldehyde (HCHO) and acetaldehyde ($CH_3CHO$) are highly reactive carbonyl compounds that are normally found in both indoor and outdoor environments. Formaldehyde is emitted by various building and insulating materials, some consumer products (e.g. disinfectants and cosmetics), carpets, fabrics and new furniture, principally if made of plywood [45,46]. In indoor environments, combustion processes, including tobacco smoking, also emit large amounts of these compounds. Acetaldehyde is also present in various consumer products such as deodorants, and in many foods and alcoholic drinks [45], which can represent emission sources in kitchens.

Indoor and outdoor formaldehyde concentrations were 7.61 ± 3.08 and 1.49 ± 0.67 µg $m^{-3}$, respectively, whilst the corresponding acetaldehyde levels were 7.94 ± 4.63 and 0.41 ± 0.36 µg $m^{-3}$ (Table 3).

**Table 3.** Comparison of carbonyl concentrations (µg $m^{-3}$) obtained in the present study with those reported for other places.

| Location | Environment | Formaldehyde | Acetaldehyde | Reference |
|---|---|---|---|---|
| Aveiro region, Portugal | Kitchens | 7.61 ± 3.08 | 7.94 ± 4.63 | Present study |
| | Outdoor | 1.49 ± 0.67 | 0.41 ± 0.36 | |
| 61 flats in Paris, France | Kitchens | 21.7 ± 1.9 | 10.1 ± 1.8 | [42] |
| Dwellings in Bari, Italy | Kitchens | 16.0 ± 8.0 | 10.7 ± 8.8 | [45] |
| | Outdoor | 4.4 ± 1.7 | 3.4 ± 2.0 | |
| 59 homes in Prince Edward Island, Canada | Not provided | 5.5–87.5 (median 29.6) | 4.4–79.1 (median 29.6) | [46] |
| Shiraz, Iran | Outdoor-summer | 15.1 ± 9.17 | 8.40 ± 4.29 | [47] |
| | Outdoor-winter | 8.57 ± 5.91 | 3.52 ± 1.69 | |

Formaldehyde levels in the kitchen of the detached house on the outskirts were found to be statistically different ($p < 0.0146$) from those in houses 1 (rural) and 2 (city centre apartment). The acetaldehyde levels of house 2 differed significantly from the values obtained in house 1 ($p = 0.0224$). Concentrations in kitchens were much higher than those observed outdoors. Statistically significant differences between the values of the indoor and outdoor environments were registered ($p = 0.0001$ for formaldehyde, $p = 0.0004$ for acetaldehyde, $\alpha = 0.05$). These carbonyls are known to be irritants of the eyes and upper airways. Formaldehyde is a known human carcinogen [48]. In the present study, its concentrations never exceeded the protection limit of 100 µg $m^{-3}$ imposed by the Portuguese legislation. Acetaldehyde was incorporated by the WHO in Group 2, which comprises pollutants of potential interest, but additional investigation would be needed before it is clear whether there is enough evidence to warrant their inclusion in the guidelines. Based on studies of short- and long-term exposure, countries such as Canada have set a maximum daily limit of 280 $\mu g\ m^{-3}$. An extended review of formaldehyde concentrations worldwide in all types of indoor environments has been compiled by Salthammer et al. [49]. In a study carried out in dwelling in Bari, Italy, indoor formaldehyde and acetaldehyde concentrations were found to be significantly higher than outdoor concentrations. No significant relation was observed by the authors between the levels of aldehydes in the kitchens and the age or restoration of the building, the time windows or balcony doors were kept open or the time the burners were kept alight [45]. These two carbonyl compounds were also assessed in three principal rooms of 61 flats in Paris [42]. Statistically, the levels monitored in the kitchens did not differ from those registered in bedrooms and living rooms.

Many of the sources that contribute to carbonyls are also emitters of other VOCs. A few of these VOCs, such as benzene, are designated by multiple authorities as human carcinogens. Short- and long-term exposures can affect many organs and cause multiple symptoms [50]. In the present study, for most compounds, all the kitchens registered indoor-to-outdoor VOC concentration ratios higher than one, proving the strong contribution of endogenous emission sources. Statistically significant differences were found between indoor and outdoor levels of toluene ($p$ = 0.0236), ethylbenzene ($p$ = 0.0397), $m$ + $p$-xylene ($p$ = 0.0273), styrene ($p$ = 0.005), $o$-xylene ($p$ = 0.0500) and α-pinene ($p$ = 0.0001). The dwelling with the most significant differences for a greater number of compounds compared to the others was the permanently occupied apartment, where an elderly woman who needs nursing care at home resides. The concentrations of VOCs in the four dwellings are within the wide range of values measured in homes of several other regions [41,43,51–53], although closer to the lower levels. The indoor concentrations of benzene, trichloroethylene, toluene, styrene and tetrachloroethylene were well below the thresholds laid down by the national regulation (Table 4).

**Table 4.** Minimum, maximum and mean values for VOC concentrations and I/O ratios, and legal limits.

| VOC | Indoor Concentration Range and Mean $\mu g\ m^{-3}$ | Indoor-to-Outdoor Ratio | Threshold by the Portuguese Legislation [54] $\mu g\ m^{-3}$ |
|---|---|---|---|
| Benzene | 0.78–3.3 (1.6) | 0.39–1.3 (0.68) | 5 |
| Ethylbenzene | 0.87–6.6 (2.4) | 1.1–7.4 (2.9) | |
| Toluene | 4.1–21.6 (9.4) | 1.1–4.6 (2.3) | 250 |
| $m$ + $p$-xylene | 2.7–20.2 (7.6) | 1.1–7.6 (3.1) | |
| $o$-Xylene | 1.0–8.9 (3.1) | 0.97–8.0 (3.0) | |
| Styrene | 0.33–1.6 (1.0) | 1.5–5.0 (3.5) | 260 |
| 1,4-Dichlorobenzene | <0.10–1.2 (0.38) | | 25 |
| Trichloroethylene | <0.10 | | |
| Tetrachloroethylene | 0.30–2.9 (0.96) | 0.56–1.2 (0.89) | 250 |
| α-Pinene | 2.9–17.4 (9.5) | 16.4–152 (71.1) | |

Indoors, benzene, toluene, ethylbenzene and xylenes (BTEX) were highly correlated with each other ($r^2$ from 0.64 to 1.0), suggesting common emission sources. Outdoors, toluene, ethylbenzene and xylenes correlated well ($r^2$ from 0.73 to 0.97), but the relationships involving benzene were weaker. However, the indoor concentrations of the various compounds did not correlate with the respective outdoor levels, indicating that the emitting sources in the kitchens are different from those observed outside. The only exceptions were tetrachlorethylene ($r^2$ = 0.95) and styrene ($r^2$ = 0.68). Tetrachlorethylene is mainly used for dry cleaning of fabrics, whereas styrene occurs naturally in small amounts in some plants and foods, such as peanuts, cinnamon, and coffee beans, although it is mostly used to make products such as food containers, rubber, plastic, carpet backing, insulation, fiberglass, pipes, and automobile parts. The highest indoor concentration was generally observed for α-pinene, followed by toluene and $m$ + $p$-xylene. α-Pinene concentrations were 16 to 152 times higher in the kitchens than outside. This monoterpene is mainly synthetised by plants and commonly incorporated as fragrance in several consumer products (e.g. cleaning agents and air fresheners). It is emitted from numerous indoor items, including furniture of wooden origin [55]. Furthermore, cooking with condiments has been reported to be an important source of terpenes in indoor environments [56]. α-Pinene was also the most abundant and frequently detected VOC in UK and Polish homes [55,57]. On average, a toluene-to-benzene (T/B) ratio of 2 was obtained for the outdoor samples, which is a typical value for traffic emissions [58]. The indoor T/B ratios were three times higher than those observed in outdoor air. Likewise, the indoor samples were characterised by $m + p$-xylene-to-benzene and ethylbenzene-to-benzene ratios of 4.9 and 1.5, on average, while the corresponding outdoor values were 1.3 and 0.4. Toluene and benzene are common constituents of gasoline. However, toluene, together with ethylbenzene and xylenes, is used in solvents, while benzene is not. Solvents are the main component of cleaning agents, coatings, paints, adhesives, etc. Thus, evaporative emissions from coated surfaces and cleaning products, among other sources, may have contributed to enhanced emissions of toluene, xylenes and ethylbenzene in the kitchens. The $m + p$-xylene-to-ethylbenzene (*mpX/E*) ratio is frequently employed as an indicator of the age of air masses at a given site [58], since $m,p$-xylene disappear more rapidly than ethylbenzene through photochemistry. Thus, a higher *mpX/E* ratio suggests fresh local emissions, whereas lower ratios are related to more photochemical activity and associated emissions from some distance. In the present study, *mpX/E* ratios presented very little variability, averaging 3.3 and 3.1 for indoor and outdoor samples, respectively. These values are in agreement with ratios of fresh in situ emissions [58].

### 3.2. *$PM_{2.5}$ Concentrations and Carbonaceous Content*

The mean indoor $PM_{2.5}$ concentration ranged from 13.8 μg $m^{-3}$, in the kitchen located in a rural area, to 30.2 μg $m^{-3}$ in the city centre apartment with permanent occupancy (Figure 1).

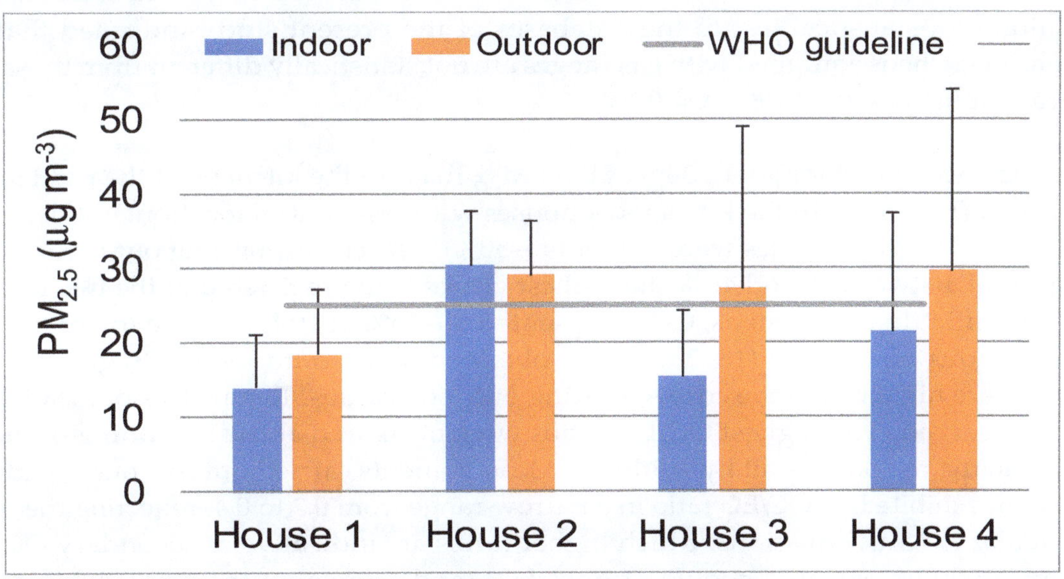


**Figure 1.** $PM_{2.5}$ concentrations monitored in kitchens and outdoor air.

The kitchen of this apartment was the one where the concentrations most often exceeded the WHO guideline value. As observed for VOCs, the $PM_{2.5}$ levels of this kitchen were found to be significantly different from those measured in any of the other dwellings ($p < 0.0262$), suggesting that concentrations increase with the occupancy rate. A mean outdoor level of 18.3 μg $m^{-3}$ was obtained in the rural area, while very close mean values were recorded in the centre and outskirts of the city (27.6–29.5 μg $m^{-3}$). Much higher $PM_{2.5}$ concentrations have been monitored in household kitchens where biomass fuels are used for cooking (Table 5).

**Table 5.** $PM_{2.5}$ concentrations measured in the present study and in kitchens of other countries using different cooking fuels or energy sources.

| Location | $PM_{2.5}$ (μg $m^{-3}$) | Cooking Fuel or Energy Source | Reference |
|---|---|---|---|
| Aveiro region, Portugal | 20.6 ± 10.9<br>17.8 ± 12.2 | gas<br>electricity | This study |
| Bhaktapur, Nepal | 630 ± 924<br>759 ± 988<br>656 ± 924<br>169 ± 207<br>101 ± 130<br>80 ± 103 | wood<br>rice husk<br>biomass mixture (wood + rice husk)<br>kerosene<br>LPG<br>electricity | [59] |
| Rural households, India | 910<br>447<br>432<br>78 | dung cakes<br>agricultural residues<br>fuel mixture (wood + dung)<br>LPG | [60] |
| Lanzhou, northwest China (heating season) | 204 ± 50<br>114 ± 39<br>107 ± 43 | coal<br>gas<br>electricity | [61] |
| Lanzhou, northwest China (non-heating season) | 213 ± 89<br>65 ± 42<br>55 ± 35 | coal<br>gas<br>electricity | [61] |

In these kitchens, it has been observed that particulate matter levels vary according to the fuel type: cow dung cakes > rice husk > agricultural residues > firewood > gas. Li et al. [61] evaluated the household concentrations of $PM_{2.5}$ among urban residents of Lanzhou, China, concluding that changing from coal to gas or electricity could result in a reduction of $PM_{2.5}$ in the kitchens by 40–70%. The application of a statistical test to the databases of the present study indicated that the $PM_{2.5}$ concentrations in kitchens equipped with gas ranges are not statistically different from those in kitchens with electrical appliances ($p = 0.486$, $\alpha = 0.05$).

Total carbon accounted for about 30% of the $PM_{2.5}$ mass in the kitchens of the rural area and city centre apartment (Figure 2). In the kitchens of houses with less central location, but near roads with intense traffic, the TC/$PM_{2.5}$ values were higher (40–50%). The corresponding outdoor mass fractions ranged, in general, between 20 and 40%, the highest values being registered at the two locations more influenced by traffic. In the kitchens, OC represented 30–35% of $PM_{2.5}$, while lower mass fractions of this carbonaceous constituent (18–23%) were obtained in the outdoor air. In general, the indoor OC/EC ratios were higher than the corresponding outdoor values. Regardless of location, ratios >2 were usually observed. The highest OC/EC ratios were measured at the beginning of the sampling campaign, when the region was hit by wildfires. Measurements carried out in a busy roadway tunnel in central Lisbon exhibited an OC/EC ratio in a narrow range from 0.3 to 0.4, reflecting the composition of fresh vehicular exhaust emissions. Much higher ratios are indicative of secondary OC formation, biomass burning emissions, and cooking fumes [62] (and references therein). Additional sources that contribute to the organic carbonaceous component of $PM_{2.5}$ in indoor air include paper and clothing fibres, microscopic specks of plastics, contaminants brought on the soles of our shoes, bacteria, skin flakes, cosmetics, cleaning products, etc. [63]. On the other hand, VOCs in indoor air react and form lower volatility reaction products. These reaction products may condense on existing particles or nucleate, producing secondary organic aerosols (SOA), which grow with time into larger particles. Surface chemistry can also be a source of indoor SOA [64].

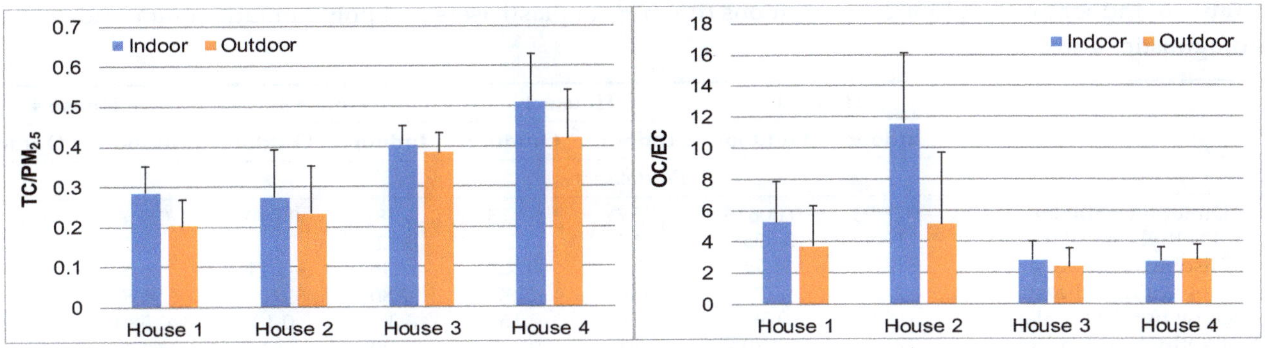


**Figure 2.** Ratios between total carbon (TC = EC + OC) and $PM_{2.5}$ and between organic carbon and elemental carbon.

$PM_{2.5}$ outdoor concentrations were weakly or moderately correlated with indoor concentrations for three of the households ($r^2$ = 0.23–0.53), while an excellent relationship ($r^2$ = 0.96) was found for the terraced house in a residential neighbourhood, near a main road with intense traffic. Similar relationships were observed for EC. The indoor and outdoor OC levels correlated well for all households ($r^2$ = 0.76–0.91), indicating common sources or formation processes. The slopes of the correlations between indoor and outdoor concentrations represent the infiltration factors, i.e., the fraction of the outdoor $PM_{2.5}$ carbonaceous component that penetrates indoors and remains suspended [65]. It was estimated that from 32% (rural house) to 74% (city centre apartment with permanent occupancy and with a window often open) of the indoor OC was infiltrated from the outside.

### *3.3. $PM_{2.5}$-Bound Polycyclic Aromatic Hydrocarbons and Plasticisers*

Eight phthalate plasticisers and one non-phthalate plasticizer [bis(2-ethylhexyl) adipate] were quantified in $PM_{2.5}$ (Table 6). This type of compounds can be found in large quantities in plastics, vinyl flooring, varnishes, coating agents, sealing compounds, industrial and natural rubber articles, and adhesives. The finishing of textiles also relies on the use of flexibilising substances to improve their feel and pliability. Plasticisers can leak out of the different products, thus escaping into the environment, making them ubiquitous. Total concentrations in the kitchens ranged from 44 to 171 ng $m^{-3}$. These values were three to 12 times higher than those detected in the outdoor air, reflecting the widespread employment of plasticised indoor materials. Total concentrations in the kitchens were found to be significantly different from those outdoors ($p$ = 0.0343). The most abundant compounds were diisobutyl phthalate and bis (2-ethylhexyl) phthalate, together with di-*n*-butyl phthalate in two of the kitchens. The latter two compounds have been listed as major plasticisers in household dust [63]. Health concerns related to phthalate ester exposures have focused primarily on cancer and reproductive effects [66]. Evidence for an association between phthalate exposure and diabetes risk and obesity was also found [67]. Moreover, it has emerged that exposure to phthalates aggravate pulmonary function and airway inflammation in asthmatic children [68].

PAHs are prevalent environmental pollutants generated primarily during the incomplete combustion of organic materials. Except in the permanently occupied housing, PAH concentrations were much higher in the outdoor air. It should be noted that in this house a greater number of meals are prepared, and gas is used for cooking. The total concentrations obtained in the kitchens were statistically different from those found outdoors ($p$ = 0.0499).

**Table 6.** Indoor and outdoor concentrations (ng $m^{-3}$) of plasticisers and polyaromatic compounds and diagnostic ratios.

| | House 1 | | House 2 | | House 3 | | House 4 | |
|---|---|---|---|---|---|---|---|---|
| | Indoor | Outdoor | Indoor | Outdoor | Indoor | Outdoor | Indoor | Outdoor |
| | | | | *Plasticisers* | | | | |
| Dimethyl phthalate | 0.0993 | 0.00616 | 1.57 | 0.115 | 0.281 | 0.0105 | 0.916 | 0.119 |
| Diethyl phthalate | 1.49 | 0.0538 | 5.55 | 1.24 | 0.825 | | 8.87 | 1.53 |
| Diisobutyl phthalate | 13.4 | 4.15 | 72.3 | 13.2 | 7.96 | 3.11 | 76.5 | 3.12 |
| Di-*n*-butyl phthalate | 1.44 | | 30.9 | 3.92 | 2.80 | | 56.8 | |
| Benzyl butyl phthalate | 0.547 | 0.394 | 2.05 | 0.160 | 0.302 | 0.135 | 0.580 | 0.132 |
| Bis(2-ethylhexyl) adipate | 7.93 | 2.70 | 6.34 | 0.847 | 3.20 | 2.18 | 6.11 | 0.985 |
| Bis(2-ethylhexyl) phthalate | 29.8 | 7.11 | 29.1 | 11.2 | 25.4 | 8.18 | 20.4 | 6.92 |
| Di-*n*-octyl phthalate | 2.95 | | 1.16 | 0.113 | 1.91 | 0.254 | 0.747 | 0.294 |
| Diisononyl phthalate | 1.31 | 0.168 | 0.211 | 0.0376 | 1.17 | 0.0601 | 0.265 | 0.101 |
| *Total* | 59.0 | 14.6 | 149 | 30.8 | 43.8 | 13.9 | 171 | 13.2 |
| | | | | *PAHs* | | | | |
| Naphthalene | 1.13 | 1.19 | 1.93 | 0.0192 | | | 1.81 | 1.22 |
| Acenaphthylene | | 0.156 | 0.00896 | | 0.0425 | 0.354 | 0.101 | 0.244 |
| Acenaphthene | | | | 0.491 | | | | |
| Fluorene | | | 0.0202 | | | 0.00557 | 0.0219 | 0.0387 |
| Phenanthrene | | 0.0540 | 0.0411 | 0.000917 | 0.00745 | 0.185 | 0.0415 | 0.233 |
| Anthracene | | 0.0132 | | 0.00579 | 0.0159 | 0.0529 | 0.0473 | 0.0402 |
| Fluoranthene | | 0.501 | 0.0484 | 0.0710 | 0.0963 | 0.666 | 0.183 | 0.602 |
| Pyrene | | 0.656 | 0.0293 | 0.0534 | 0.0109 | 0.718 | 0.188 | 0.662 |
| *p*-Terphenyl | | 0.0229 | | 0.0110 | | 0.0155 | | 0.0168 |
| Retene | 0.106 | 5.26 | 0.568 | 1.31 | 0.117 | 0.621 | 0.136 | 0.581 |
| Benzo[a]anthracene | 0.0387 | 0.780 | 0.0705 | 0.0981 | 0.0604 | 1.13 | 0.194 | 0.731 |
| Chrysene | 0.0612 | 1.11 | 0.0950 | 0.162 | 0.0643 | 1.64 | 0.330 | 1.30 |
| Benzo[b]fluoranthene | 0.319 | 1.86 | 0.297 | 0.329 | 0.671 | 2.43 | 0.830 | 1.77 |
| 7,12-Dimethylbenz[a]anthracene | 0.164 | 1.60 | 0.486 | 0.448 | 0.0910 | 0.298 | 0.0933 | 0.190 |
| Benzo[k]fluoranthene | 0.277 | 1.83 | 0.261 | 0.280 | 0.715 | 2.24 | 0.813 | 1.80 |
| Benzo[e]pyrene | 0.258 | 1.37 | 0.288 | 0.304 | 0.556 | 1.71 | 0.665 | 1.28 |
| Benzo[a]pyrene | 0.202 | 1.36 | 0.128 | 0.157 | 0.627 | 1.89 | 0.668 | 1.15 |
| Perylene | 0.108 | 0.719 | 0.0806 | 0.0920 | 0.301 | 0.875 | 0.314 | 0.528 |
| Indeno[1,2,3-cd]pyrene | 0.533 | 1.72 | 0.345 | 0.332 | 0.633 | 1.68 | 1.05 | 1.50 |
| Dibenzo[a,h]anthracene | 0.0749 | 0.245 | 0.0424 | 0.0389 | 0.0924 | 0.222 | 0.122 | 0.180 |
| Benzo[g,h,i]perylene | 0.460 | 1.47 | 0.359 | 0.345 | 0.547 | 1.52 | 1.05 | 1.41 |
| *Total* | 3.73 | 21.9 | 5.10 | 4.55 | 4.65 | 18.3 | 8.65 | 15.5 |
| | | | | *Ratios between PAHs* | | | | |
| BaA/(BaA + Chry) | 0.39 | 0.41 | 0.43 | 0.38 | 0.48 | 0.41 | 0.37 | 0.36 |
| IP/(IP + BghiP) | 0.54 | 0.54 | 0.49 | 0.49 | 0.54 | 0.53 | 0.50 | 0.52 |
| BaP/(BaP + BghiP) | 0.36 | 0.48 | 0.45 | 0.47 | 0.50 | 0.53 | 0.39 | 0.48 |
| BaP/BghiP | 0.44 | 0.92 | 0.36 | 0.46 | 1.2 | 1.2 | 0.64 | 0.82 |

Empty cells mean below detection limit or of the same order of the blanks; BaA—Benzo[a]anthracene; Chry—Chrysene; IP—Indeno[1,2,3-cd]pyrene; BghiP—Benzo[g,h,i]perylene; BaP—Benzo[a]pyrene.

The highest PAH levels were obtained in the house under the influence of wildfires. In the outdoor air of this dwelling, the retene concentration deserves to be highlighted. This alkylated phenanthrene has been described as the most abundant polyaromatic in particulate matter samples from several wildfire events [69]. An overwhelming proportion of retene has also been found in the organic extracts of $PM_{2.5}$ from the combustion of vegetal charcoal in barbecue grills [70]. More recently, it has been detected in non-exhaust particles resulting from tyre wear [71]. Retene in tyre-related samples may originate from the natural waxes and resins added as softeners and extenders to rubbers. In addition to biomass burning, this traffic non-exhaust source may justify the detection of retene in the outdoor air, which in part penetrates inside the buildings. In all samples, high PAH molecular weights with $\geq$ 4 rings dominated over lighter compounds, indicating prevalence of pyrogenic with respect to petrogenic sources.

Concentration ratios between PAHs have been frequently used as diagnostic tools to infer their sources [69,70,72]. In the present study, regardless of the location, BaA/(BaA + Chry) ratios around 0.4

were obtained, revealing a mixed contribution from petrogenic sources, cooking fumes and biomass burning emissions. Also, irrespective of the sampling site, a rather constant IP/(IP + BghiP) ratio around 0.5 was observed. Values $\geq 0.5$ have been linked to wildfires, coal combustion and residential wood burning, while emissions from petroleum combustion are characterised by much lower ratios. Values between 0.4 and 0.5 have been described as typical of cooking emissions [73]. In contrast, the BaP/BghiP ratio showed some variations. The indoor ratios were always lower than those observed outdoors. The highest ratios were obtained in the suburban detached house, which is close to a charcoal grilled chicken restaurant without fume removal or scrubbing system. BaP/BghiP ratios > 1.2 have been pointed out as typical of both wildfires and coal combustion, whilst values around 0.4–0.5 are characteristic of vehicle emissions [69,70,72].

### 3.4. $PM_{2.5}$ Morphological Characteristics

SEM images are widely used in the study of atmospheric particle morphology, and can directly show the particle size, shape, aggregation characteristics, composition, and even sources. The individual particle details could contribute to establish pollution tracers emitted by specific sources in future studies. The filters used in this study are made of quartz fibres with different diameters in a tree-dimensional filtration substrate. A blank/clean quartz filter (Figure 3a) was analysed by SEM in order to compare its microstructure with that of the filters on which the particles were collected. The particulate material was mainly composed of soot aggregates, fly ash particles and mineral particles, which mainly derive from combustion and dust.

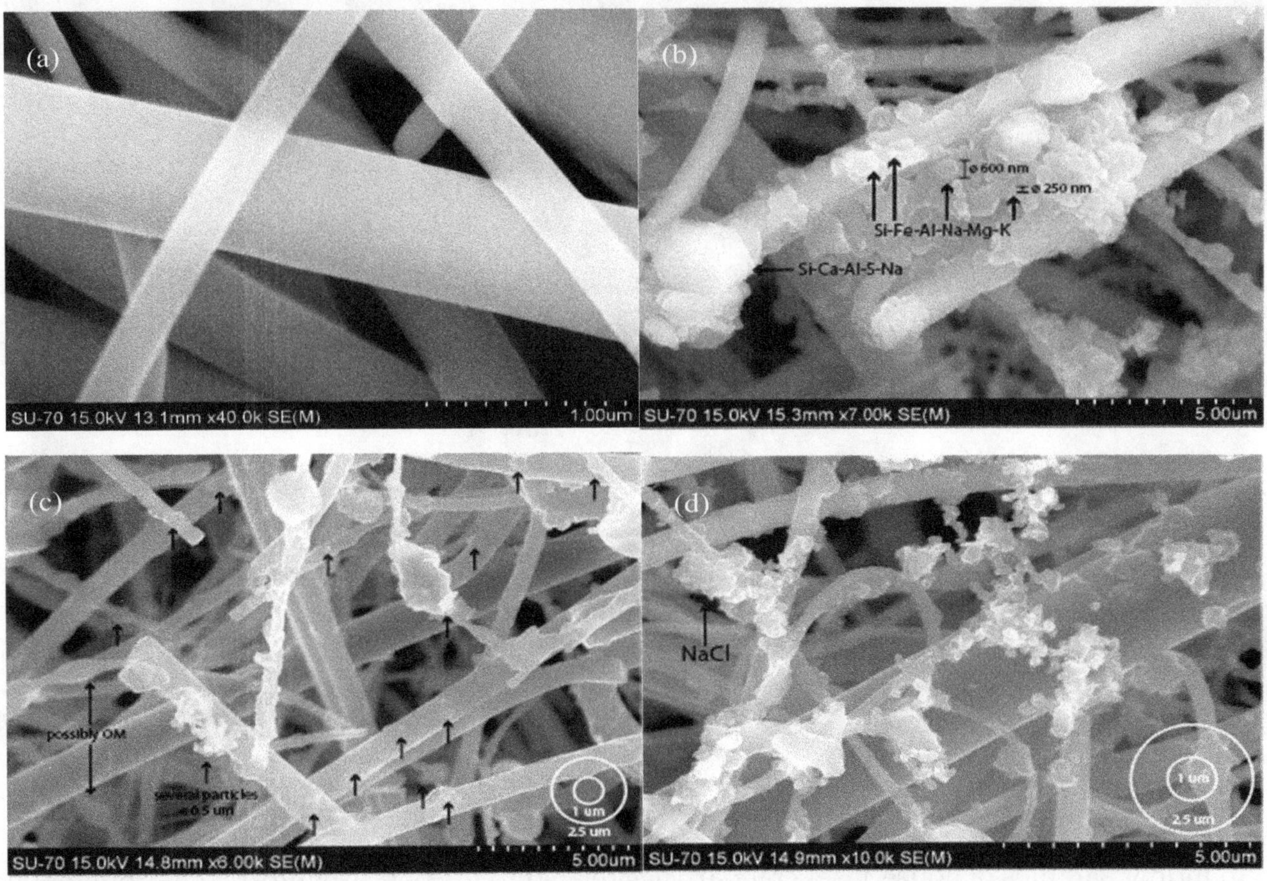


**Figure 3.** SEM imagens of (**a**) control blank quartz filter with 1.0 μm scale; (**b**–**d**) collected particles.

Figure 3b shows the diverse types of materials found in the outdoor environment of house 1 (rural) between 22 and 24 October 2017. Black arrows (Figure 3c) indicate some of the $PM_{2.5}$ present in different depths of the quartz fibre filter and the size of the circles on the bottom right corner represents the aerodynamic diameter cut-off for $PM_1$ and $PM_{2.5}$ in a sample collected in the kitchen of house 2 (city centre apartment). The kitchen samples from house 3 (Figure 3d) also reveal a mix composition of different materials, reflecting the contribution from diverse sources, e.g. cooking activities and particle resuspension.

In outdoor samples, carbonaceous particles represent a significant amount of the total particulates, being soot aggregates the dominant carbonaceous material (Figure 4a,b). These soot masses are formed by ultrafine aggregates of spherical particles with nanometric variable sizes (50–200 nm). Previous studies suggest that these types of particles are formed during combustion processes, e.g. biomass, coal, and diesel [74]. The EDS analysis revealed carbon, oxygen and sulphur peaks typical from combustion. Indoor particles from households 1 and 2 also show soot aggregates with lower sulphur content, which are likely associated with the use of burner gas hobs in these two kitchens. Additionally, silicate and iron plerospheres and cenospheres (Figure 4c,d) fly ashes, with diameter $\leq 2$ μm, were also found in outdoor samples, with Fe-Si-Cu-Al-Ca variable composition. In the outdoor sample of house 1 collected from 16 to 18 October, soot materials and fly ashes were more abundant than in other filters from the same location, possibly due to the occurrence of wildfires in those days in the nearby forests.

**Figure 4.** SEM images of soot aggregates with branching structures (**a**,**b**) and with cenosphere fly ash (**c**,**d**) in outdoor samples.

In indoor samples of house 1, several particles related to the peri urban/rural environment were found. Brochosomes (Figure 5a) are spherical honeycomb like particles, composed of proteins and lipids < 1 μm in diameter with which the leafhoppers (family Cicadellidae) coat themselves [75]. Also, in all houses, kitchen salt (NaCl) particles with dimensions <2.5 μm were abundantly found (Figures 3c and 5b).

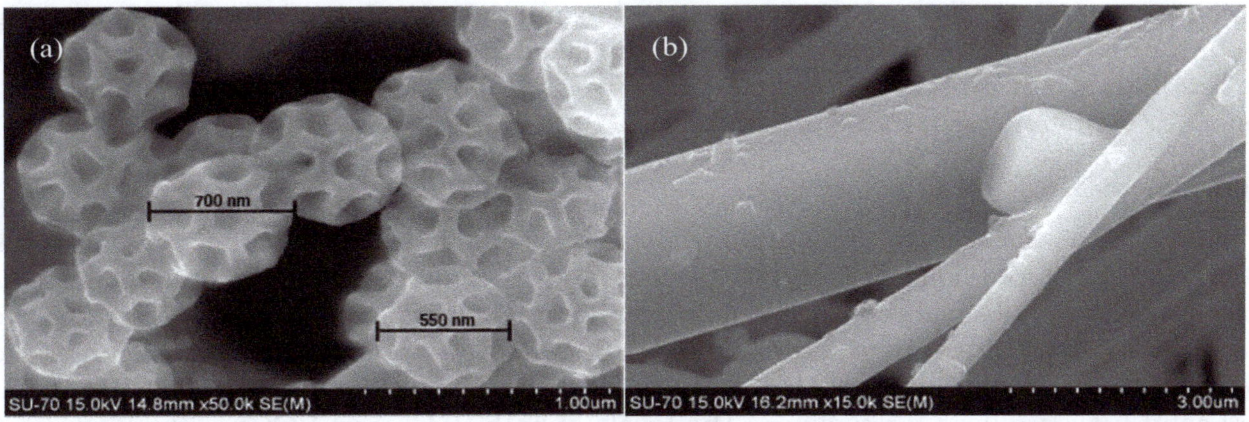


**Figure 5.** SEM imagens of (**a**) brochosomes, and (**b**) salt (NaCl) particles in indoor samples.

In summary, $PM_{2.5}$ was not only comprised of irregularly shaped agglomerate particles but also contained spherical, elongated, and flocculent particles. It is known that spherical particles and soot aggregates can enable the fine particles to easily adsorb toxic and harmful substances, such as heavy metals, volatile organic contaminants, and semivolatile organic pollutants. The observation of many particles in the ultrafine mode, including in the nanoscale size range, is relevant from the point of view of health. In addition to being able to penetrate deeply into the airways, these particles have a high adhesion surface to adsorb various chemical constituents, resulting in an enhanced complexity and toxicity.

### 3.5. Cancer and Non-Cancer Risks

Regardless of household, the hazard quotients associated with VOC inhalation were always below 1, indicating a low probability of non-cancer effects (Figure 6). The total hazard index ranged from 0.40 to 0.64. As observed in previous works [41,43], formaldehyde and acetaldehyde were the compounds that contributed most to the total risk, accounting for 25–63% and 32–70% of HI, respectively. The global excess lifetime cancer risk varied between $2.7 \times 10^{-5}$ and $4.7 \times 10^{-5}$. Thus, CR was lower than the USEPA guideline of $1.0 \times 10^{-4}$, but not negligible ($>1 \times 10^{-6}$). The major contribution to CR came, once again, from formaldehyde (59–81%) and acetaldehyde (8–35%). The highest risks were obtained in households located near roads with more intense traffic.

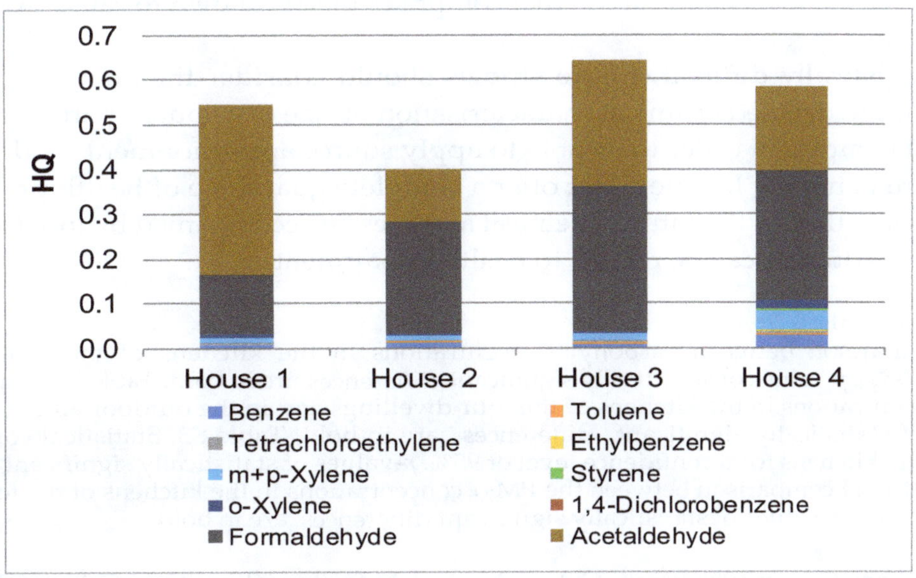


**Figure 6.** Hazard quotients (HQ) for individual VOCs for the four dwellings.

The cancer risks associated with domestic exposure to PAHs through the inhalation pathway ranged from $1.1 \times 10^{-7}$ to $3.5 \times 10^{-7}$, which can be taken as negligible. Benzo[a]pyrene accounted for more than a half (51–61%) of the total cancer risk, followed by dibenzo[a,h]anthracene and indeno[1,2,3-cd]pyrene with shares of 10–19% and 7–13%, respectively.

## 4. Conclusions

Particulate matter ($PM_{2.5}$), VOCs and carbonyls were monitored in the indoor and outdoor air of modern kitchens of four houses. Except for benzene and tetrachloroethylene, all pollutants presented indoor-to-outdoor ratios higher than 1, demonstrating the contribution of domestic emission sources. Concentrations were lower than the thresholds stipulated in the legislation or recommended by international organisations and well below the values reported for kitchens in developing countries where solid fuels are used for cooking. The levels of both $PM_{2.5}$ and VOCs in the kitchen of the permanently occupied home were significantly different from those observed in the other dwellings. Thus, it seems that a higher frequency of activities associated with full-time occupancy is the most determining factor for air quality. Carbonaceous constituents represented about 30–50% of the $PM_{2.5}$ mass, with the highest mass fractions recorded in houses closest to high traffic routes. It was estimated that from 32% to 74% of the indoor OC penetrates from outdoors. OC/EC ratios were higher indoors than outdoors, always surpassing 2, and peaked when the region was plagued by wildfires. In outdoor samples, pherospheres and cenospheres fly ashes composed of Fe-Si-Cu-Al-Ca were abundant, while ultrafine soot aggregates represented the dominant carbonaceous material. Soot aggregates with lower sulphur content were also found in kitchens with burner gas hobs. Salt and mineral particles from soil resuspension were observed in all kitchens. Brochosomes were only detected in the kitchen of the rural house. Although the particle levels were found to be statistically different in only one of the dwellings, the $PM_{2.5}$ morphology indicated the presence of particles with distinct properties in kitchens with gas cooking appliances compared to those equipped with electric hobs.

Irrespective of the type of house, a low probability of non-cancer effects due to inhalation of VOCs was estimated. The global excess lifetime cancer risk was lower than the USEPA guideline of $1 \times 10^{-4}$ but was higher than $1 \times 10^{-6}$, so it cannot be considered negligible. Formaldehyde and acetaldehyde were the compounds that contributed most to the total cancer and non-cancer risks in the indoor environments. The cancer risk associated with residential exposure to particle-bound PAHs via inhalation was found to be insignificant. However, it is necessary to bear in mind that the morphological analysis revealed the presence of numerous ultrafine particles, including nanometric variable sizes, with a complex composition that comprises metals known to cause oxidative stress and other health hazards.

Although logistically difficult, future studies should consider the analysis of other gaseous pollutants and a more detailed chemical characterisation of size distributed particulate material for a larger number of samples in order to be able to apply source apportionment models. The fact that concentrations are generally low does not offer a complete guarantee of health protection. For this reason, it is advisable that, in the future, chemical analyses be accompanied by in vitro toxicity testing to assess which constituents can be related to health impairment.

**Supplementary Materials:**
S1: Statistical comparison between carbonyl concentrations in the kitchens of the four dwellings for a confidence level of 95%. *p*-values of statistically significant differences are in bold, Table S2: Statistical comparison between VOC concentrations in the kitchens of the four dwellings and in the outdoor air for a confidence level of 95%. *p*-values of statistically significant differences are in bold, Table S3: Statistical comparison of VOC concentrations in the kitchens for a confidence level of 95%. *p*-values of statistically significant differences are in bold, Table S4: Statistical comparison between the $PM_{2.5}$ concentrations in the kitchens of the four dwellings for a confidence level of 95%. *p*-values of statistically significant differences are in bold.

**Author Contributions:** Conceptualisation, C.A., T.N., and M.C.; sampling, C.A., T.N., and M.E.; weighting and carbon analyses, A.R.O. and T.N.; SEM analysis, C.C. and F.R.; GC-MS analyses, A.V.; VOC and carbonyl determinations, E.V., S.M.A. and M.C.; writing of original draft, C.A.; review of original draft, C.A. in collaboration with all co-authors. All authors have read and agreed to the published version of the manuscript.

**Acknowledgments:** We thank all homeowners who have allowed the installation of noisy sampling equipment in their kitchens and helped to change the filters.

## References


1. Marć, M.; Śmiełowska, M.; Namiesnik, J.; Zabiegała, B. Indoor air quality of everyday use spaces dedicated to specific purposes—A review. *Environ. Sci. Pollut. Res.* **2017**, *25*, 2065–2082. [CrossRef]
2. Huboyo, H.S.; Tohno, S.; Cao, R. Indoor $PM_{2.5}$ characteristics and CO concentration related to water-based and oil-based cooking emissions using a gas stove. *Aerosol Air Qual. Res.* **2011**, *11*, 401–411. [CrossRef]
3. Torkmahalleh, M.A.; Gorjinezhad, S.; Unluevcek, H.S.; Hopke, P.K. Review of factors impacting emission/concentration of cooking generated particulate matter. *Sci. Total. Environ.* **2017**, *586*, 1046–1056. [CrossRef] [PubMed]
4. Ambade, B. A critical comparative study of indoor air pollution from household cooking fuels and its effect on health. *Orient. J. Chem.* **2016**, *32*, 473–480. [CrossRef]
5. Chakraborty, D.; Mondal, N.K.; Datta, J.K. Indoor pollution from solid biomass fuel and rural health damage: A micro-environmental study in rural area of Burdwan, West Bengal. *Int. J. Sustain. Built Environ.* **2014**, *3*, 262–271. [CrossRef]
6. Huboyo, H.S.; Tohno, S.; Lestari, P.; Mizohata, A.; Okumura, M. Characteristics of indoor air pollution in rural mountainous and rural coastal communities in Indonesia. *Atmos. Environ.* **2014**, *82*, 343–350. [CrossRef]
7. Rajkumar, S.; Clark, M.L.; Young, B.N.; Benka-Coker, M.L.; Bachand, A.M.; Brook, R.D.; Nelson, T.L.; Volckens, J.; Reynolds, S.J.; L'Orange, C.; et al. Exposure to household air pollution from biomass-burning cookstoves and HbA1c and diabetic status among Honduran women. *Indoor Air* **2018**, *28*, 768–776. [CrossRef] [PubMed]
8. Das, I.; Jagger, P.; Yeatts, K. Biomass cooking fuels and health outcomes for women in Malawi. *Ecohealth* **2016**, *14*, 7–19. [CrossRef]
9. GBD 2015 Risk Factors Collaborators. Global, regional, and national comparative risk assessment of 79 behavioural, environmental and occupational, and metabolic risks or clusters of risks, 1990–2015: A systematic analysis for the Global Burden of Disease Study 2015. *Lancet* **2016**, *388*, 1659–1724. [CrossRef]
10. Lacey, F.; Henze, D.K.; Lee, C.J.; van Donkelaar, A.; Martin, R.V. Transient climate, and ambient health impacts due to national solid fuel cookstove emissions. *Proc. Natl. Acad. Sci. USA* **2017**, *114*, 1269–1274. [CrossRef]
11. Sehgal, M.; Rizwan, S.A.; Krishnan, A. Disease burden due to biomass cooking-fuel-related household air pollution among women in India. *Glob. Heal. Action* **2014**, *7*, 25326. [CrossRef] [PubMed]
12. Gioda, A.; Tonietto, G.B.; de Leon, A.P. Exposure to the use of firewood for cooking in Brazil and its relationship with the health problems of the population. *Cienc. Saude Coletiva* **2019**, *24*, 3079–3088. [CrossRef] [PubMed]
13. Sana, A.; Meda, N.; Badoum, G.; Kafando, B.; Bouland, C. Primary cooking fuel choice and respiratory health outcomes among women in charge of household cooking in Ouagadougou, Burkina Faso: Cross-sectional study. *Int. J. Environ. Res. Public Heal.* **2019**, *16*, 1040. [CrossRef]
14. World Health Organization (WHO). *WHO Guidelines for Indoor Air Quality: Household Fuel Combustion*; World Health Organization: Geneva, Switzerland, 2014.
15. Zhao, Y.; Zhao, B. Emissions of air pollutants from Chinese cooking: A literature review. *Build. Simul.* **2018**, *11*, 977–995. [CrossRef]
16. Sharma, D.; Jain, S. Impact of intervention of biomass cookstove technologies and kitchen characteristics on indoor air quality and human exposure in rural settings of India. *Environ. Int.* **2019**, *123*, 240–255. [CrossRef]

17. Jiang, R.; Bell, M. A comparison of particulate matter from biomass-burning rural and non-biomass-burning urban households in Northeastern China. *Environ. Heal. Perspect.* **2008**, *116*, 907–914. [CrossRef] [PubMed]
18. Singh, S.; Gupta, G.P.; Kumar, B.; Kulshrestha, U. Comparative study of indoor air pollution using traditional and improved cooking stoves in rural households of Northern India. *Energy Sustain. Dev.* **2014**, *19*, 1–6. [CrossRef]
19. Chowdhury, Z.; Le, L.T.; al Masud, A.; Chang, K.C.; Alauddin, M.; Hossain, M.; Zakaria, A.; Hopke, P.K. Quantification of indoor air pollution from using cookstoves and estimation of its health effects on adult women in Northwest Bangladesh. *Aerosol Air Qual. Res.* **2012**, *12*, 463–475. [CrossRef]
20. Dasgupta, S.; Martin, P.; Samad, H.A. Lessons from rural Madagascar on improving air quality in the kitchen. *J. Environ. Dev.* **2015**, *24*, 345–369. [CrossRef]
21. Mukhopadhyay, R.; Sambandam, S.; Pillarisetti, A.; Jack, D.; Mukhopadhyay, K.; Balakrishnan, K.; Vaswani, M.; Bates, M.N.; Kinney, P.L.; Arora, N.; et al. Cooking practices, air quality, and the acceptability of advanced cookstoves in Haryana, India: An exploratory study to inform large-scale interventions. *Glob. Heal. Action* **2012**, *5*, 1–13. [CrossRef] [PubMed]
22. Gautam, S.K.; Suresh, R.; Sharma, V.P.; Sehgal, M. Indoor air quality in the rural India. *Manag. Environ. Qual. Int. J.* **2013**, *24*, 244–255. [CrossRef]
23. Bhargava, A.; Khanna, R.; Bhargava, S.; Kumar, S. Exposure risk to carcinogenic PAHs in indoor air during biomass combustion whilst cooking in rural India. *Atmos. Environ.* **2004**, *38*, 4761–4767. [CrossRef]
24. Naeher, L.P.; Leaderer, B.P.; Smith, K.R. Particulate matter and carbon monoxide in Highland Guatemala: Indoor and outdoor levels from traditional and improved wood stoves and gas stoves. *Indoor Air* **2000**, *10*, 200–205. [CrossRef] [PubMed]
25. Wangchuk, T.; He, C.; Knibbs, L.D.; Mazaheri, M.; Morawska, L. A pilot study of traditional indoor biomass cooking and heating in rural Bhutan: Gas and particle concentrations and emission rates. *Indoor Air* **2016**, *27*, 160–168. [CrossRef] [PubMed]
26. Chartier, R.; Phillips, M.; Mosquin, P.L.; Elledge, M.; Bronstein, K.; Nandasena, S.; Thornburg, V.; Thornburg, J.; Rodes, C. A comparative study of human exposures to household air pollution from commonly used cookstoves in Sri Lanka. *Indoor Air* **2016**, *27*, 147–159. [CrossRef]
27. Röllin, H.B.; Mathee, A.; Bruce, N.; Levin, J.; von Schirnding, Y.E.R. Comparison of indoor air quality in electrified and un-electrified dwellings in rural South African villages. *Indoor Air* **2004**, *14*, 208–216. [CrossRef]
28. Chen, Y.; Du, W.; Shen, G.; Zhuo, S.; Zhu, X.; Shen, H.; Huang, Y.; Su, S.; Lin, N.; Pei, L.; et al. Household air pollution and personal exposure to nitrated and oxygenated polycyclic aromatics (PAHs) in rural households: Influence of household cooking energies. *Indoor Air* **2016**, *27*, 169–178. [CrossRef]
29. Colbeck, I.; Nasir, Z.A.; Ali, Z. Characteristics of indoor/outdoor particulate pollution in urban and rural residential environment of Pakistan. *Indoor Air* **2010**, *20*, 40–51. [CrossRef]
30. Mccord, A.I.; Stefanos, S.A.; Tumwesige, V.; Lsoto, D.; Meding, A.H.; Adong, A.; Schauer, J.J.; Larson, R.A. The impact of biogas and fuelwood use on institutional kitchen air quality in Kampala, Uganda. *Indoor Air* **2017**, *27*, 1067–1081. [CrossRef]
31. Siddiqui, A.R.; Lee, K.; Bennett, D.; Yang, X.; Brown, K.H.; Bhutta, Z.A.; Gold, E.B. Indoor carbon monoxide and $PM_{2.5}$ concentrations by cooking fuels in Pakistan. *Indoor Air* **2009**, *19*, 75–82. [CrossRef]
32. Giwa, S.O.; Nwaokocha, C.N.; Odufuwa, B.O. Air pollutants characterization of kitchen microenvironments in southwest Nigeria. *Build. Environ.* **2019**, *153*, 138–147. [CrossRef]
33. Muralidharan, V.; Sussan, T.; Limaye, S.; Koehler, K.; Williams, D.L.; Rule, A.M.; Juvekar, S.; Breysse, P.N.; Salvi, S.; Biswal, S. Field testing of alternative cookstove performance in a rural setting of Western India. *Int. J. Environ. Res. Public Heal.* **2015**, *12*, 1773–1787. [CrossRef]
34. Less, B.; Mullen, N.; Singer, B.; Walker, I. Indoor air quality in 24 California residences designed as high-performance homes. *Sci. Technol. Built Environ.* **2015**, *21*, 14–24. [CrossRef]
35. Perrino, C.; Tofful, L.; Canepari, S. Chemical characterization of indoor and outdoor fine particulate matter in an occupied apartment in Rome, Italy. *Indoor Air* **2015**, *26*, 558–570. [CrossRef] [PubMed]
36. Mullen, N.A.; Li, J.; Russell, M.L.; Spears, M.; Less, B.D.; Singer, B.C. Results of the California Healthy Homes Indoor Air Quality Study of 2011-2013: Impact of natural gas appliances on air pollutant concentrations. *Indoor Air* **2015**, *26*, 231–245. [CrossRef]

37. Liu, Y.; Misztal, P.K.; Xiong, J.; Tian, Y.; Arata, C.; Weber, R.J.; Nazaroff, W.W.; Goldstein, A.H. Characterizing sources and emissions of volatile organic compounds in a northern California residence using space- and time-resolved measurements. *Indoor Air* **2019**, *29*, 630–644. [CrossRef]
38. Lee, S.-C.; Li, W.-M.; Ao, C.-H. Investigation of indoor air quality at residential homes in Hong Kong—Case study. *Atmos. Environ.* **2002**, *36*, 225–237. [CrossRef]
39. World Health Organization (WHO). Air Quality Guidelines. *Risks Hazard. Wastes* **2011**, 379. [CrossRef]
40. Cassee, F.R.; Héroux, M.-E.; Gerlofs-Nijland, M.E.; Kelly, F.J. Particulate matter beyond mass: Recent health evidence on the role of fractions, chemical constituents, and sources of emission. *Inhal. Toxicol.* **2013**, *25*, 802–812. [CrossRef] [PubMed]
41. Fang, L.; Norris, C.; Johnson, K.; Cui, X.; Sun, J.; Teng, Y.; Tian, E.; Xu, W.; Li, Z.; Mo, J.; et al. Toxic volatile organic compounds in 20 homes in Shanghai: Concentrations, inhalation health risks, and the impacts of household air cleaning. *Build. Environ.* **2019**, *157*, 309–318. [CrossRef]
42. Clarisse, B.; Laurent, A.; Seta, N.; le Moullec, Y.; el Hasnaoui, A.; Momas, I. Indoor aldehydes: Measurement of contamination levels and identification of their determinants in Paris dwellings. *Environ. Res.* **2003**, *92*, 245–253. [CrossRef]
43. Cheng, Z.; Li, B.; Yu, W.; Wang, H.; Zhang, T.; Xiong, J.; Bu, Z. Risk assessment of inhalation exposure to VOCs in dwellings in Chongqing, China. *Toxicol. Res.* **2018**, *7*, 59–72. [CrossRef] [PubMed]
44. Alves, C.A.; Vicente, E.D.; Vicente, A.M.P.; Rienda, I.C.; Tomé, M.; Querol, X.; Amato, F. Loadings, chemical patterns, and risks of inhalable road dust particles in an Atlantic city in the north of Portugal. *Sci. Total Environ.* **2020**, *737*, 139596. [CrossRef]
45. Lovreglio, P.; Carrus, A.; Iavicoli, S.; Drago, I.; Persechino, B.; Soleo, L. Indoor formaldehyde and acetaldehyde levels in the province of Bari, South Italy, and estimated health risk. *J. Environ. Monit.* **2009**, *11*, 955. [CrossRef] [PubMed]
46. Gilbert, N.L.; Guay, M.; Miller, J.D.; Judek, S.; Chan, C.C.; Dales, R.E. Levels and determinants of formaldehyde, acetaldehyde, and acrolein in residential indoor air in Prince Edward Island, Canada. *Environ. Res.* **2005**, *99*, 11–17. [CrossRef] [PubMed]
47. Delikhoon, M.; Fazlzadeh, M.; Sorooshian, A.; Baghani, A.N.; Golaki, M.; Ashournejad, Q.; Barkhordari, A. Characteristics and health effects of formaldehyde and acetaldehyde in an urban area in Iran. *Environ. Pollut.* **2018**, *242*, 938–951. [CrossRef]
48. Swenberg, J.A.; Moeller, B.C.; Lu, K.; Rager, J.E.; Fry, R.C.; Starr, T.B. Formaldehyde carcinogenicity research: 30 years and counting for mode of action, epidemiology, and cancer risk assessment. *Toxicol. Pathol.* **2013**, *41*, 181–189. [CrossRef] [PubMed]
49. Salthammer, T.; Mentese, S.; Marutzky, R. Formaldehyde in the indoor environment. *Chem. Rev.* **2010**, *110*, 2536–2572. [CrossRef] [PubMed]
50. IAQ-SFRB. Indoor Air Quality (IAQ) Volatile Organic Compounds. Scientific Findings Resource Bank (IAQ-SFRB). Available online: https://iaqscience.lbl.gov/voc-summary (accessed on 6 April 2020).
51. Rovelli, S.; Cattaneo, A.; Fazio, A.; Spinazzè, A.; Borghi, F.; Campagnolo, D.; Dossi, C.; Cavallo, D.M. VOC's measurements in residential buildings: Quantification via thermal desorption and assessment of indoor concentrations in a case-study. *Atmosphere* **2019**, *10*, 57. [CrossRef]
52. Chin, J.-Y.; Godwin, C.; Parker, E.; Robins, T.; Lewis, T.; Harbin, P.; Batterman, S. Levels and sources of volatile organic compounds in homes of children with asthma. *Indoor Air* **2014**, *24*, 403–415. [CrossRef]
53. Li, Y.; Cakmak, S.; Zhu, J. Profiles, and monthly variations of selected volatile organic compounds in indoor air in Canadian homes: Results of Canadian national indoor air survey 2012–2013. *Environ. Int.* **2019**, *126*, 134–144. [CrossRef]
54. *Ordinance No. 353-A/2013 (Portaria no. 353-A/2013)—Regulamento de Desempenho Energético dos Edifícios de Comércio e Serviços (RECS)—Requisitos de Ventilação e Qualidade do ar Interior*; Ministério do Ambiente Ordenamento do Território e Energia da Saúde e da Solidariedade Emprego e Segurança Social: Lisbon, Portugal, 2013.
55. Król, S.; Namieśnik, J.; Zabiegała, B. $\alpha$-Pinene, 3-carene and d-limonene in indoor air of Polish apartments: The impact on air quality and human exposure. *Sci. Total. Environ.* **2014**, 985–995. [CrossRef] [PubMed]
56. Klein, F.; Farren, N.J.; Bozzetti, C.; Daellenbach, K.R.; Kilic, D.; Kumar, N.K.; Pieber, S.M.; Slowik, J.G.; Tuthill, R.N.; Hamilton, J.F.; et al. Indoor terpene emissions from cooking with herbs and pepper and their secondary organic aerosol production potential. *Sci. Rep.* **2016**, *6*, 36623. [CrossRef]

57. Wang, C.M.; Barratt, B.; Carslaw, N.; Doutsi, A.; Dunmore, R.E.; Ward, M.W.; Lewis, A.C. Unexpectedly high concentrations of monoterpenes in a study of UK homes. *Environ. Sci. Process. Impacts* **2017**, *19*, 528–537. [CrossRef]
58. Bretón, J.G.C.; Bretón, R.M.C.; Ucan, F.V.; Baeza, C.B.; Fuentes, M.D.L.L.E.; Chi, M.P.U.; Marrón, M.R.; Pacheco, J.A.M.; Guzmán, A.R.; Chi, M.P.U. Characterization and sources of aromatic hydrocarbons (BTEX) in the atmosphere of two urban sites located in Yucatan Peninsula in Mexico. *Atmosphere* **2017**, *8*, 107. [CrossRef]
59. Pokhrel, A.; Bates, M.N.; Acharya, J.; Valentiner-Branth, P.; Chandyo, R.K.; Shrestha, P.S.; Raut, A.; Smith, K.R. $PM_{2.5}$ in household kitchens of Bhaktapur, Nepal, using four different cooking fuels. *Atmos. Environ.* **2015**, *113*, 159–168. [CrossRef]
60. Sidhu, M.K.; Khaiwal, R.; Mor, S.; John, S. Household air pollution from various types of rural kitchens and its exposure assessment. *Sci. Total. Environ.* **2017**, *586*, 419–429. [CrossRef] [PubMed]
61. Li, T.; Cao, S.; Fan, D.; Zhang, Y.; Wang, B.; Zhao, X.; Leaderer, B.P.; Shen, G.; Zhang, Y.; Duan, X. Household concentrations and personal exposure of $PM_{2.5}$ among urban residents using different cooking fuels. *Sci. Total. Environ.* **2016**, *548*, 6–12. [CrossRef]
62. Alves, C.A.; Duarte, M.; Nunes, T.; Moreira, R.; Rocha, S. Carbonaceous particles emitted from cooking activities in Portugal. *Glob. Nest J.* **2014**, *16*, 412–420.
63. Vicente, E.D.; Vicente, A.; Nunes, T.; Calvo, A.; Blanco-Alegre, C.; Oduber, F.; Castro, A.; Fraile, R.; Amato, F.; Alves, C. Household dust: Loadings and $PM_{10}$-bound plasticizers and polycyclic aromatic hydrocarbons. *Atmosphere* **2019**, *10*, 785. [CrossRef]
64. Weschler, C.J.; Carslaw, N. Indoor Chemistry. *Environ. Sci. Technol.* **2018**, *52*, 2419–2428. [CrossRef] [PubMed]
65. Othman, M.; Latif, M.T.; Matsumi, Y. The exposure of children to $PM_{2.5}$ and dust in indoor and outdoor school classrooms in Kuala Lumpur City Centre. *Ecotoxicol. Environ. Saf.* **2019**, *170*, 739–749. [CrossRef] [PubMed]
66. Kay, V.R.; Bloom, M.S.; Foster, W.G. Reproductive and developmental effects of phthalate diesters in males. *Crit. Rev. Toxicol.* **2014**, *44*, 467–498. [CrossRef] [PubMed]
67. Radke, E.G.; Galizia, A.; Thayer, K.A.; Cooper, G.S. Phthalate exposure and metabolic effects: A systematic review of the human epidemiological evidence. *Environ. Int.* **2019**, *132*, 104768. [CrossRef] [PubMed]
68. Kim, Y.-M.; Kim, J.; Cheong, H.-K.; Jeon, B.-H.; Ahn, K. Exposure to phthalates aggravates pulmonary function and airway inflammation in asthmatic children. *PLoS ONE* **2018**, *13*, e0208553. [CrossRef] [PubMed]
69. Vicente, A.; Calvo, A.; Fernandes, A.P.; Nunes, T.; Monteiro, C.; Pio, C.; Alves, C.A. Hydrocarbons in particulate samples from wildfire events in central Portugal in summer 2010. *J. Environ. Sci.* **2017**, *53*, 122–131. [CrossRef] [PubMed]
70. Vicente, E.; Vicente, A.; Evtyugina, M.; Carvalho, R.L.; Tarelho, L.A.C.; Oduber, F.; Alves, C.A. Particulate, and gaseous emissions from charcoal combustion in barbecue grills. *Fuel Process. Technol.* **2018**, *176*, 296–306. [CrossRef]
71. Alves, C.A.; Vicente, A.; Calvo, A.; Baumgardner, D.; Amato, F.; Querol, X.; Pio, C.; Gustafsson, M. Physical and chemical properties of non-exhaust particles generated from wear between pavements and tyres. *Atmos. Environ.* **2020**, *224*, 117252. [CrossRef]
72. Alves, C.A. Characterisation of solvent extractable organic constituents in atmospheric particulate matter: An overview. *Anais Acad. Brasil. Ciênc.* **2008**, *80*, 21–82. [CrossRef]
73. Zhang, N.; Han, B.; He, F.; Xu, J.; Zhao, R.; Zhang, Y.; Bai, Z. Chemical characteristic of $PM_{2.5}$ emission and inhalational carcinogenic risk of domestic Chinese cooking. *Environ. Pollut.* **2017**, *227*, 24–30. [CrossRef]
74. Huang, Y.; Wang, L.; Zhang, S.; Zhang, M.; Wang, J.; Cheng, X.; Li, T.; He, M.; Ni, S. Source apportionment and health risk assessment of air pollution particles in eastern district of Chengdu. *Environ. Geochem. Heal.* **2020**, 1–13. [CrossRef]
75. Schroeder, T.B.H.; Houghtaling, J.; Wilts, B.D.; Mayer, M. It is not a bug; it is a feature: Functional materials in insects. *Adv. Mater.* **2018**, *30*, e1705322. [CrossRef] [PubMed]

# Permissions

All chapters in this book were first published by MDPI; hereby published with permission under the Creative Commons Attribution License or equivalent. Every chapter published in this book has been scrutinized by our experts. Their significance has been extensively debated. The topics covered herein carry significant findings which will fuel the growth of the discipline. They may even be implemented as practical applications or may be referred to as a beginning point for another development.

The contributors of this book come from diverse backgrounds, making this book a truly international effort. This book will bring forth new frontiers with its revolutionizing research information and detailed analysis of the nascent developments around the world.

We would like to thank all the contributing authors for lending their expertise to make the book truly unique. They have played a crucial role in the development of this book. Without their invaluable contributions this book wouldn't have been possible. They have made vital efforts to compile up to date information on the varied aspects of this subject to make this book a valuable addition to the collection of many professionals and students.

This book was conceptualized with the vision of imparting up-to-date information and advanced data in this field. To ensure the same, a matchless editorial board was set up. Every individual on the board went through rigorous rounds of assessment to prove their worth. After which they invested a large part of their time researching and compiling the most relevant data for our readers.

The editorial board has been involved in producing this book since its inception. They have spent rigorous hours researching and exploring the diverse topics which have resulted in the successful publishing of this book. They have passed on their knowledge of decades through this book. To expedite this challenging task, the publisher supported the team at every step. A small team of assistant editors was also appointed to further simplify the editing procedure and attain best results for the readers.

Apart from the editorial board, the designing team has also invested a significant amount of their time in understanding the subject and creating the most relevant covers. They scrutinized every image to scout for the most suitable representation of the subject and create an appropriate cover for the book.

The publishing team has been an ardent support to the editorial, designing and production team. Their endless efforts to recruit the best for this project, has resulted in the accomplishment of this book. They are a veteran in the field of academics and their pool of knowledge is as vast as their experience in printing. Their expertise and guidance has proved useful at every step. Their uncompromising quality standards have made this book an exceptional effort. Their encouragement from time to time has been an inspiration for everyone.

The publisher and the editorial board hope that this book will prove to be a valuable piece of knowledge for researchers, students, practitioners and scholars across the globe.

# List of Contributors

**Thao Thi Thu Nguyen, Yasuhiro Kambayashi, Enoch Olando Anyenda, Hiromasa Tsujiguchi, Masami Kitaoka, Hiroki Asakura, Daisuke Hori, Yuri Hibino and Hiroyuki Nakamura**
Department of Environmental Health and Preventive Medicine, Graduate School of Medical Sciences, Kanazawa University, 13-1 Takara-machi, Kanazawa 920-8640, Japan

**Tomomi Higashi**
Department of Hygiene, Graduate School of Medical Sciences, Kanazawa University, 13-1 Takara-machi, Kanazawa 920-8640, Japan

**Yoshimasa Michigami**
Environment Preservation Center, Kanazawa University, Kakuma-machi, Kanazawa 920-1192, Japan

**Johsuke Hara**
Respiratory Medicine, Cellular Transplantation Biology, Graduate School of Medical Sciences, Kanazawa University, 13-1 Takara-machi, Kanazawa 920-8640, Japan

**Masaki Fujimura**
Respiratory Medicine, Cellular Transplantation Biology, Graduate School of Medical Sciences, Kanazawa University, 13-1 Takara-machi, Kanazawa 920-8640, Japan
Respiratory Medicine, National Hospital Organization Nanao Hospital, 3-1 Mattou-machi Yabe, Nanao, Ishikawa 926-8531, Japan

**Tadashi Konoshita**
Third Department of Internal Medicine, Fukui University School of Medicine, Eiheiji, Fukui 910-1193, Japan

**Anthony Skipper, Clement G. Yedjou and Paul B. Tchounwou**
Molecular Toxicology Research Laboratory, NIH-Center for Environmental Health, College of Science, Engineering and Technology, Jackson State University, 1400 Lynch Street, Jackson, MS 39217, USA

**Jennifer N. Sims**
Department of Pathology, Beth Israel Deaconess Medical Center, Harvard Medical School, 330 Bookline Avenue, Boston, MA 02215, USA

**Zhouping Liu, Qiaofen Zhang, Tiqian Han, Yanfei Ding, Junwei Sun, Feijuan Wang and Cheng Zhu**
College of Life Sciences, China Jiliang University, Hangzhou 310018, China

**Gehendra Mahara, Chao Wang, Qin Xu, Fangfang Huang, Lixin Tao, Jin Guo, Kai Cao, Liu Long, Qi Gao and Xiuhua Guo**
Department of Epidemiology and Biostatistics, School of Public Health, Capital Medical University, Beijing 100069, China
Beijing Municipal Key Laboratory of Clinical Epidemiology, Beijing 100069, China

**Da Huo and Quanyi Wang**
Institute for Infectious Disease & Endemic Disease Control, Beijing Center for Disease Prevention & Control (CDC), Beijing Center for Preventive Medical Research, Beijing 100069, China

**Jagadish K. Chhetri**
Department of Geriatrics, XuanWu Hospital of Capital Medical University, Beijing 100069, China

**Wei Wang**
Department of Epidemiology and Biostatistics, School of Public Health, Capital Medical University, Beijing 100069, China
Beijing Municipal Key Laboratory of Clinical Epidemiology, Beijing 100069, China
Systems and Intervention Research Centre for Health, School of Medical Sciences, Edith Cowan University, Perth 6027, Australia

**Jun-Hyun Kim, Chanam Lee and Wonmin Sohn**
Department of Landscape Architecture and Urban Planning; Texas A&M University, College Station, TX 77843, USA

**Hui-Ching Wu**
Department of Medical Sociology and Social Work, Chung Shan Medical University, Taichung 402367, Taiwan
Social Service Section, Chung Shan Medical University Hospital, Taichung 402367, Taiwan

**Ming-Hseng Tseng**
Department of Medical Informatics, Chung Shan Medical University, Taichung 402367, Taiwan

**Chuan-Chao Lin**
School of Medicine, Chung Shan Medical University, Taichung 402367, Taiwan
Department of Physical Medicine and Rehabilitation, Chung Shan Medical University Hospital, Taichung 402367, Taiwan

**Marina M. S. Cabral Pinto, Manuela Inácio and Eduardo Ferreira da Silva**
Geobiotec Research Centre, Department of Geosciences, University of Aveiro, 3810-193 Aveiro, Portugal

**Agostinho Almeida and Luísa Diniz**
Laboratory of Applied Chemistry, Department of Chemical Sciences, Faculty of Pharmacy, LAQV/ REQUIMTE, Porto University, 4050-313 Porto, Portugal

**Orquídia Neves**
CERENA, DECivil, Instituto Superior Técnico, University of Lisbon, 1049-001 Lisbon, Portugal

**Pedro A. Dinis**
MARE—Marine and Environmental Sciences Centre, Department of Earth Sciences, University of Coimbra, 3030-790 Coimbra, Portugal

**Maria Orquídia Neves and José Marques**
Department of Civil Engineering, Architecture and Georesources, CERENA (Centro de Recursos Naturais e Ambiente), Instituto Superior Técnico, University of Lisbon, 1049-001 Lisbon, Portugal

**Hans G.M. Eggenkamp**
Department of Petrology and Mineral Resources, Eberhard-Karls University of Tübingen, Schnarrenbergstraße 94-96, 72076 Tübingen, Germany

**Amit Kumar**
School of Hydrology and Water Resources, Nanjing University of Information Science and Technology, Nanjing 210044, China

**Amit Kumar, Aftab A. Shabnam and Gangavarapu Subrahmanyam**
Central Muga Eri Research and Training Institute, Central Silk Board, Jorhat, Assam 785000, India

**Cabral-Pinto M.M.S.**
Geobiotec Research Centre, Department of Geosciences, University of Aveiro, 3810-193 Aveiro, Portugal

**Ashish K. Chaturvedi**
Water Management (Agriculture) Division, Centre for Water Resources Development and Management, Kozhikode, Kerala 673571, India

**Raju Mondal**
Central Sericultural Germplasm Resources Centre (CSGRC), Central Silk Board, Ministry of Textiles, Thally Road, Hosur, Tamil Nadu 635109, India

**Dipak Kumar Gupta**
ICAR-Central Arid Zone Research Institute Regional Research Station Pali Marwar, Rajasthan 342003, India

**Sandeep K. Malyan**
Institute of Soil, Water and Environmental Sciences, Agricultural Research Organization (ARO), The Volcani Center, Rishon LeZion 7505101, Israel

**Smita S. Kumar**
Department of Environment science, J.C. Bose University of Science & Technology, YMCA, NH-2, Sector-6, Mathura Road, Faridabad, Haryana 121006, India

**Shakeel A. Khan**
Centre for Environment Science and Climate Resilient Agriculture, ICAR-Indian Agricultural Research Institute, New Delhi 110012, India

**Krishna K. Yadav**
Institute of Environment and Development Studies, Bundelkhand University, Kanpur Road, Jhansi 284128, India

**Bianca Gerardo, Joana Nogueira, Mário R. Simões and Sandra Freitas**
Center for Research in Neuropsychology and Cognitive and Behavioral Intervention (CINEICC), Faculty of Psychology and Educational Sciences (FPCEUC), Univ Coimbra, 3000-115 Coimbra, Portugal
Psychological Assessment and Psychometrics Laboratory (PsyAssessmentLab), Faculty of Psychology and Educational Sciences (FPCEUC), Univ Coimbra, 3000-115 Coimbra, Portugal

**Paula Pinto**
Psychological Assessment and Psychometrics Laboratory (PsyAssessmentLab), Faculty of Psychology and Educational Sciences (FPCEUC), Univ Coimbra, 3000-115 Coimbra, Portugal

**Marina Cabral Pinto**
Geobiotec Research Centre, Department of Geosciences, University of Aveiro, 3810-193 Aveiro, Portugal

**Edgar Pinto**
LAQV/REQUIMTE, Laboratory of Applied Chemistry, Faculty of Pharmacy, University of Porto, 4050-313 Porto, Portugal
Department of Environmental Health, School of Health, P.Porto, CISA/Research Center in Environment and Health, 4200-072 Porto, Portugal

**Paula Marinho-Reis**
Geobiotec Research Centre, Department of Geosciences, University of Aveiro, 3810-193 Aveiro, Portugal
Departamento de Ciências da Terra, Instituto de Ciências da Terra, Polo da Universidade do Minho, Campus de Gualtar, 4710-057 Braga, Portugal

**Paula I. Moreira**
Center for Neuroscience and Cell Biology (CNC), University of Coimbra, 3004-504 Coimbra, Portugal
Institute of Physiology, Faculty of Medicine, University of Coimbra, 3000-548 Coimbra, Portugal

**Ângela Cerqueira, Denise Terroso and Cristina Sequeira**
GeoBioTec Research Centre, Department of Geosciences, University of Aveiro, 3810-193 Aveiro, Portugal

**Carla Marina Bastos**
GeoBioTec Research Centre, Department of Geosciences, University of Aveiro, 3810-193 Aveiro, Portugal
Exatronic, Lda, 3800-373 Aveiro, Portugal

**Paula Tilley**
CIISA - Centre for Interdisciplinary Research in animal Health, Faculty of Veterinary Medicine, University of Lisbon, 1300-477 Lisbon, Portugal

**Abraham Addo-Bediako and Karabo Malakane**
Department of Biodiversity, University of Limpopo, Sovenga 0727, South Africa

**Pedro Dinis**
MARE - Marine and Environmental Sciences Centre, Department of Earth Sciences, University of Coimbra, 3030-790 Coimbra, Portugal

**Amílcar Armando**
Universidade Metodista de Angola, Rua Nossa Senhora da Muxima 10, Caixa Postal 6739, Luanda, Angola

**João Pratas**
Geosciences Center, Department of Earth Sciences, University of Coimbra, Rua Sílvio Lima, Univ. Coimbra - Pólo II, 3030-790 Coimbra, Portugal

**Célia Alves, Ana Vicente, Ana Rita Oliveira, Estela Vicente, Teresa Nunes, Mário Cerqueira and Margarita Evtyugina**
Centre for Environmental and Marine Studies (CESAM), Department of Environment, University of Aveiro, 3810-193 Aveiro, Portugal

**Carla Candeias and Fernando Rocha**
Geobiosciences, Geotechnologies and Geoengineering Research Centre (GeoBioTec), Department of Geosciences, University of Aveiro, 3810-193 Aveiro, Portugal

**Susana Marta Almeida**
Centre for Nuclear Sciences and Technologies (C2TN), Instituto Superior Técnico, University of Lisbon, Estrada Nacional 10, 2695-066 Bobadela, Portugal

# Index

www.ingramcontent.com/pod-product-compliance
Lightning Source LLC
Chambersburg PA
CBHW081416220926
51613CB00010B/250
* 9 7 8 1 6 3 9 8 7 7 7 9 9 *